Hyperbolic Differential Operators and Related Problems

PURE AND APPLIED MATHEMATICS

A Program of Monographs, Textbooks, and Lecture Notes

LECTURE NOTES IN PURE AND APPLIED MATHEMATICS

Additional Volumes in Preparation

Hyperbolic Differential Operators and Related Problems

edited by

Vincenzo Ancona
Università degli Studi di Firenze
Florence, Italy

Jean Vaillant
Université Pierre et Marie Curie, Paris VI
Paris, France

CRC Press
Taylor & Francis Group
Boca Raton London New York

CRC Press is an imprint of the
Taylor & Francis Group, an **informa** business

First published 2003 Marcel Dekker, Inc.

Published 2019 by CRC Press
Taylor & Francis Group
6000 Broken Sound Parkway NW, Suite 300
Boca Raton, FL 33487-2742

© 2003 by Taylor & Francis Group, LLC
CRC Press is an imprint of Taylor & Francis Group, an Informa business

First issued in paperback 2019

No claim to original U.S. Government works

ISBN 13: 978-0-367-44677-2 (pbk)
ISBN 13: 978-0-8247-0963-1 (hbk)

**Visit the Taylor & Francis Web site at
http://www.taylorandfrancis.com**

**and the CRC Press Web site at
http://www.crcpress.com**

Library of Congress Cataloging-in-Publication Data
A catalog record for this book is available from the Library of Congress.

Preface

The papers collected in this volume are concerned with hyperbolic problems, or problems the methods of which are related to hyperbolic techniques.

T. NISHITANI introduces a notion of nondegenerate characteristic for systems of linear partial differential equations of general order. He shows that nondegenerate characteristics are stable under hyperbolic perturbations, and he proves that if the coefficients of the system are real analytic and all characteristics are nondegenerate then the Cauchy problem for the system is well posed in the class of smooth functions.

K. KAJITANI studies a class of operators that generalize the linear hyperbolic operators, introducing the notion of *time function*, and proving the well-posedness of the Cauchy problem in the class of C^∞ functions.

The Cauchy problem is also the subject of the paper by A. BOVE and C. BERNARDI; they state some results for a class of linear hyperbolic operators with double characteristics, not effectively hyperbolic. In particular they prove well-posedness in the C^∞ class under a geometric condition and a Levi condition, and well-posedness in the Gevrey class under more general assumptions.

For a linear system whose principal part is hyperbolic and whose coefficients depend only on time, H. YAMAHARA establishes necessary and sufficient conditions for well-posedness in the Gevrey class, whatever the lower order terms are.

L. MENCHERINI and S. SPAGNOLO consider a first order hyperbolic system in two variables whose coefficients depend only on time; they define the notion of *pseudosymmetry* for matrix symbols of order zero, and determine the Gevrey class where the Cauchy problem is well-posed, according to the type of pseudosymmetry of the principal matrix symbol.

The 2-phase Goursat problem has been solved by means of Bessel functions; here J. CARVALHO E SILVA considers the 3-phase Goursat problem, using instead some hypergeometric functions in four variables. He also discusses the general problem, pointing out that the main difficulties are due to the lack of results on special functions.

The Stricharz inequality for the classical linear wave equation has been generalized by M. REISSIG and Y.G. WANG to the case of time-dependent coefficients: the coefficient is the product of an increasing factor and an oscillatory factor. The interaction was studied by the authors; in the present paper they extend the result to a one-dimensional system describing thermoelasticity.

The noncharacteristic, nonlinear Cauchy problem is the subject of the paper by M. TSUJI. The classical solution has some singularities, so that the problem arises of studying the extension of the solution beyond the singularities. The author constructs a Lagrangian solution in the cotangent bundle, getting a multivalued classical solution; then he explores how to construct a reasonable univalued solution with singularities.

Y. CHOQUET considers the Einstein equations. By a suitable choice of the gauge, (for instance, an a priori hypothesis on coordinates choice) she obtains a hyperbolic system in the sense of Leray-Ohya, well-posed in the Gevrey class of index 2. She studies old and new cases where the system is strictly hyperbolic and well-posed in the C^∞ class.

Necessary and sufficient conditions for the Cauchy-Kowalevskaya theorem on systems of linear PDEs were given by Matsumoto and Yamahara; on the other hand, Nagumo constructed a local solution, unique, for a higher order scalar Kowalevskian operator, whose coefficients are analyitic in x and continuous in t. Here W. MATSUMOTO, M. MURAI and T. NAGASE show that for a system whose coefficients are analytic in x and

C^∞ in t the above condition of Matsumoto and Yamahara is necessary and sufficient also when the order on $\partial/\partial t$ is one, the order on $\partial/\partial x$ is two, and the rank of the second order part is at most one.

B.W. SCHULZE and N. TARKHANOV construct a general calculus of pseudodifferential operators on a smoothly stratified space, with local cone bundle neighborhood, including ellipticity and the Fredholm property.

M. DREHER and I. WITT propose edge Sobolev spaces for the investigation of weak hyperbolicity for linear and semilinear operators; propagation of singularities is stated.

For the surface waves of water K. O. Friedrichs gave a systematic procedure to obtain the Airy equations from the Euler equations; a rigorous mathematical approach was more recently given by T. Kano in the analytic case. Here T. KANO and S. MIKI develop the theory in the Lagrange coordinate system. The wave equation in shallow water and the Boussineq equation, via Korteweg-de Vries, are obtained as approximate equations in Lagrangian coordinates.

For certain equations of Schrödinger type, J. TAKEUCHI states necessary and sufficient conditions for the Cauchy problem to be well-posed in H^∞; he uses a time independent L_2 symmetrization, with loss of regularity.

D. GOURDIN studies a 2-evolution operator in the sense of Petrosky; subject to the condition that the real roots of the principal polynomial with constant coefficients have constant multiplicity. He finds sufficient conditions for the well-posedness of the Cauchy problem in Sobolev spaces. Some generalizations are also given.

K. KAJITANI investigates the Gevrey smoothing effects of the solution to the Cauchy problem for Schrödinger- type equations; the decay of initial data is related to the Gevrey class with respect to the space variables of the solutions.

The metaplectic representation allows the construction of the solutions of the Schrödinger equation for the quadratic hamiltonians. M de GOSSON is able to obtain the result more generally for any physical hamiltonian.

F. COLOMBINI and C. GRAMMATICO consider the problem of the strong uniqueness of the solution (in a neighborhood of the origin) in $\mathbf{R}^n$, for particular fourth order elliptic operators flat at the origin. In the second order case, the known result in $\mathbf{R}^2$ is obtained in $\mathbf{R}^n$. In the case of a product of some model second order operators in $\mathbf{R}^2$ with Gevrey coefficients strong uniqueness is obtained under a condition on the Gevrey index, related to the form of the operator.

A sharp condition on the modulus of continuity was obtained by Tarama for an elliptic operator of second order to have the local uniqueness property. D. DEL SANTO shows that this condition is necessary by constructing a nonuniqueness example.

For certain holomorphic operators with polynomial coefficients Y. HAMADA constructs some solutions whose domain of holomorphy has an exterior point. The results are based on the earier work by Hamada, Leray, Takeuchi, as well as Darboux, Halphen and Chazy.

The remaining papers contain more theoretical results.

P. LAUBIN presents some known results and new advances on the topology of spaces of holomorphic functions in an open subset of a Frechet space; he gives a projective characterization of holomorphic germs using seminorms whose form is similar to the one appearing in the Whitney extension theorem for C^∞ functions.

Let Y be a smooth submanifold of a C^∞ manifold X; a distribution u in the complement of Y has the property that the closure of its wave front in the cotangent bundle of X is orthogonal to the tangent bundle of Y. M.K.V. MURTHY describes its analytic behavior in a neighborhood of Y by the notion of *microlocal scaling degree* along Y, and obtains a result similar to the Hörmander theorem for homogeneous distributions.

A. DEBIARD and B. GAVEAU in their paper apply stochastic methods to determine the ground state of an atomic molecular system. The system is represented by a Hamiltonian operator acting on a Hilbert space. A lower bound for the ground state is obtained using the Feynman-Kac formula and the special homogeneity property of the Coulomb potential.

L.S. SCHULMAN raises the difficulty of conceiving that a single dynamical system could contain subsystems, in significant mutual contact, possessing opposite thermodynamics arrows of time. By examining possible cosmological justification for the usual arrow of time he finds that a consistent way to establish such justification is by giving symmetric boundary conditions at two (cosmologically remote) times and seeking irreversible behavior in between. Other boundary conditions, modeling shorter periods in the evolution of the cosmos, can be found that allow the simultaneous existence of two thermodynamic arrows, notwithstanding moderate interaction between the two systems.

Vincenzo Ancona
Jean Vaillant

Contents

Contributors

Enrico Bernardi   University of Bologna, Bologna, Italy

Antonio Bove   University of Bologna, Bologna, Italy

Jaime Carvalho e Silva   Universidade de Coimbra, Coimbra, Portugal

Yvonne Choquet-Bruhat   Université de Paris 6, Paris, France

F. Colombini   Università di Pisa, Pisa, Italy

Maurice de Gosson   Blekinge Institute of Technology, Karlskrona, Sweden, and University of Colorado at Boulder, Boulder, Colorado, U.S.A.

A. Debiard   Laboratoire Analyse et Physique Mathematique, Université Pierre et Marie Curie, Paris, France

Daniele Del Santo   Università di Trieste, Trieste, Italy

Michael Dreher   University of Tsukuba, Tsukuba, Japan

Bernard Gaveau   Laboratoire Analyse et Physique Mathématique, Université Pierre et Marie Curie, Paris, France

Daniel Gourdin   Université de Paris 6, Paris, France

C. Grammatico   Università di Bologna, Bologna, Italy

Yûsaku Hamada   Kyoto, Japan

Kunihiko Kajitani   University of Tsukuba, Tsukuba, Japan

Tadayoshi Kano   University of Osaka, Toyonaka, Japan

P. Laubin†   University of Liège, Liège, Belgium

Waichiro Matsumoto   Ryukoku University, Otsu, Japan

Lorenzo Mencherini   Università di Firenze, Florence, Italy

Sae Miki   University of Osaka, Toyonaka, Japan

Minoru Murai   Ryukoku University, Otsu, Japan

M. K. Venkatesha Murthy   Università di Pisa, Pisa, Italy

---

†Deceased.

Takaaki Nagase   Ryukoku University, Otsu, Japan

Tatsuo Nishitani   Osaka University, Osaka, Japan

Michael Reissig   TU Bergakademie Freiberg, Freiberg, Germany

L. S. Schulman   Clarkson University, Potsdam, New York, U.S.A.

B. W. Schulze   Universität Potsdam, Potsdam, Germany

Sergio Spagnolo   Università di Pisa, Pisa, Italy

Jiro Takeuchi   Science University of Tokyo, Hokkaido, Japan

N. Tarkhanov   University of Potsdam, Potsdam, Germany

Mikio Tsuji   Kyoto Sangyo University, Kyoto, Japan

Ya-Guang Wang   Shanghai Jiao Tong University, Shanghai, P.R. China

Ingo Witt   University of Potsdam, Potsdam, Germany

Hideo Yamahara   Osaka Electro-Communication University, Osaka, Japan

The conference in honor of Jean Vaillant

BERNARD GAVEAU
Laboratoire Analyse et Physique Mathématique,
Université Pierre et Marie Curie, Paris, France

Since his thesis in 1964 prepared under the direction of J. Leray and A. Lichnerowicz the main theme of the mathematical work of Jean Vaillant has been the study of systems of hyperbolic or holomorphic partial differential equations. The basic example of an hyperbolic equation is the wave equation which is the mathematical description of wave propagation at finite velocity, as, for example, the propagation of small disturbances in fluids (the sound) or of electromagnetic waves in vacuum. Examples of hyperbolic systems include the Maxwell system for the propagation of electromagnetic waves, the Dirac system for the propagation of spinors and Einstein equations in general relativity. The wave equation is the fundamental example of a strictly hyperbolic equation, for which the propagation velocities are different and do not vanish. An approximation of the solutions of a strictly hyperbolic equation is the high frequency approximation or geometrical optics approximation : sound or light propagates essentially along the trajectories of the Hamilton Jacobi equation associated to the partial differential equation. One can say, in a rather unprecise manner, that « singularities are propagated along bicharacteristics », which, a posteriori, justifies the use of geometrical optics, the laws of reflexion and refraction. On the other hand, interference and diffraction phenomena show that sound or light can be described as fields which can be added, rather than particles, but again a good approximation of these phenomena is the propagation along bicharacteristics at least in the simplest situations. Dirac system is an example of a system with multiple characteristics with constant multiplicities. Maxwell system in a non isotropic medium, like a crystal, is an hyperbolic system with multiple characteristics, but their multiplicity is non constant : The velocities of propagation depends of the direction of propagation, but for special directions, some velocities may coincide. In this situation, the approximation of geometrical optics is no more valid : The propagation along bicharacteristics (or rays) is not a good description or approximation of the phenomenon, and indeed this can be checked experimentally. A light ray falling on certain crystals, is, in general refracted along a certain direction. Nevertheless for special incidence angles, corresponding to the geometry of the crystal, the ray is refracted, not along another ray, but on a whole conical surface. Until the end of 19th century, this experiment was the only proof of the electromagnetic nature of light, because all the other light propagation phenomena could be described by a wave equation, without the use of the complete Maxwell system (see [1], [2]).

1

In the beginning of the 1960's, strictly hyperbolic equations with simple characteristics (the velocities of propagation are distinct and non zero), are well understood (see [1], [3]). Around that time, Jean Vaillant begins a systematic study of equations or systems of equations which are non strict. In his thesis ([4], [5]), he introduces the notion of localization with respect to a factor of the characteristic determinant of a system with constant coefficients and he relates this notion to the equation of propagation along the bicharacteristics. This seems to be the first attempt to apply the method of localization, in particular using invariant factors. Following the article of Gärding, Kotake, Leray (Problème de Cauchy VI, [6]), J. Vaillant defines a new invariant associated to systems with double characteristics and gives an application to the Goursat problem [7], and to the localization for systems with variables coefficients and double characteristics : This is the first example of a Levi condition in this setting. J. Vaillant relates also the difference of the subcharacteristic polynomial, and the second coefficient of $\exp(-i\omega\varphi) \, P\,(x, D_x) \exp(i\omega\varphi)$ to the Lie derivative of the volume form along the bicharacteristics [8]. This is an important invariant, because it is well known that the existence and the regularity of the solution depend, in degenerate cases not only of the principal symbol but also on the lower order terms of the operator. This result was rederived by Duistermaat and Hörmander.

In 1973-74, J. Vaillant constructs the asymptotic expansion of the solution of an hyperbolic systems with characteristics of variable multiplicities [9]. He defines the localization of an hyperbolic system at a multiple points with application to conical refraction [9]. This work will be extended in 1978, when he constructs the parametrix for the Cauchy problem with multiple characteristics [10], in relation with the invariants of the system.

In [11], J. Vaillant studies the symmetrisation of localized hyperbolic systems and defines the notion of « reduced dimension » : The property of symmetry is proved in the case of a maximal reduced dimension. These last few years, he has continued to study the symmetric of strongly hyperbolic systems, in particular with T. Nishitani. If the reduced dimension of a system of rank m is not less than $\dfrac{m(m+1)}{2} - 2$, a constant coefficient systems is symmetrizable ([7], [8]). For a system with non constant coefficients, if at any point the reduced dimension is not less than $\dfrac{m(m+1)}{2} - 2$, the system is also symmetrizable with a regular symmetrizer [18]. Recently he has determined the multiple points according to the reduced dimension [19].

In 1982, in collaboration with D. Schiltz and C. Wagschal [12], J. Vaillant has studied the ramification of the Cauchy problem for a system in involution with triple characteristics. This problem reduces to the question of the singularities of integrals of holomorphic forms depending of parameters, on chains depending also of parameters. The problem is to determine the singularities of these integrals with respect to the parameters. The first systematic work in this direction was the article of J. Leray [13] in the algebraic case. J. Vaillant studies the ramification in the general holomorphic case, using a grassmann boundle [14]. Since 1987, J. Vaillant has started the problem of the classification of systems with constant multiplicities : definition of invariant Levi conditions, relations to the Cauchy problem in the C^∞ and Gevrey classes. For any system, he obtains systematically the Levi conditions [15], [16].

J. Vaillant has founded a research group and a seminar, which he has maintained, for more than thirty years, independent of fashions « mots d'ordre » and which survives in difficult conditions. Freedom of thought, which is a necessary condition for any creative work, is paid a very high price. Creation, scientific or artistic, cannot be judged according to economic or social criteria, measured in monetary values. Research is not a collective activity. The highly mysterious activity of thought can only be a personal activity. During all his career, Jean Vaillant, following the example of Leray, has tried to defend by his attitudes and his work, the values of scientific creation and intellectual independence.

For more than forty years, J. Vaillant has developed many collaborations with his Japanese and Italian friends, in particular Y. Hamada, Y. Ohya, K. Kajitani, T. Nishitani and S. Spagnolo, F. Colombini, A. Bove and E. Bernardi. He has also developed many european collaborations and he has created a european network of belgian, french, italian, and portugese universities. All his friends know that they can rely on his help and his advises.

acknowledgment : We are very grateful to Anne Durrande, Evelyne Guilloux and Maryse Loiseau for their help during the preparation of this conference. We also thank the Maison Européenne des Technologies, in particular Madame Muller for her help.

B. Gaveau
Laboratoire Analyse et Physique Mathématique
14 avenue Félix Faure
75015 PARIS

4

[1]   Courant-Hilbert : Methods of Mathematical Physics Vol II
        Interscience 1962

[2]   M. Born & E. Wolf : Principles of optics
        Pergamon Press 1980

[3]   J. Leray : Hyperbolic differential equations
        Lectures notes, Princeton 1950

[4]   J. Vaillant : Sur les discontinuités du tenseur de courbure en théorie
        d'Einstein-Schrödinger
        CR Acad Sci Paris – 10 juillet 1961, 30 octobre 1961, 15 janvier 1962

[5]   J. Vaillant : Caractéristiques multiples et bicaractéristiques des systèmes d'équations
        aux dérivées partielles linéaires et à coefficients constants
        Annales Institut Fourier 15 (1965) et 16 (1966)

[6]   L. Gärding, T. Kotake, J. Leray : Uniformation et développement asymptotique de la
        solution du problème de Cauchy linéaire à données holomorphes ; analogie avec la
        théorie des ondes asymptotiques et approchées (Problème de Cauchy I bis et VI)
        Bull. Sci. Math. France 92 1964, 263-361.

[7]   J. Vaillant : Données de Cauchy portées par une caractéristique double : rôle des
        bicaractéristiques
        J. Maths Pures et Appliquées 47 (1968), 1-40

[8]   J. Vaillant : Dérivée de Lie de la forme élément de volume le long des
        bicaractéristiques et polynôme sous-caractéristique de Gärding-Kotake-Leray
        CR Acad Sci Paris -10 mars 1969

[9]   J. Vaillant : Solutions asymptotiques d'un système à caractéristiques de multiplicité
        variable
        J. Maths Pures et Appliquées 53 (1974), 71-98

[10]   R. Berzin, J. Vaillant – Paramétrix à caractéristiques multiples
        Bull. Sci. Math 102 (1978), 287-294

[11]   J. Vaillant : Symétrisation de matrices localisées
        Annali della Scuola Normale Superiore di Pisa. Ser. IV, 5 (1978), 405-427

[12]   D. Schiltz, J. Vaillant, C. Wagschal : Problème de Cauchy ramifié
        J. Math. Pures et appliquées (1982)

[13]   J. Leray : Un complément au théorème de N.Nilsson sur les intégrales de formes
        différentielles à support singulier algébrique

Bull. Soc. Math. Fr 95 (1967), 313-374

[14]   J. Vaillant : Ramifications d'intégrales holomorphes
            J. Math. Pures et Appliquées 65 (1986), 343-402

[15]   J. Vaillant : Conditions d'hyperbolicité pour les systèmes
            Bull. Sci Math 114 (1990), 243-328

        Conditions de Levi   in Travaux en cours 48. Herman (1994)

        Analytic hyperbolic systems
            Pitman, Research Notes in Mathematics 349 (1996), 209-229

        Conditions invariantes sur les systèmes d'équations aux dérivées partielles et
        problème de Cauchy
            in complex Analysis and microlocal analysis, RIMS Kokyuroku 1090,
        p. 131-142
            Kyoto University (1999)

        Invariants des systèmes d'opérateurs différentiels et sommes formelles
        asymptotiques
            Japanese J. Math. 15 (1999), 1-153

[16]   J. Vaillant, G. Taglialatela : Conditions invariantes d'hyperbolicité et réduction des
        systèmes
            Bull. Sci. Math. 120 (1996), 19-97

[17]   J. Vaillant : Symétrie des opérateurs hyperboliques 4×4 et dimension réduite
            Annali della Scuola Normale Superiore di Pisa 29 (2000), p. 839-890.

[18]   J. Vaillant, T. Nishitani : Smoothy symmetrizable systems and the reduced
         dimensions
            Tsukuba Journal Vol. 25, n° 1, juin 2001, p. 165-177.

[19]   J. Vaillant, H. Delquié : Dimension réduite et valeurs propres multiples d'une matrice
        diagonalisable 4×4
            Bull. Sci. Math. 124 – 4 (2000).

Hyperbolic systems with nondegenerate characteristics

Tatsuo Nishitani

Department of Mathematics, Osaka University, Machikaneyama 1-16, Toyonaka Osaka, 560-0043, Japan

1   Introduction

In this note we discuss the stability of non degenerate characteristics of hyperbolic systems of general order under hyperbolic perturbations. We also study the well posedness of the Cauchy problem for hyperbolic systems of general order with non degenerate characteristics.

For first order systems we have introduced non degenerate characteristics in [7], [8]. According to this definition, simple characteristics are non degenerate and non degenerate double characteristics coincide with those studied in [2], [3], [4]. We adapt this definition for higher order systems with an obvious modification. For a system of general order, in a standard manner, we can associate a first order system and we prove in section 2 that non degenerate characteristics of the original system are also non degenerate characteristics of the same order for the corresponding first order system and vice versa (Proposition 2.1).

In section 3 we prove that one can not remove non degenerate characteristics by hyperbolic perturbations: any hyperbolic system which is sufficiently close to a hyperbolic system with a non degenerate characteristic must have a non degenerate characteristic of the same order. In particular, near a hyperbolic system with a non degenerate multiple characteristic, there is no strictly hyperbolic system. Moreover we show that, near a non degenerate characteristic of order r, the characteristics of order r form a smooth manifold of codimension $r(r+1)/2$ (Theorem 3.1). The same result was proved for analytic first order systems in [7] and for systems with non degenerate double characteristics in [2].

In section 4, the well posedness of the Cauchy problem for hyperbolic systems with non degenerate characteristics is discussed. We prove that a hyperbolic system of which every characteristic is non degenerate is smoothly symmetrized and hence the Cauchy problem is C^∞ well posed for arbitrary lower order terms (Theorem 4.1, Theorem 4.2). This generalizes a result in [2] (see also [4]) where the same result was proved for hyperbolic systems with constant coefficients with non degenerate double characteristics.

In the last section we restrict our considerations to 2×2 first order hyperbolic systems with constsnt coefficients with n independent variables. We show that if $n \le 3$

then such a system is a limit of strictly hyperbolic systems (Proposition 5.1). Contrary if $n > 3$, by [5] there is no strictly hyperbolic 2×2 system.

2   Non degenerate characteristics

Let $P(x)$ be a $m \times m$ matrix valued smooth function defined near $\bar{x} \in \mathbf{R}^n$. We assume that $P(x)$ is a polynomial in x_1 so that

$$(2.1) \qquad P(x) = \sum_{j=0}^{q} A_j(x')x_1^{q-j}$$

where $x' = (x_2, ..., x_n)$. We say that $P(x)$ is hyperbolic with respect to $\theta = (1, 0, ..., 0) \in \mathbf{R}^n$ if $\det A_0(x') \neq 0$ near $x' = \bar{x}'$ and

$$(2.2) \qquad \det P(x + \lambda\theta) = 0 \Longrightarrow \lambda \text{ is real.}$$

We call $\bar{x}$ is a characteristic of order r if

$$(2.3) \qquad \partial_x^\alpha (\det P)(\bar{x}) = 0, \quad \forall |\alpha| < r, \quad \partial_x^\alpha (\det P)(\bar{x}) \neq 0, \quad \exists |\alpha| = r.$$

Following [7], [8], we introduce non degenerate characteristics. To do so we first define the localization of $P(x)$ at a multiple characteristic $\bar{x}$ verifying

$$(2.4) \qquad \operatorname{Ker} P(\bar{x}) \cap \operatorname{Im} P(\bar{x}) = \{0\}.$$

Let $\dim \operatorname{Ker} P(\bar{x}) = r$ and let $v_1, ..., v_r$ be a basis for $\operatorname{Ker} P(\bar{x})$. Taking (2.4) into account we can choose linear forms $\ell_1, ..., \ell_r$ so that

$$\ell_i(\operatorname{Im} P(\bar{x})) = 0, \quad \ell_i(v_j) = \delta_{ij}$$

where δ_{ij} is the Kronecker's delta. Then we define a $r \times r$ matrix $P_{\bar{x}}(x)$ by

$$(2.5) \qquad (\ell_i(P(\bar{x} + \mu x)v_j))_{1 \leq i,j \leq r} = \mu[P_{\bar{x}}(x) + O(\mu)].$$

It is easy to see that the definition of $P_{\bar{x}}(x)$ is independent of the choice of $v_1, ..., v_r$, that is

$$P_{\bar{x}}(x)v = \sum_{j=1}^{r} [\lim_{\mu \downarrow 0} \mu^{-1}\ell_j(P(\bar{x} + \mu x)v)]v_j$$

is a well defined map from $\operatorname{Ker} P(\bar{x})$ to $\operatorname{Ker} P(\bar{x})$. We denote

$$(2.6) \qquad P_{\bar{x}} = \{P_{\bar{x}}(x) \mid x \in \mathbf{R}^n\} \subset M(r; \mathbf{R})$$

which is a subspace of $M(r; \mathbf{R})$, the space of all real $r \times r$ matrices. We first show

Lemma 2.1 *Assume (2.2) and let $\bar{x}$ be a characteristic verifying (2.4) with $\dim \operatorname{Ker} P(\bar{x}) = r$. Then we have*

$$(2.7) \qquad \det P(\bar{x} + \mu x) = \mu^r[c \det P_{\bar{x}}(x) + O(\mu)] \quad \text{with } c \neq 0.$$

Assume further that $\det P_{\bar{x}}(x) \neq 0$ then

$$(2.8) \qquad \det P_{\bar{x}}(\theta) \neq 0,$$
$$(2.9) \qquad \det P_{\bar{x}}(x + \lambda\theta) = 0 \Longrightarrow \lambda \text{ is real for all } x \in \mathbf{R}^n.$$

Proof: In view of (2.4) we can choose a non singular constant matrix T so that

$$T^{-1}P(\bar{x})T = \begin{pmatrix} 0 & 0 \\ 0 & G \end{pmatrix}$$

where G is a non singular $(m-r) \times (m-r)$ matrix. With $\tilde{P}(x) = T^{-1}P(x)T$ we write

$$\tilde{P}(\bar{x} + \mu x) = \tilde{P}(\bar{x}) + \mu \hat{P}(x) + O(\mu^2).$$

Denoting

$$\hat{P}(x) = \begin{pmatrix} \hat{P}_{11}(x) & \hat{P}_{12}(x) \\ \hat{P}_{21}(x) & \hat{P}_{22}(x) \end{pmatrix}$$

it is clear $\tilde{P}_{\bar{x}}(x) = \hat{P}_{11}(x)$ which follows from the definition. Since $\tilde{P}_{\bar{x}}(x) = P_{\bar{x}}(x)$ we have

(2.10) $$P_{\bar{x}}(x) = \hat{P}_{11}(x).$$

Note that

(2.11) $$\det P(\bar{x} + \mu x) = \det \tilde{P}(\bar{x} + \mu x) = \mu^r[\det G \; \det \hat{P}_{11}(x) + O(\mu)]$$

which shows the first assertion. To prove the second assertion suppose that $\det P_{\bar{x}}(\theta) = 0$ so that $\det P(\bar{x} + \mu\theta) = o(\mu^r)$ by (2.7). Since $\det P(x)$ is hyperbolic in the sense (2.2) it follows that

$$[\partial_x^\alpha \det P](\bar{x}) = 0, \quad \forall |\alpha| \leq r.$$

This implies $\det P_{\bar{x}}(x) = 0$ which is a contradiction. We turn to the third assertion. Since

$$\det P(\bar{x} + \mu(x + \lambda\theta)) = \mu^r[c \; \det P_{\bar{x}}(x + \lambda\theta) + O(\mu)]$$

if $\det P_{\bar{x}}(x + \lambda\theta) = 0$ has a non real root λ, then taking $\mu \neq 0$ sufficiently small the equation

$$c \; \det P_{\bar{x}}(x + \lambda\theta) + O(\mu) = 0$$

admits a non real root. This contradicts (2.2).

We now generalize the notion of non degenerate characteristics for $P(x)$ in (2.1) which is defined in [7], [8] for the case $q = 1$, making an obvious modification.

DEFINITION 2.1: We say that $\bar{x}$ is a non degenerate characteristic (of order r) of $P(x)$ if the following conditions are verified.

(2.12) $$\text{Ker}P(\bar{x}) \cap \text{Im}P(\bar{x}) = \{0\},$$

(2.13) $$\dim P_{\bar{x}} = r(r+1)/2, \quad r = \dim \text{Ker}P(\bar{x}),$$

(2.14) $$\det P_{\bar{x}}(\theta) \neq 0, \quad P_{\bar{x}}(\theta)^{-1}P_{\bar{x}}(x) \text{ is diagonalizable for every } x \in \mathbf{R}^n.$$

Simple characteristics verify (2.12)-(2.14) with $r = 1$ and hence non degenerate. Non degenerate double characteristics have a special feature:

Lemma 2.2 *Assume that* $\dim \mathrm{Ker} P(\bar{x}) = 2$ *and* $\mathrm{Ker} P(\bar{x}) \cap \mathrm{Im} P(\bar{x}) = \{0\}$. *Then a double characteristic* $\bar{x}$ *is non degenerate if and only if the rank of the Hessian of* $P(x)$ *at* $\bar{x}$ *is maximal, that is 3.*

Proof: Assume that $\mathrm{rank} \mathrm{Hess}_{\bar{x}} P = 3$. From Lemma 2.1 and the assumption it is clear that $\det P_{\bar{x}}(\theta) \neq 0$. Hence, by Lemma 2.1 again, $P_{\bar{x}}(\theta)^{-1} P_{\bar{x}}(x)$ has only real eigenvalues for every x. From Lemma 4.1 in [6] there exists a constant 2×2 matrix T such that $T^{-1}(P_{\bar{x}}(\theta)^{-1} P_{\bar{x}}(x))T$ is symmetric for every x so that one can write

$$(2.15) \qquad\qquad T^{-1}(P_{\bar{x}}(\theta)^{-1} P_{\bar{x}}(x))T$$
$$= \phi_1(x) \begin{pmatrix} 1 & 0 \\ 0 & 1 \end{pmatrix} + \phi_2(x) \begin{pmatrix} 1 & 0 \\ 0 & -1 \end{pmatrix} + \phi_3(x) \begin{pmatrix} 0 & 1 \\ 1 & 0 \end{pmatrix}$$

and obviously $P_{\bar{x}}(\theta)^{-1} P_{\bar{x}}(x)$ is diagonalizable for every x. Since the rank of the quadratic form $\det(P_{\bar{x}}(\theta)^{-1} P_{\bar{x}}(x))$ is 3 and hence $\phi_i(x)$, $i = 1, 2, 3$ are linearly independent. Thus it is clear that $\dim P_{\bar{x}} = 3$. Conversely assume $\dim P_{\bar{x}} = 3$ and $\det P_{\bar{x}}(\theta) \neq 0$. Since $P_{\bar{x}}(\theta)^{-1} P_{\bar{x}}(x)$ has only real eigenvalues for every x, (2.15) still holds. Then it follows that ϕ_i are linearly independent. Hence $\mathrm{rank} \mathrm{Hess}_{\bar{x}} P = 3$ by (2.7) again. This proves the assertion.

REMARK 2.1: Assume that $q = 1$ and $A_1(x')$ is symmetric in (2.1). Then (2.12) and (2.14) are always verified.

REMARK 2.2: By definition, the order of non degenerate characteristics never exceed m, the size of the matrix whatever q is.

To study $P(x)$ we consider the following $mq \times mq$ matrix valued function

$$\mathcal{P}(x) = x_1 + \begin{pmatrix} 0 & -I & & & \\ 0 & 0 & -I & & \\ & & & \ddots & \\ & & & & -I \\ A_q(x') & \cdots & & \cdots & A_1(x') \end{pmatrix} = x_1 + \mathcal{A}(x')$$

where I is the identity matrix of order m. It is clear that

$$(2.16) \qquad\qquad \det \mathcal{P}(x) = \det P(x).$$

Then the condition (2.2) implies that all eigenvalues of $\mathcal{A}(x')$ are real, equivalently

$$(2.17) \qquad\qquad \text{all eigenvalues of } \mathcal{P}(x) \text{ are real.}$$

Our aim in this section is to prove

Proposition 2.1 *Let* $\bar{x}$ *be a non degenerate characteristic of order* r *of* $P(x)$. *Then* $\bar{x}$ *is also a non degenerate characteristic of order* r *for* $\mathcal{P}(x)$ *and vice versa.*

Proof: Assume that $\bar{x}$ is a non degenerate characteristic of order r of $P(x)$. We first check

(2.18) $$\frac{\partial P}{\partial x_1}(\bar{x})\,(\mathrm{Ker}\,P(\bar{x})) = \mathrm{Ker}\,P(\bar{x}).$$

Let $v_1,...,v_r$ be a basis for $\mathrm{Ker}\,P(\bar{x})$ and take ℓ_i so that $\ell_i(\mathrm{Im}\,P(\bar{x})) = 0$ and $\ell_i(v_j) = \delta_{ij}$. Then by definition we have

$$P_{\bar{x}}(x) = \left(\ell_i[(\sum_{k=1}^{n} \frac{\partial P}{\partial x_k}(\bar{x})x_k)v_j] \right) = \sum_{k=1}^{n} \left(\ell_i[(\frac{\partial P}{\partial x_k}(\bar{x}))v_j] \right) x_k.$$

Hence

$$P_{\bar{x}}(\theta) = \left(\ell_i[(\frac{\partial P}{\partial x_1}(\bar{x}))v_j] \right).$$

Then $\det P_{\bar{x}}(\theta) \neq 0$ implies that

$$\mathrm{Ker}\,P(\bar{x}) \subset \frac{\partial P}{\partial x_1}(\bar{x})(\mathrm{Ker}\,P(\bar{x}))$$

and hence the result.

We note that

$$\mathrm{Ker}\,\mathcal{P}(x) = \{{}^t(u, x_1 u, ..., x_1^{q-1}u) \mid u \in \mathrm{Ker}\,P(x)\}$$

and $\dim\mathrm{Ker}\,\mathcal{P}(\bar{x}) = r$ of course. We next describe $\mathrm{Im}\,\mathcal{P}(x)$. Write

$$\phi_k(x) = \sum_{j=0}^{q-k} A_j(x')x_1^{q-j-k}$$

then it is easy to see that

$$\mathrm{Im}\,\mathcal{P}(x) = \{{}^t(w^{(1)}, ..., w^{(q-1)}, P(x)v - \sum_{k=1}^{q-1} \phi_k(x)w^{(k)}) \mid w^{(1)}, ..., w^{(q-1)}, v \in \mathbf{R}^m\}.$$

We first show that

(2.19) $$\mathrm{Ker}\,\mathcal{P}(\bar{x}) \cap \mathrm{Im}\,\mathcal{P}(\bar{x}) = \{0\}.$$

Let ℓ be a linear form on $\mathbf{R}^{mq}$. Writing $v = {}^t(v^{(1)}, ..., v^{(q)}) \in \mathbf{R}^{mq}$ one can write

$$\ell(v) = \sum_{j=1}^{q} \ell^{(j)}(v^{(j)})$$

where $\ell^{(j)}$ are linear forms on $\mathbf{R}^m$. Assume $\ell(\mathrm{Im}\,\mathcal{P}(\bar{x})) = 0$. This implies that

(2.20) $$\ell^{(j)}(\cdot) = \ell^{(q)}(\phi_j(\bar{x})\cdot), \quad 1 \leq j \leq q-1, \quad \ell^{(q)}(\mathrm{Im}\,P(\bar{x})) = 0.$$

Assume moreover $\ell(\mathrm{Ker}\,\mathcal{P}(\bar{x})) = 0$ so that

(2.21) $$\sum_{j=1}^{q} \ell^{(j)}(\bar{x}_1^{j-1}u) = \sum_{j=1}^{q-1} \ell^{(q)}(\bar{x}_1^{j-1}\phi_j(\bar{x})u) + \ell^{(q)}(\bar{x}_1^{q-1}u)$$

$$= \ell^{(q)}(\sum_{j=1}^{q-1} \bar{x}_1^{j-1}\phi_j(\bar{x})u + \bar{x}_1^{q-1}u) = 0.$$

From this, noting the identity

$$\sum_{j=1}^{q-1} x_1^{j-1}\phi_j(x) + x_1^{q-1} = \frac{\partial P}{\partial x_1}(x)$$

one gets

(2.22)
$$\ell^{(q)}(\frac{\partial P}{\partial x_1}(\bar{x})u) = 0, \quad \forall u \in \mathrm{Ker}P(\bar{x}).$$

From (2.18) it follows that $\ell^{(q)}(\mathrm{Ker}P(\bar{x})) = 0$. Since $\mathrm{Ker}P(\bar{x}) + \mathrm{Im}P(\bar{x}) = \mathbf{R}^m$ then (2.20) shows that $\ell^{(q)} = 0$ and hence $\ell = 0$. This proves that

(2.23)
$$\mathrm{Ker}\mathcal{P}(\bar{x}) + \mathrm{Im}\mathcal{P}(\bar{x}) = \mathbf{R}^{mq}$$

and hence (2.19).

We next examine (2.13), (2.14) for $\mathcal{P}(x)$. Let $U = {}^t(u, \bar{x}_1 u, ..., \bar{x}_1^{q-1}u) \in \mathrm{Ker}\mathcal{P}(\bar{x})$ where $u \in \mathrm{Ker}P(\bar{x})$. Consider $\mathcal{P}(x)U$:

$$\mathcal{P}(x)U = {}^t((x_1 - \bar{x}_1)u, (x_1 - \bar{x}_1)\bar{x}_1 u, ..., (x_1 - \bar{x}_1)\bar{x}_1^{q-2}u, *)$$
$$= {}^t(w^{(1)}, w^{(2)}, ..., w^{(q-1)}, *)$$

where the last component $*$ is

$$P(\bar{x}_1, x')u + (x_1\bar{x}_1^{q-1} - \bar{x}_1^q)u$$
$$= P(x)u + [P(\bar{x}_1, x') - P(x_1, x')]u + \bar{x}_1^{q-1}(x_1 - \bar{x}_1)u.$$

Now it is easy to see that this is equal to

(2.24)
$$P(x)u - \sum_{k=1}^{q-1} \phi_k(\bar{x})w^{(k)} + O((x_1 - \bar{x}_1)^2).$$

Let ℓ be a linear form on $\mathbf{R}^{mq}$ with $\ell(\mathrm{Im}\mathcal{P}(\bar{x})) = 0$. From (2.24) it follows that

$$\ell(\mathcal{P}(x)U) = \sum_{j=1}^{q-1} \ell^{(j)}(w^{(j)}) + \ell^{(q)}(P(x)u - \sum_{k=1}^{q-1} \phi_k(\bar{x})w^{(k)}) + O((x_1 - \bar{x}_1)^2)$$

(2.25)
$$= \ell^{(q)}(P(x)u) + O((x_1 - \bar{x}_1)^2)$$

by (2.20). Let us take $U_j = {}^t(u_j, \bar{x}_1 u_j, ..., \bar{x}_1^{q-1}u_j) \in \mathrm{Ker}\mathcal{P}(\bar{x})$ where $\{u_j\}$ is a basis for $\mathrm{Ker}P(\bar{x})$ and write

$$\frac{\partial P}{\partial x_1}(\bar{x})u_j = \sum_{k=1}^{r} a_{jk}u_k, \quad A = (a_{jk}).$$

Take $\tilde{\ell}_i$ so that

$$\tilde{\ell}_i(\mathrm{Im}P(\bar{x})) = 0, \quad \tilde{\ell}_i(u_j) = \delta_{ij}.$$

Let us take $\ell_i^{(q)}$

$$\ell_i^{(q)} = \sum_{k=1}^{r} b_{ik}\tilde{\ell}_k, \quad B = (b_{ik}) = {}^t A^{-1}$$

so that

$$\ell_i^{(q)}(\frac{\partial P}{\partial x_1}(\bar{x})u_j) = \sum_{k=1}^r b_{ik} \sum_{p=1}^r a_{jp}\tilde{\ell}_k(u_p) = \delta_{ij}.$$

We now define linear forms $\ell_i(\cdot) = {}^t(\ell_i^{(q)}(\phi_1(\bar{x})\cdot), ..., \ell_i^{(q)}(\phi_{q-1}(\bar{x})\cdot), \ell_i^{(q)}(\cdot))$ on $\mathbf{R}^{mq}$ then we have

(2.26) $\ell_i(\text{Im}\mathcal{P}(\bar{x})) = 0, \quad \ell_i(U_j) = \delta_{ij}$

as observed above. From (2.25) it follows that

$$\ell_i(\mathcal{P}(\bar{x} + \mu x)U_j) = \ell_i^{(q)}(P(\bar{x} + \mu x)u_j) + O(\mu^2)$$

$$= \sum_{k=1}^r b_{ik}\tilde{\ell}_k(P(\bar{x} + \mu x)u_j) + O(\mu^2)$$

$$= \mu[BP_{\bar{x}}(x) + O(\mu)].$$

Since $B = ({}^tA)^{-1} = P_{\bar{x}}(\theta)^{-1}$ we conclude that

(2.28) $\mathcal{P}_{\bar{x}}(x) = P_{\bar{x}}(\theta)^{-1}P_{\bar{x}}(x).$

Since $\mathcal{P}_{\bar{x}}(\theta) = I$ then (2.13) and (2.14) for $\mathcal{P}_{\bar{x}}(x)$ follow immediately.

Conversely assume (2.23). Let $\ell^{(q)}$ be a linear form on $\mathbf{R}^m$ such that $\ell^{(q)}(\text{Im}P(\bar{x})) = 0$, $\ell^{(q)}(\text{Ker}P(\bar{x})) = 0$ and define $\ell^{(j)}$, $1 \leq j \leq q-1$ by (2.20). Then we have $\ell(\text{Im}\mathcal{P}(\bar{x})) = 0$ and moreover (2.20) shows $\ell(\text{Ker}\mathcal{P}(\bar{x})) = 0$ and hence $\ell = 0$ by (2.23). Thus we have $\ell^{(q)} = 0$ which proves $\text{Ker}P(\bar{x}) + \text{Im}P(\bar{x}) = \mathbf{R}^m$ and hence (2.12). To check (2.13), (2.14) for $P(x)$ we note that $\text{Ker}P(\bar{x}) \cap \text{Im}P(\bar{x}) = \{0\}$ implies that

$$u \in \text{Ker}P(\bar{x}), \frac{\partial P}{\partial x_1}(\bar{x})u \in \text{Im}P(\bar{x}) \Longrightarrow u = 0.$$

Hence we have (2.18) again and thus (2.28). Then the rest of the proof is clear.

3   Stability of non degenerate characteristics

In this section we discuss the stability of non degenerate characteristics under hyperbolic perturbations.

Theorem 3.1 *Assume that $P(x)$ is a $m \times m$ real matrix valued smooth function of the form (2.1) verifying (2.2) in a neighborhood U of $\bar{x}$ and let $\bar{x}$ be a non degenerate characteristic of order r of P. Let $\tilde{P}(x)$ be another $m \times m$ real matrix valued smooth function of the form (2.1) verifying (2.2) which is sufficiently close to $P(x)$ in C^{q+2}, then $\tilde{P}(x)$ has a non degenerate characteristic of the same order close to $\bar{x}$. Moreover, near $\bar{x}$, the characteristics of order r are non degenerate and they form a smooth manifold of codimension $r(r+1)/2$. In particular, near $\bar{x}$ the set of characteristics of order r of $P(x)$ itself consists of non degenerate ones which form a smooth manifold of codimension $r(r+1)/2$.*

To prove Theorem 3.1, taking Proposition 2.1 into account, we study $P(x)$ of the form

(3.1) $$P(x) = x_1 + P^\#(x')$$

where we assume that

(3.2) $$\det P(x) = 0 \Longrightarrow x_1 \text{ is real near } x' = \bar{x}'.$$

This is equivalent to say that all eigenvalues of $P^\#(x')$ are real. We extend the stability result for P of the form (3.1) which is proved for real analytic $P(x)$, $\tilde{P}(x)$ in the case $r = m$ in [7] and in [2] when $\bar{x}$ is a non degenerate double characteristic.

Proposition 3.1 *Assume that $P(x)$ is a $m \times m$ real matrix valued smooth function verifying (3.2) and $\bar{x}$ is a non degenerate characteristic of order r of $P(x)$. Let $\tilde{P}(x)$ be another $m \times m$ real matrix valued smooth function of the form (3.1) verifying (3.2) which is sufficiently close to $P(x)$ in C^2. Then $\tilde{P}(x)$ has a non degenerate characteristic of the same order close to $\bar{x}$. Moreover, near $\bar{x}$, the characteristics of order r of $\tilde{P}(x)$ are non degenerate and form a smooth manifold of codimension $r(r+1)/2$. In particular, the characteristics of order r of $P(x)$ itself consists of non degenerate ones which form a smooth manifold of codimension $r(r+1)/2$.*

Proof: We first show that the proof is reduced to the case that P and $\tilde{P}$ are $r \times r$ matrix valued function. Without restrictions we may assume that $\bar{x} = 0$. As in the previous section, we take T so that one has

$$T^{-1}P(0)T = \begin{pmatrix} 0 & 0 \\ 0 & G \end{pmatrix}$$

where G is non singular. Denote $T^{-1}P(x)T$ and $T^{-1}\tilde{P}T$ by $P(x)$ and $\tilde{P}(x)$ again. Writing

$$P(x) = \begin{pmatrix} P_{11}(x) & P_{12}(x) \\ P_{21}(x) & P_{22}(x) \end{pmatrix}$$

we have

(3.3) $$P_{11}(x) = x_1 + \sum_{j=2}^{n} A_j x_j + O(|x|^2) = P_0(x) + O(|x|^2).$$

From the assumption $P_0(x)$ is diagonalizable for every x and $\{I, A_2, ..., A_n\}$ span a $r(r+1)/2$ dimensional subspace in $M(r; \mathbf{R})$. By Lemma 2.1 all eigenvalues of $P_0(x)$ are real then one can apply Theorem 3.4 in [9] and conclude that: there is a constant matrix S such that

$$S^{-1}(x_1 + \sum_{j=2}^{n} A_j x_j)S = x_1 + \sum_{j=2}^{n} \tilde{A}_j x_j$$

where $\tilde{A}_j$ are symmetric and $\{I, \tilde{A}_1, ..., \tilde{A}_n\}$ span $M^s(r; \mathbf{R})$, the space of all $r \times r$ real symmetric matrices. We still denote

$$\begin{pmatrix} S^{-1} & 0 \\ 0 & I \end{pmatrix} P(x) \begin{pmatrix} S & 0 \\ 0 & I \end{pmatrix}, \quad \begin{pmatrix} S^{-1} & 0 \\ 0 & I \end{pmatrix} \tilde{P}(x) \begin{pmatrix} S & 0 \\ 0 & I \end{pmatrix}$$

by $P(x)$ and $\tilde{P}(x)$ again so that writing

$$P(x) = \begin{pmatrix} P_{11}(x) & P_{12}(x) \\ P_{21}(x) & P_{22}(x) \end{pmatrix}$$

we may assume that

(3.4)
$$P_{11}(x) = x_1 I + \sum_{j=2}^{n} A_j x_j + O(|x|^2)$$

where

(3.5)
$$\{I, A_1, ..., A_n\} \text{ span } M^s(r; \mathbf{R}).$$

Let $\{F_1, F_2, ..., F_k\}$, $F_1 = I$ be a basis for $M^s(r; \mathbf{R})$ where $k = r(r+1)/2$. Writing

$$x_1 I + \sum_{j=2}^{n} A_j x_j = \sum_{j=1}^{k} F_j \ell_j(x)$$

we make a linear change of coordinates $\tilde{x}_j = \ell_j(x)$, $j = 1, ..., n$ so that denoting $x_j = \tilde{x}_j$, $1 \le j \le k$ again and $(\tilde{x}_{k+1}, ..., \tilde{x}_n) = (y_1, ..., y_l)$ we have

(3.6)
$$P_{11}(x, y) = \sum_{j=1}^{k} F_j x_j + O((|x| + |y|)^2).$$

Note that the coefficient of x_1 in $\tilde{P}_{11}(x, y)$ is the identity matrix I. We now prepare the next lemma.

Lemma 3.3 *Let $P(x)$ be a $m \times m$ matrix valued C^∞ function defined near $x = 0$. With a blocking*

$$P(0) = \begin{pmatrix} A_{11} & A_{12} \\ A_{21} & A_{22} \end{pmatrix}$$

we assume that A_{11} and A_{22} has no common eigenvalue. Then there is $\epsilon = \epsilon(A_{11}, A_{22}) > 0$ such that if $\|A_{21}\| + \|A_{12}\| < \epsilon$ then one can find a smooth matrix $T(x)$ defined in $|x| < \epsilon$ such that

$$T(x)^{-1} P(x) T(x) = \begin{pmatrix} \hat{P}_{11}(x) & 0 \\ 0 & \hat{P}_{22}(x) \end{pmatrix}$$

where $T(x) = I + T_1(x)$ and $\|T_1(0)\| \to 0$ as $\|A_{21}\| + \|A_{12}\| \to 0$.

Proof: We first show that there are G_{12}, G_{21} such that

(3.7)
$$\begin{pmatrix} A_{11} & A_{12} \\ A_{21} & A_{22} \end{pmatrix} \begin{pmatrix} I & G_{12} \\ G_{21} & I \end{pmatrix} = \begin{pmatrix} I & G_{12} \\ G_{21} & I \end{pmatrix} \begin{pmatrix} A_{11} + X_{11} & 0 \\ 0 & A_{22} + X_{22} \end{pmatrix}$$

provided $\|A_{12}\| + \|A_{21}\|$ is small. The equation (3.7) is written as

$$\begin{pmatrix} A_{11} + A_{12}G_{21} & A_{11}G_{12} + A_{12} \\ A_{21} + A_{22}G_{21} & A_{21}G_{12} + A_{22} \end{pmatrix} = \begin{pmatrix} A_{11} + X_{11} & G_{12}A_{22} + G_{12}X_{22} \\ G_{21}A_{11} + G_{21}X_{11} & A_{22} + X_{22} \end{pmatrix}.$$

This gives $A_{12}G_{21} = X_{11}$, $A_{21}G_{12} = X_{22}$. Plugging these relations into the remaining two equations, we have

$$A_{12} + A_{11}G_{12} = G_{12}A_{22} + G_{12}A_{21}G_{12}$$
$$A_{21} + A_{22}G_{21} = G_{21}A_{11} + G_{21}A_{12}G_{21}.$$

Let us set

$$F_1(G_{12}, G_{21}, A_{12}, A_{21}) = G_{12}A_{22} - A_{11}G_{12} + G_{12}A_{21}G_{12} - A_{12}$$
$$F_2(G_{12}, G_{21}, A_{12}, A_{21}) = G_{21}A_{11} - A_{22}G_{12} + G_{21}A_{12}G_{21} - A_{21}$$

then the equations become

(3.8)
$$\begin{cases} F_1(G_{12}, G_{21}, A_{12}, A_{21}) = 0 \\ F_2(G_{12}, G_{21}, A_{12}, A_{21}) = 0. \end{cases}$$

It is well known that
$$\frac{\partial(F_1, F_2)}{\partial(G_{12}, G_{21})}(0, 0, 0, 0)$$
is non singular if A_{11} and A_{22} have no common eigenvalue. Then by the implicit function theorem there exist smooth $G_{12}(A_{12}, A_{21})$ and $G_{21}(A_{12}, A_{21})$ defined in $\|A_{12}\| + \|A_{21}\| < \delta$ with $G_{12}(0, 0) = 0$, $G_{21}(0, 0) = 0$ verifying (3.8). This proves the assertion.

We look for $T(x)$ in the form

$$T(x) = T_0 + T_1(x), \quad T_0(x) = \begin{pmatrix} I & G_{12} \\ G_{21} & I \end{pmatrix}, \quad T_1(0) = 0.$$

The equation which is verified by $T(x)$ is:

(3.9)
$$(P_0 + P_1(x))(T_0 + T_1(x)) = (T_0 + T_1(x))(\tilde{P}_0 + \tilde{P}_1(x))$$

where $P_0 = P(0)$, $P_0 T_0 = T_0 \tilde{P}_0$ and

$$\tilde{P}_1(x) = \begin{pmatrix} \tilde{P}_{11}(x) & 0 \\ 0 & \tilde{P}_{22}(x) \end{pmatrix}.$$

Recall that

$$\tilde{P}_0 = \begin{pmatrix} A_{11} + A_{12}G_{21} & 0 \\ 0 & A_{22} + A_{21}G_{12} \end{pmatrix}, \quad P_1(x) = \begin{pmatrix} P_{11}(x) & P_{12}(x) \\ P_{21}(x) & P_{22}(x) \end{pmatrix}.$$

Look for $T_1(x)$ in the form

$$T_1(x) = \begin{pmatrix} 0 & T_{12}(x) \\ T_{21}(x) & 0 \end{pmatrix}.$$

Equating the off diagonal entries of both sides of (3.9) we get

(3.10)
$$\begin{cases} A_{11}T_{12} + P_{12}(x) + P_{11}(x)G_{12} + P_{11}(x)T_{12} \\ \quad = (G_{12} + T_{12})\tilde{P}_{22}(x) + T_{12}(A_{22} + A_{21}G_{12}) \\ A_{22}T_{21} + P_{21}(x) + P_{22}(x)G_{21} + P_{22}(x)T_{21} \\ \quad = (G_{21} + T_{21})\tilde{P}_{11}(x) + T_{21}(A_{11} + A_{12}G_{21}). \end{cases}$$

On the other hand, equating the diagonal entries of both sides we have

(3.11)
$$\begin{cases} \tilde{P}_{11}(x) = A_{12}T_{21} + P_{11}(x) + P_{12}(x)(G_{21} + T_{21}) \\ \tilde{P}_{22}(x) = A_{21}T_{12} + P_{22}(x) + P_{21}(x)(G_{12} + T_{12}). \end{cases}$$

Plugging (3.11) into (3.10) we obtain

$$F_1(T_{12}, x) = A_{11}T_{12} - T_{12}(A_{22} + A_{21}G_{12}) + P_{11}(x)G_{12} + P_{12}(x) + P_{11}(x)T_{12}$$
$$-(G_{12} + T_{12})(A_{21}T_{12} + P_{21}(x)[G_{12} + T_{12}] + P_{22}(x)) = 0$$

and

$$F_2(T_{21}, x) = A_{22}T_{21} - T_{21}(A_{11} + A_{12}G_{21}) + P_{22}(x)G_{21} + P_{21}(x) + P_{22}(x)T_{12}$$
$$-(G_{21} + T_{21})(A_{12}T_{21} + P_{12}(x)[G_{21} + T_{21}] + P_{11}(x)) = 0.$$

Since

$$F_1(T_{12}, 0) = A_{11}T_{12} - T_{12}A_{22}, \quad F_2(T_{21}, 0) = A_{22}T_{21} - T_{21}A_{11}$$

when $A_{21} = 0$, $A_{12} = 0$, $x = 0$, it is clear that

$$\frac{\partial F_1}{\partial T_{12}}(0,0), \quad \frac{\partial F_2}{\partial T_{21}}(0,0)$$

are non singular if $\|A_{12}\| + \|A_{21}\|$ is small. Then by the implicit function theorem there exist smooth $T_{12}(x)$ and $T_{21}(x)$ with $T_{12}(0) = 0$, $T_{21}(0) = 0$ such that

$$F_1(T_{12}(x), x) = 0, \quad F_2(T_{21}(x), x) = 0.$$

This proves the assertion.

We return to the proof of Proposition 3.1. Since $\tilde{P}(x, y)$ is sufficiently close to $P(x, y)$ and

$$P(0,0) = \begin{pmatrix} 0 & 0 \\ 0 & G \end{pmatrix}, \quad \det G \neq 0$$

one can apply Lemma 3.3 to $\tilde{P}(x, y)$ and find $G(x, y)$ such that

(3.12)
$$G(x, y)^{-1}\tilde{P}(x, y)G(x, y) = \begin{pmatrix} \tilde{P}_{11}(x, y) & 0 \\ 0 & \tilde{P}_{22}(x, y) \end{pmatrix}.$$

Denote $G(x, y)^{-1}P(x, y)G(x, y)$ and $G(x, y)^{-1}\tilde{P}(x, y)G(x, y)$ by $P(x, y)$ and $\tilde{P}(x, y)$ again. We summarize our arguments in

Proposition 3.2 *Assume that P_{orig} and $\tilde{P}_{orig}$ verify the assumption in Proposition 3.1. Then we may assume that P_{orig} and $\tilde{P}_{orig}$ have the form*

$$\tilde{P}(x, y) = \begin{pmatrix} \tilde{P}_{11}(x, y) & 0 \\ 0 & \tilde{P}_{22}(x, y) \end{pmatrix}, \quad P(x, y) = \begin{pmatrix} P_{11}(x, y) & P_{12}(x, y) \\ P_{21}(x, y) & P_{22}(x, y) \end{pmatrix}$$

with

$$P_{11}(x,y) = \sum_{j=1}^{k} A_j x_j + \sum_{j=1}^{l} B_j y_j + R(x,y), \quad R(x,y) = O((|x|+|y|)^2)$$

where the following properties are verified: for any neighborhood U of the origin there is a neighborhood $W \subset U$ of the origin such that for any $\epsilon > 0$ one can find $\tilde{\epsilon} > 0$ so that if $|\tilde{P}_{orig} - P_{orig}|_{C^2(U)} < \tilde{\epsilon}$ then we have

(3.13)
$$|\tilde{P}_{11}(x,y) - P_{11}(x,y)|_{C^2(W)} < \epsilon,$$

(3.14)
$$|\sum_{j=1}^{k} A_j x_j + \sum_{j=1}^{l} B_j y_j - \sum_{j=1}^{k} F_j x_j| < C\epsilon(|x|+|y|).$$

Moreover one has

$$\det(\lambda + \tilde{P}_{11}(x,y)) = 0 \Longrightarrow \lambda \text{ is real.}$$

Proof: Since $P(x,y)$ and $\tilde{P}(x,y)$ are obtained from P_{orig} and $\tilde{P}_{orig}$ by a smooth change of basis and a linear change of coordinates then (3.13) is clear. Let us recall

$$G(x,y) = \begin{pmatrix} I & G_{12}(x,y) \\ G_{21}(x,y) & I \end{pmatrix}$$

which verifies (3.12) where $\|G_{12}(0,0)\| + \|G_{21}(0,0)\|$ becomes as small as we please if $\tilde{\epsilon}$ is small. Hence $G(x,y)$ is enough close to the identity and then (3.14) follows from (3.6). Note that

$$\det(\lambda + \tilde{P}_{orig}) = \det(\lambda + \tilde{P}_{11}(x,y))\det(\lambda + \tilde{P}_{22}(x,y)).$$

Then the last assertion follows immediately.

We proceed to the next step. Write

(3.15)
$$\tilde{P}_{11}(x,y) = \tilde{P}_{11}(0,y) + (\tilde{\phi}_j^i(x,y))_{1 \leq i,j \leq r}.$$

Let us denote

$$\tilde{\phi}_j^i(x,y) = \phi_j^i(x) + t_j^i(x,y)$$

where

$$\sum_{j=1}^{k} F_j x_j = (\phi_j^i(x))_{1 \leq i,j \leq r}.$$

Lemma 3.4 *Assume that $|\tilde{P}_{11}(x,y) - P_{11}(x,y)|_{C^2(W)} < \epsilon$ and $\{(x,y) \mid |x|, |y| < \epsilon\} \subset W$. Then for $|x|, |y| < \epsilon$ we have*

$$|t_j^i(x,y)| < C|x|, \quad |\partial_{x_\mu} t_j^i(x,y)| < C\epsilon, \quad \mu = 1,...,k.$$

Proof: Write

$$\tilde{P}_{11}(x,y) = \tilde{P}_{11}(0,y) + \sum_{j=1}^{k} \tilde{A}_j(y)x_j + \tilde{R}(x,y), \quad \tilde{R}(x,y) = O(|x|^2)$$

so that

(3.16)
$$T = (t_j^i(x,y)) = \sum_{j=1}^{k} \tilde{A}_j(y)x_j - \sum_{j=1}^{k} F_j x_j + \tilde{R}(x,y).$$

Noting $\partial_{x_j}\tilde{P}_{11}(0,y) = \tilde{A}_j(y)$, $\partial_{x_j}P_{11}(0,y) = A_j + \partial_{x_j}R(0,y)$ and

$$|\partial_{x_j}R(0,y)| \le C|y| \le C\epsilon \quad \text{if } |y| < \epsilon$$

with C independent of $\tilde{P}$, one gets

(3.17)
$$|\tilde{A}_j(y) - A_j| \le C\epsilon \quad \text{if } |y| < \epsilon.$$

Now it is clear that
(3.18)
$$|\tilde{A}_j(y) - F_j| \le C'\epsilon \quad \text{if } |y| < \epsilon$$

because of (3.14) and (3.17). On the other hand from

$$P_{11}(0,y) = \sum_{j=1}^{l} B_j y_j + R(0,y)$$

and (3.14) it follows that

$$|P_{11}(0,y)| \le C\epsilon|y| + C|y|^2 \le C\epsilon|y| \quad \text{if } |y| < \epsilon.$$

Moreover $|\tilde{P}_{11}(0,y) - P_{11}(0,y)|_{C^2(W)} < \epsilon$ shows

(3.19)
$$|\tilde{P}_{11}(0,y)| < \epsilon + C\epsilon|y| < C'\epsilon \quad \text{if } |y| < \epsilon.$$

We now estimate $T(x,y) = (t_j^i(x,y))$ and $\partial_{x_j}T(x,y)$. Note that $|\partial_{x_j}\tilde{R}(x,y)| \le C|x|$ since $\partial_{x_j}\tilde{R}(0,y) = 0$ and $|\partial_x^\alpha \tilde{R}(x,y)| \le C$ for $|\alpha| = 2$ with C independent of $\tilde{P}$. Then by (3.16) and (3.18) one sees

(3.20)
$$|T(x,y)| \le C\epsilon|x| + C|x|^2 \le C'\epsilon|x| \quad \text{if } |x| < \epsilon,$$
$$|\partial_{x_j}T(x,y)| \le C\epsilon + C|x| \le C'\epsilon \quad \text{if } |x|, |y| < \epsilon$$

which proves the assertion.

Let us study the map

$$\Phi : B_a \ni x \mapsto (\tilde{\phi}_j^i(x,y))_{i \ge j} \in \mathbf{R}^k$$

where $B_a = \{x \in \mathbf{R}^k \mid |x| \le a\}$. Let

$$A : \mathbf{R}^k \ni x \mapsto (\phi_j^i(x))_{i \ge j} \in \mathbf{R}^k$$

which is a linear transformation on $\mathbf{R}^k$. Since $\phi_j^i(x)$, $i \geq j$ are linearly independent, A is non singular. From Lemma 3.4 one can choose $a > 0$ so that

$$|A^{-1}\Phi'_x(x,y) - I| < 1/2 \quad \text{if } |x|, |y| < a.$$

Let us write $\tilde{P}_{11}(0,y) = (b_j^i(y))$ and we apply the implicit function theorem to

$$\tilde{\phi}_j^i(x,y) = \theta_j^i - b_j^i(y), \quad i \geq j$$

where $\theta \in \mathbf{R}^k$, $y \in \mathbf{R}^k$. Then there exist $a_1 > 0$, $a_2 > 0$ and a smooth $g(y,\theta)$ defined in $\theta \in B_{a_1}$, $y \in B_{a_2}$ such that

(3.21) $$\tilde{\phi}_j^i(g(y,\theta),y) = \theta_j^i - b_j^i(y), \quad i \geq j.$$

Note that
(3.22) $$|g(y,\theta)| < C\epsilon$$

if $|y|, |\theta| < \epsilon$ because of (3.19). Set

(3.23) $$(\psi_j^i(y,\theta)) = \tilde{P}_{11}(g(y,\theta),y)$$

then from (3.15) and (3.21) it follows that

(3.24) $$\psi_j^i(y,\theta) = \theta_j^i, \quad i \geq j.$$

Let us write

$$\psi_j^i(y,\theta) = c_j^i(y) + \chi_j^i(y,\theta)$$

where $c_j^i(y) = \psi_j^i(y,0)$ and $\chi_j^i(y,\theta) = O(|\theta|)$. We show that $c_q^p(y) = 0$ for $p < q$. This, together with (3.24), implies
(3.25) $$\tilde{P}_{11}(g(y,0),y) = 0.$$

Let us put $h(\lambda) = \det(\lambda + \tilde{P}_{11}(g(y,\theta),y))$. From Proposition 3.2 it follows that $h(\lambda) = 0$ implies λ is real. Take $\theta_j^i = 0$ for $i \geq j$ unless $(i,j) = (q,p)$. Then one has

$$h(\lambda) = \lambda^{m-2}[\lambda^2 - \theta_p^q(c_q^p(y) + \chi_q^p(y,\theta))].$$

If $c_q^p(y) \neq 0$ then $h(\lambda) = 0$ has a non real root for sufficiently small θ because $\chi_q^p(y,0) = 0$. This is a contradiction and hence the assertion.

Here we extend Proposition 3.1 in [7]. Let

$$F(x) = \sum_{j=1}^k F_j x_j$$

where $F_1 = I$ and $\{F_1, ..., F_k\}$ be a basis for $M^s(m; \mathbf{R})$ and hence $k = m(m+1)/2$.

Proposition 3.3 *Assume that $P(x)$ is a $m \times m$ real matrix valued smooth function defined in a neighborhood of the origin of $\mathbf{R}^n$. Assume that all eigenvalues of $P(x)$ are real and*

(3.26) $$\sum_{j=1}^k \frac{\partial P}{\partial x_j}(0)x_j$$

is sufficiently close to $F(x)$. Then there is a $\delta > 0$ such that $P(x)$ is diagonalizable for every x with $|x| < \delta$.

Proof: It is enough to repeat the proof of Proposition 3.1 in [7] with a slight modification. Let us write

$$P(\omega + x) = P(\omega) + Q(x, \omega)$$

so that $Q(0, \omega) = 0$. For $T \in O(m)$, an orthogonal matrix of order m we consider

$$^tTP(\omega + x)T = {}^tTP(\omega)T + {}^tTQ(x, \omega)T$$
$$= P^T(\omega) + Q^T(x, \omega).$$

Denoting $Q^T(x, \omega) = (\phi_j^i(x, \omega; T))_{1 \leq i, j \leq m}$ we show that there exist a $\delta > 0$ and a neighborhood W of the origin of $\mathbf{R}^k$ such that with $x = (x_a, x_b)$, $x_a = (x_1, ..., x_k)$, $x_b = (x_{k+1}, ..., x_n)$

$$W \ni x_a \mapsto (\phi_j^i(x, \omega; T))_{i \geq j} \in \mathbf{R}^k$$

is a diffeomorphism from W into $\{y \in \mathbf{R}^k \mid |y| < \delta\}$ for every $T \in O(m)$ and every x_b, ω with $|x_b|, |\omega| < \delta$. To see this we write

$$Q(x, \omega) = P(x + \omega) - P(\omega)$$
$$= \sum_{j=1}^{k} F_j x_j + \sum_{j=1}^{k} \left(\frac{\partial P}{\partial x_j}(\omega) - F_j \right) x_j + \sum_{j=k+1}^{n} \frac{\partial P}{\partial x_j}(\omega) x_j + \tilde{R}(x, \omega)$$
$$= \sum_{j=1}^{k} F_j x_j + R(x, \omega)$$

then it is clear that for any $\epsilon > 0$ one can find $\delta' > 0$ such that

(3.27) $$\|R(x, \omega)\| \leq \epsilon |x|$$

if $|x|, |\omega| < \delta'$ and (3.26) is sufficiently close to $F(x)$. Let us study

$$Q^T(x, \omega) = \sum_{j=1}^{k} F_j^T x_j + R^T(x, \omega)$$
$$= \sum_{j=1}^{k} \ell_j(x_a; T) F_j + R^T(x, \omega)$$

where $\ell_j(x_a; T)$ are linear in x_a. Since $O(m) \subset \mathbf{R}^{m^2}$ is compact it is clear that we have

$$\left| \frac{\partial(\ell_1, ..., \ell_k)}{\partial(x_1, ..., x_k)}(x_a; T) \right| \geq c > 0$$

with some $c > 0$ for every $T \in O(m)$. In view of (3.27), taking $\epsilon > 0$ so small we conclude that

$$\left| \frac{\partial((\phi_j^i)_{i \geq j})}{\partial(x_a)}(0, 0, 0; T) \right| \geq c' > 0$$

with some $c' > 0$ for every $T \in O(m)$. By the implicit function theorem and the compactness of $O(m)$ there exists a smooth $x_a(y_a, x_b, \omega; T)$ defined in $|y_a|, |x_b|, |\omega| < \delta$. and $T \in O(m)$ such that

$$\phi_j^i(x_a(y_a, x_b, \omega; T), x_b, \omega; T) = y_j^i \quad \text{for} \quad i \geq j$$

where we have set $y_a = (y_j^i)_{i \geq j} \in \mathbf{R}^k$. This proves the assertion.

We now show that $P(\omega)$ is diagonalizable for every $\omega \in \mathbf{R}^n$ with $|\omega| < \delta$. Take $T \in O(m)$ so that

(3.28)
$$P^T(\omega) = (\bigoplus_{i=1}^s \lambda_i I_{r_i}) + (A_{ij})_{1 \leq i, j \leq s}$$

where $\{\lambda_i\}$ are different from each other and A_{ij} are $r_i \times r_j$ matrices such that $A_{ij} = 0$ if $i > j$ and A_{ii} are upper triangular with zero diagonals. Let us set

$$J = \bigcup_{p=1}^{s-1} \{(i, j) \mid r_p < i \leq m, r_{p-1} < j \leq r_p\}$$

where $r_0 = 0$. As observed above one can take $((y_j^i)_{i \geq j}, x_b)$ as a new system of local coordinates around the origin of $\mathbf{R}^n$. Denote

$$y_{II} = (y_j^i)_{(i,j) \in J}, \quad y_a = (y_I, y_{II})$$

and, putting $y_{II} = 0$, $x_b = 0$, consider

$$\det(\lambda + P(\omega + x)) = \det(\lambda + P^T(\omega + x)) = \prod_{i=1}^s \det(\lambda + K_i(y_I, \omega; T))$$

where

$$K_i(y_I, \omega; T) = \lambda_i I_{r_i} + A_{ii} + (\phi_q^p(y_I, \omega; T))_{s_{i-1} < p, q \leq s_i}$$

with $s_i = r_1 + \cdots + r_i$, $s_0 = 0$. Note that we have

$$\phi_q^p(y_I, \omega; T) = y_q^p \quad \text{if} \quad p \geq q, \quad \phi_q^p(0, \omega; T) = 0.$$

Hence applying the same arguments as in the proof of Lemma 3.4 in [7] we get $A_{ii} = 0$. Then from Lemma 3.5 in [7] it follows that $P^T(\omega)$ is diagonalizable. Since ω, $|\omega| < \delta$ is arbitrary we get the assertion.

We now prove that near $(0, 0)$ the characteristics of order r of $\tilde{P}(x, y)$ form a smooth manifold given by $x = g(y, 0)$. Let $(\bar{x}, \bar{y})$ be a characteristic of order r of $\tilde{P}(x, y)$ close to $(0, 0)$. Then it is clear that $(\bar{x}, \bar{y})$ is a characteristic of the same order for $\tilde{P}_{11}(x, y)$ because $\det \tilde{P}_{22}(x, y) \neq 0$ near $(0, 0)$. Recalling that $\tilde{P}_{11}(x, y)$ has the form

$$\tilde{P}_{11}(x, y) = x_1 + \tilde{P}_{11}^{\#}(x^{\#}, y), \quad x^{\#} = (x_2, ..., x_k)$$

we see that $\det \tilde{P}_{11}(x_1, \bar{x}^{\#}, \bar{y}) = (x_1 - \bar{x}_1)^r$ and hence

$$\det(\lambda + \tilde{P}_{11}(\bar{x}, \bar{y})) = \lambda^r.$$

Thus the zero is an eigenvalue of order r of $\tilde{P}_{11}(\bar{x}, \bar{y})$. On the other hand Proposition 3.2 gives

(3.29)
$$\left| \frac{\partial \tilde{P}_{11}}{\partial x_j}(0) - F_j \right|, \quad \left| \frac{\partial \tilde{P}_{11}}{\partial y_j}(0) \right| < C\epsilon.$$

Then one can apply Proposition 3.3 to conclude that $\tilde{P}(\bar{x}, \bar{y})$ is diagonalizable. This shows that

$$\tilde{P}_{11}(\bar{x}, \bar{y}) = 0$$

and hence $\theta_j^i = 0$, $i \geq j$. Then one gets $\bar{x} = g(\bar{y}, 0)$.

Finally we show that the characteristics $(g(y, 0), y)$ are non degenerate. From (3.25) we have

$$\tilde{P}(g(y, 0), y) = \begin{pmatrix} 0 & 0 \\ 0 & \tilde{P}_{22}(g(y, 0), y) \end{pmatrix}$$

and hence

(3.30) $$\operatorname{Ker} \tilde{P}(g(y, 0), y) \cap \operatorname{Im} \tilde{P}(g(y, 0), y) = \{0\}.$$

It is also clear that $\tilde{P}_{(g(y,0),y)}(x, y)$ is given by

$$\sum_{j=1}^{k} \frac{\partial \tilde{P}_{11}}{\partial x_j}(g(y, 0), y) x_j + \sum_{j=1}^{l} \frac{\partial \tilde{P}_{11}}{\partial y_j}(g(y, 0), y) y_j.$$

On the other hand since $|\tilde{P}_{11} - P_{11}|_{C^2(W)} < \epsilon$ it follows from Proposition 3.3 and (3.22) that

(3.31) $$\left| \frac{\partial \tilde{P}_{11}}{\partial x_j}(g(y, 0), y) - F_j \right|, \quad \left| \frac{\partial \tilde{P}_{11}}{\partial y_j}(g(y, 0), y) \right| < C\epsilon$$

if $|y| < \epsilon$. This clearly shows that

(3.32) $$\dim \tilde{P}_{(g(y,0),y)} = \frac{r(r+1)}{2}.$$

To finish the proof, taking $\tilde{P}_{(g(y,0),y)}(\theta) = I$ into account, it is enough to show that $\tilde{P}_{(g(y,0),y)}(x, y)$ is diagonalizable for every (x, y). Note that from Lemma 2.1 all eigenvalues of $\tilde{P}_{(g(y,0),y)}(x, y)$ are real. Then from Proposition 3.3 and (3.31) it follows that $\tilde{P}_{(g(y,0),y)}(x, y)$ is diagonalizable for every (x, y) near $(0, 0)$ and hence for all (x, y).

4 Well posedness of the Cauchy problem

Let us study a system

(4.1) $$P(x, D) = \sum_{|\alpha| \leq q} A_\alpha(x) D^\alpha, \quad D_j = \frac{1}{i} \frac{\partial}{\partial x_j}$$

where $A_\alpha(x)$ are $m \times m$ matrix valued smooth function defined in a neighborhood Ω of the origin of $\mathbf{R}^n$. We assume that $x_1 = const.$ are non characteristic and without restrictions we may assume that

(4.2) $$A_{(q,0,\ldots,0)}(x) = I$$

the identity matrix of order m. We are concerned with the following Cauchy problem:.

(4.3) $$\begin{cases} P(x, D)u = f, & \operatorname{supp} f \subset \{x_1 \geq 0\} \\ \operatorname{supp} u \subset \{x_1 \geq 0\} \end{cases}$$

Let $P_q(x, \xi)$ be the principal symbol of $P(x, D)$:

$$P_q(x, \xi) = \sum_{|\alpha|=q} A_\alpha(x)\xi^\alpha$$

and we assume that

(4.4) $\det P_q(x, \xi) = 0 \Longrightarrow \xi_1$ is real $\forall x \in \Omega, \forall \xi' = (\xi_2, ..., \xi_n) \in \mathbf{R}^{n-1}$.

We prove the following result which extends those in [2], [4].

Theorem 4.1 *Assume that every characteristic over $(0, \xi')$, $|\xi'| = 1$ of $P_q(x, \xi)$ is at most double and non degenerate. Then the Cauchy problem for $P(x, D)$ is C^∞ well posed near the origin for arbitrary lower order terms. Moreover if $\tilde{P}(x, D)$ is another system of the form (4.1) verifying (4.4) with the principal symbol $\tilde{P}_q(x, \xi) = \sum_{|\alpha|=q} \tilde{A}_\alpha(x)\xi^\alpha$ of which $\tilde{A}_\alpha(x)$ are sufficiently close to $A_\alpha(x)$ in $C^2(\Omega)$ for $|\alpha| = q$ then the Cauchy problem for $\tilde{P}(x, D)$ is C^∞ well posed near the origin for arbitrary lower order terms.*

Assuming the analyticity of the coefficients we have

Theorem 4.2 *Assume that $A_\alpha(x)$, $|\alpha| = q$ are real analytic in Ω and every characteristic of $P_q(x, \xi)$ over (x, ξ'), $|\xi'| = 1$ is non degenerate. Then the Cauchy problem for $P(x, D)$ is C^∞ well posed near the origin for arbitrary lower order terms.*

The proof is very simple. We reduce the Cauchy problem for $P(x, D)$ to that for a first order system $\mathcal{P}(x, D)$. Taking the invariance of non degeneracy of characteristics proved in Proposition 3.1, it suffices to apply the previous results in [6] and [7] which assert the existence of a smooth symmetrizer $S(x, D')$ for $\mathcal{P}(x, D)$ defined near the origin.

     Let us write

(4.5) $P(x, D)u = D_1^q u + \sum_{j=2}^n A_j(x, D')D_1^{q-j}u = f$.

Put

$$u^{(k)} = \langle D' \rangle^{q-k} D_1^{k-1} u, \quad k = 1, ..., q$$

where $\langle D' \rangle^2 = 1 + \sum_{j=2}^n D_j^2$. Then (4.5) is reduced to

$$D_1 U + \begin{pmatrix} 0 & -I & & & \\ 0 & 0 & -I & & \\ & & & \ddots & \\ & & & & -I \\ A_q^{\#}(x, D') & & \cdots & & A_1^{\#}(x, D') \end{pmatrix} \langle D' \rangle U = F$$

where $U = {}^t(u^{(1)}, ..., u^{(q)})$, $F = {}^t(0, ..., 0, f)$ and

$$A_j^{\#}(x, D') = A_j(x, D')\langle D' \rangle^{-j}.$$

Let us denote by $A_j^0(x, \xi')$ the principal symbol of $A_j^\#(x, \xi')$ and set

$$(4.6) \qquad \mathcal{A}(x, \xi') = \begin{pmatrix} 0 & -I & & & \\ 0 & 0 & -I & & \\ & & & \ddots & \\ & & & & -I \\ A_q^0(x, \xi') & & \cdots & & A_1^0(x, \xi') \end{pmatrix}$$

Fix $(0, \bar{\xi}')$, $|\bar{\xi}'| = 1$. Let $(0, \lambda_i, \bar{\xi}')$, $i = 1, ..., p$, be characteristics of $\xi_1 + \mathcal{A}(x, \xi')$ where $(0, \lambda_i, \bar{\xi}')$ are at most double and non degenerate and λ_i are different from each other. Then there exists a smooth $T(x, \xi')$ defined near $(0, \bar{\xi}')$, homogeneous of degree 0, such that

$$T(x, \xi')^{-1} \mathcal{A}(x, \xi') T(x, \xi') = \mathcal{A}_1(x, \xi') \oplus \cdots \oplus \mathcal{A}_p(x, \xi')$$

where $(0, \lambda_i, \bar{\xi}')$ is a non degenerate characteristic of $\mathcal{P}^{(i)}(x, \xi) = \xi_1 + \mathcal{A}_i(x, \xi')$. From Proposition 3.1 it follows that the characteristic set of $\det \mathcal{P}^{(i)}(x, \xi)$ near $(0, \lambda_i, \bar{\xi}')$ is a smooth manifold of codimension 3 through $(0, \lambda_i, \bar{\xi}')$. Then one can apply Lemma 3.1 in [6] to get a smooth $\mathcal{S}^{(i)}(x, \xi')$ defined near $(0, \bar{\xi}')$, homogeneous of degree 0 such that

$$\mathcal{S}^{(i)}(x, \xi')^{-1} \mathcal{A}^{(i)}(x, \xi') \mathcal{S}^{(i)}(x, \xi')$$

is symmetric. This proves that $\mathcal{A}(x, \xi')$ is smoothly symmetrizable near $(0, \bar{\xi}')$ by $\mathcal{S}^{(1)}(x, \xi') \oplus \cdots \oplus \mathcal{S}^{(p)}(x, \xi')$. By the usual argument of partition of unity one can prove that there is a smooth $\mathcal{S}(x, \xi')$ which symmetrizes $\mathcal{A}(x, \xi')$. Thus the Cauchy problem for $\mathcal{P}(x, D)$ is C^∞ well posed for arbitrary lower order terms and hence so is for $P(x, D)$.

Let $\tilde{A}_\alpha(x)$ be sufficiently close to $A_\alpha(x)$ for $|\alpha| = q$ in $C^2(\Omega)$. Let $\tilde{\mathcal{A}}(x, \xi')$ be defined by (4.6) from $\tilde{P}_q(x, \xi')$. Then it is clear that $\tilde{\mathcal{A}}(x, \xi')$ is sufficiently close to $\mathcal{A}(x, \xi')$ in $C^2(\Omega \times \{1/2 \le |\xi'| \le 2\})$. To prove the last assertion it suffices to show that every characteristic of $\tilde{\mathcal{P}}(x, \xi) = \xi_1 + \tilde{\mathcal{A}}(x, \xi')$ over $(0, \bar{\xi}')$ is non degenerate and at most double. Let $(0, \lambda, \bar{\xi}')$ be a characteristic of $\mathcal{P}(x, \xi) = \xi_1 + \mathcal{A}(x, \xi')$. If $(0, \lambda, \bar{\xi}')$ is simple then the characteristic of $\tilde{\mathcal{P}}(x, \xi)$ close to $(0, \lambda, \bar{\xi}')$ is also simple. If $(0, \lambda, \bar{\xi}')$ is a non degenerate double characteristic, then from Proposition 3.1 it follows that the characteristic of $\tilde{\mathcal{P}}(x, \xi)$ near $(0, \lambda, \bar{\xi}')$ is simple except in a manifold of codimension 3 consisting of non degenerate double characteristics. This proves that the characteristic of $\tilde{\mathcal{P}}(x, \xi)$ close to $(0, \lambda, \bar{\xi}')$ is non degenerate and at most double. This completes the proof of Theorem 4.1.

To prove Theorem 4.2, it suffices to apply Theorem 1.1 in [7]:

Proposition 4.1 ([7]) *Assume that $\mathcal{A}(x, \xi')$ is a $m \times m$ matrix valued real analytic function defined near $(0, \bar{\xi}')$ and $(0, \lambda, \bar{\xi}')$ is a non degenerate characteristic for $\xi_1 + \mathcal{A}(x, \xi')$ of order m. Then there exists a real analytic $\mathcal{S}(x, \xi')$ defined near $(0, \bar{\xi}')$ such that*

$$\mathcal{S}(x, \xi')^{-1} \mathcal{A}(x, \xi') \mathcal{S}(x, \xi')$$

is symmetric.

5   A remark

As proved in Theorem 3.1, if

$$(5.1) \qquad P(x,\xi) = \sum_{j=0}^{q} A_j(x,\xi')\xi_1^{q-j}, \quad A_0 = I$$

has a non degenerate multiple characteristic $(\bar{x},\bar{\xi})$ then near P there is no strictly hyperbolic system of the form (5.1). When $q = 1$ more detailed facts are known (see [5], [1]). Let P be a first order $(q = 1)$ system with constant coefficients:

$$(5.2) \qquad P(\xi) = \xi_1 - \sum_{j=2}^{n} A_j\xi_j$$

where A_j are $m \times m$ constant matrices. We always assume that

$$(5.3) \qquad \text{all eigenvalues of } \sum_{j=2}^{n} A_j\xi_j \text{ are real for every } \xi' = (\xi_2, ..., \xi_n).$$

Then from [5] $P(\xi)$ can not be strictly hyperbolic if $n > 3$ and $m \equiv 2$ modulo 4. Contrary to this if $q = 2$ it is clear that for any n and m there exist strictly hyperbolic systems.

In this section we study 2×2 first order systems with constant coefficients: from the result refered above if $n > 3$ then $P(\xi)$ can not be strictly hyperbolic. On the other hand we have

Proposition 5.1 *Let $m = 2$ and $P(\xi)$ verify (5.3). If $n \leq 3$ then $P(\xi)$ can be approximated by strictly hyperbolic systems: there are strictly hyperbolic $P_\epsilon(\xi)$ of the form (5.2) converging to $P(\xi)$ as $\epsilon \to 0$.*

Proof: Since the assertion is clear for $n = 2$ we may assume that $n = 3$. We first consider the case that $P(\xi)$ is diagonalizable for every ξ. From Lemma 4.1 in [6] there exists a constant matrix T such that

$$T^{-1}P(\xi)T = \xi_1 - \tilde{A}_2\xi_2 - \tilde{A}_3\xi_3$$

is symmetric for every ξ. Take a basis

$$I = \begin{pmatrix} 1 & 0 \\ 0 & 1 \end{pmatrix}, \quad A = \begin{pmatrix} 1 & 0 \\ 0 & -1 \end{pmatrix}, \quad B = \begin{pmatrix} 0 & 1 \\ 1 & 0 \end{pmatrix}$$

for $M^s(2; \mathbf{R})$ and write with $\xi' = (\xi_2, \xi_3)$

$$\begin{aligned} T^{-1}P(\xi)T &= \xi_1 - \{\phi_1(\xi')I + \phi_2(\xi')A + \phi_3(\xi')B\} \\ &= (\xi_1 - \phi_1(\xi'))I - \phi_2(\xi')A - \phi_3(\xi')B. \end{aligned}$$

If ϕ_2 and ϕ_3 are linearly independent then $P(\xi)$ itself is strictly hyperbolic because

$$\det P(\xi) = (\xi_1 - \phi_1(\xi'))^2 - \phi_2(\xi')^2 - \phi_3(\xi')^2.$$

If $\phi_2 = \phi_3 = 0$ then taking $\psi_2(\xi')$ and $\psi_3(\xi')$ which are linearly independent we define $P_\epsilon(\xi)$ as

$$T^{-1}P_\epsilon(\xi)T = (\xi_1 - \phi_1(\xi'))I - \epsilon\psi_2(\xi')A - \epsilon\psi_3(\xi')B.$$

Assume that $\phi_2 \neq 0$ and $\phi_3 = k\phi_2$ with some constant $k \neq 0$. Take $\psi_3(\xi')$ so that ϕ_2 and ψ_3 are linearly independent and define $P_\epsilon(\xi)$ by

$$T^{-1}P_\epsilon(\xi)T = (\xi_1 - \phi_1(\xi'))I - (\phi_2 + \epsilon\psi_3)A - k(\phi_2 - \epsilon k^{-2}\psi_3)B.$$

Then one has

$$\det P_\epsilon(\xi) = (\xi_1 - \phi_1)^2 - (1 + k^2)\phi_2^2 - \epsilon^2(1 + k^{-2})\psi_3^2$$

and hence $P_\epsilon(\xi)$ is strictly hyperbolic which converges to $P(\xi)$ as $\epsilon \to 0$. If $k = 0$ then it suffices to take P_ϵ so that

$$T^{-1}P_\epsilon T = (\xi_1 - \phi_1)I - \phi_2 A - \epsilon\psi_3 B.$$

The proof for the remaining case is just a repetition.

We turn to the case that $P(\xi)$ is not diagonalizable for some $\xi = \omega$, $\omega \neq 0$. Choose a constant matrix T so that

$$T^{-1}P(\omega)T = \begin{pmatrix} \lambda & 1 \\ 0 & \lambda \end{pmatrix}.$$

With $T^{-1}P(\xi)T = (\ell_{ij}(\xi))$ we show that $\ell_{11}(\xi)$, $\ell_{22}(\xi)$ and $\ell_{21}(\xi)$ are linearly dependent. Suppose this were not true then we could choose ξ so that $\ell_{11}(\xi) = \ell_{22}(\xi)$, $1 + \ell_{12}(\xi) > 0$ and $\ell_{21}(\xi) < 0$. This shows that $T^{-1}P(\omega + \xi)T = T^{-1}P(\omega)T + (\ell_{ij}(\xi))$ would have a non real eigenvalue which contradicts (5.3). Then one has

(5.4) $$\alpha\ell_{11}(\xi') + \beta\ell_{22}(\xi') + \gamma\ell_{21}(\xi') = 0.$$

We first assume $\gamma \neq 0$ so that

$$\ell_{21} = a(\ell_{11} - \ell_{22})$$

with some $a \neq 0$ because ℓ_{21} contains no ξ_1. Let us consider

$$\begin{pmatrix} 1 & 0 \\ -a & 1 \end{pmatrix} \begin{pmatrix} \ell_{11} & \ell_{12} \\ \ell_{21} & \ell_{22} \end{pmatrix} \begin{pmatrix} 1 & 0 \\ a & 1 \end{pmatrix}$$

$$= \begin{pmatrix} \ell_{11} + a\ell_{12} & \ell_{12} \\ -a^2\ell_{12} & \ell_{22} - a\ell_{12} \end{pmatrix} = (TS)^{-1}P(\xi)TS.$$

Note that $\ell_{12} \neq 0$ for $(TS)^{-1}P(\xi)TS$ is not diagonalizable at ω. Let us write

$$(TS)^{-1}P(\xi)TS = (\xi_1 + \ell(\xi'))I + \begin{pmatrix} \phi(\xi') & \ell_{12}(\xi') \\ -a^2\ell_{12}(\xi') & -\phi(\xi') \end{pmatrix}$$

where

(5.5) $$\phi(\xi')^2 - a^2\ell_{12}(\xi')^2 \geq 0$$

by (5.3). Note that $\phi \neq 0$ because $\ell_{12} \neq 0$. Then one can write $\ell_{12} = k\phi$ with some k because of (5.5) and hence

$$(TS)^{-1}P(\xi)TS = (\xi_1 + \ell(\xi'))I + \begin{pmatrix} 1 & k \\ -a^2k & -1 \end{pmatrix}\phi(\xi')$$

where $1 - a^2k^2 \geq 0$ by (5.5). If

(5.6) $\begin{pmatrix} 1 & k \\ -a^2k & -1 \end{pmatrix}$

is diagonalizable then so is $P(\xi)$ for every ξ contradicting the assumption. If (5.6) is not diagonalizable and necessarily $1 - a^2k^2 = 0$ one can take U so that

$$U^{-1}\begin{pmatrix} 1 & k \\ -a^2k & -1 \end{pmatrix}U = \begin{pmatrix} 0 & 1 \\ 0 & 0 \end{pmatrix}.$$

Take ψ so that ψ and ϕ are linearly independent and define $P_\epsilon(\xi)$ by

(5.7) $(TSU)^{-1}P_\epsilon(\xi)TSU = (\xi_1 + \ell(\xi'))I + \epsilon\psi\begin{pmatrix} 1 & 0 \\ 0 & -1 \end{pmatrix} + \phi\begin{pmatrix} 0 & 1 \\ \epsilon & 0 \end{pmatrix}.$

It is clear that $P_\epsilon(\xi)$ is strictly hyperbolic and converges to $P(\xi)$ as $\epsilon \to 0$ because

$$\det P_\epsilon(\xi) = (\xi_1 + \ell(\xi'))^2 - \epsilon\phi(\xi')^2 - \epsilon^2\psi(\xi')^2.$$

If $\gamma = 0$ then we have $\ell_{11} = \ell_{22}$ from (5.4). Hence one can write

$$T^{-1}P(\xi)T = (\xi_1 + \ell_{11}(\xi'))I + \begin{pmatrix} 0 & \ell_{12}(\xi') \\ \ell_{21}(\xi') & 0 \end{pmatrix}.$$

Since $\ell_{12}(\xi')\ell_{21}(\xi') \geq 0$ by (5.3), assuming $\ell_{12} \neq 0$ we can write

$$\ell_{21} = k\ell_{12}$$

with some constant $k \geq 0$ so that

$$T^{-1}P(\xi)T = (\xi_1 + \ell_{11}(\xi'))I + \ell_{12}(\xi')\begin{pmatrix} 0 & 1 \\ k & 0 \end{pmatrix}.$$

If $k > 0$ then $P(\xi)$ is diagonalizable for every ξ and contradicts the assumption. If $k = 0$ it suffices to define $P_\epsilon(\xi)$ by the right-hand side of (5.7) with $\ell = \ell_{11}$ and $\phi = \ell_{12}$. The proof for the case $\ell_{21} \neq 0$ is just a repetition. Thus we complete the proof.

References

[1] S. FRIEDLAND, J. ROBBIN AND J. SYLVESTER, *On the crossing rule*, Comm. Pure Appl. Math. **37** (1984), 19-37.

[2] L.HÖRMANDER, *Hyperbolic systems with double characteristics*, Comm. Pure Appl. Math. **46** (1993), 261-301.

[3] F.JOHN, *Algebraic conditions for hyperbolicity of systems of partial differential equations*, Comm. Pure Appl. Math. **31** (1978), 89-106.

[4] F.JOHN, *Addendum to : Algebraic conditions for hyperbolicity of systems of partial differential equations*, Comm. Pure Appl. Math. **31** (1978), 787-793.

[5] P.D.LAX, *The multiplicity of eigenvalues*, Bull. Amer. Math. Soc. **6** (1982), 213-214.

[6] T.NISHITANI, *On strong hyperbolicity of systems*, in Research Notes in Mathematics, **158**, pp. 102-114, 1987.

[7] T.NISHITANI, *Symmetrization of hyperbolic systems with non degenerate characteristics*, J. Func. Anal. **132** (1995), 92-120.

[8] T.NISHITANI, *Stability of symmetric systems under hyperbolic perturbations*, Hokkaido Math. J. **26** (1997), 509-527.

[9] T.NISHITANI, *Symmetrization of hyperbolic systems with real constant coefficients*, Scuola Norm. Sup. Pisa **21** (1994), 97-130.

The Cauchy problem for hyperbolic operators dominated by the time function

Kunihiko Kajitani

Institute of Mathematics, University of Tsukuba,
Tsukuba, Ibaraki 305, Japan

In Honor of Jean Vaillant

0   Introduction

This note is devoted to the C^∞ well posedness of the Cauchy problem for a hyperbolic operator which are a generalization of the effectively hyperbolic operator. The result in this note is a joint work with S. Wakabayashi and K. Yagdjian and the detail of the proof will be appeared in [8]. Let

$$P(t,x,D_t,D_x) := \sum_{j+|\alpha|\leq m} a_{j,\alpha}(t,x)D_x^\alpha D_t^j,$$

be a partial diferential operator with smooth coefficients $a_{j,\alpha}(t,x) \in \mathcal{B}^\infty(\mathbf{R}^{n+1})$. We consider the Cauchy problem for P

$$\begin{cases} P(t,x,D_t,D_x)u(t,x) = f(t,x), & t \in [0,T], \ x \in \mathbf{R}^n, \\ D_t^j u(0,x) = u_j(x), & j = 0,\ldots,m-1, \ x \in \mathbf{R}^n. \end{cases} \tag{0.1}$$

For the principal symbol $P_m(t,x,\lambda,\xi)$ of the operator P defined by

$$P_m(t,x,\lambda,\xi) := \sum_{j+|\alpha|=m} a_{j,\alpha}(t,x)\xi^\alpha\lambda^j$$

we assume that for all $t \in \mathbf{R}$, $x \in \mathbf{R}^n$, $\lambda \in \mathbf{R}$, $\xi \in \mathbf{R}^n$, the following representation with the real-valued functions $\lambda_j(t,x,\xi)$, $j = 1,\ldots,m$, and with an integer $d \geq 2$ and a nonnegative Lipschitz continuous function $\varphi(t,x)$ and $\varphi_0(t,x) \in B^1(\mathbf{R}^{n+1})$ satisfying $\partial_t\varphi_0 \neq 0$ if $\varphi_0 = 0$,

$$\begin{cases} P_m(t,x,\lambda,\xi) = \prod_{j=1}^m (\lambda - \lambda_j(t,x,\xi)), \\ |\lambda_j(t,x,\xi) - \lambda_k(t,x,\xi)| \geq \varphi(t,x)^d|\xi|, \ j \neq k, \\ \varphi(t,x) \geq |\varphi_0(t,x)|, \end{cases} \tag{0.2}$$

31

holds and $\varphi(t,x)$ does not vanish outside a compact set in R^{n+1}. Thus the operator $P(t,x,D_t,D_x)$ is a hyperbolic operator with the characteristics $\lambda_j(t,x,\xi)$, $j = 1, \ldots, m$, of variable multiplicity in a compact set and strictly hyperbolic outside a compact set. It is well-known that the lower order terms of the operators with multiple characteristics play crucial role in the well-posedness of the Cauchy problem (see, for example, [2], [9]). Therefore we make an assumption

$$|\partial_t^k \partial_x^\beta a_{j,\alpha}(t,x)| \leq C_{k,\beta}\varphi(t,x)^{d(s-j)-(m-s)-k-|\beta|}, \quad j+|\alpha| = s, \ 1 \leq s \leq m, \qquad (0.3)$$

for $(t,x) \in R^{n+1}$, where $\varphi(t,x)$ is given in (0.2). These kind of conditions for the coefficients with $s \leq m-1$ are called *Levi conditions*. To describe a propagation phenomena in the Cauchy problem we denote

$$\lambda_{max} := \sup_{\substack{j=1,\ldots,m, \\ (t,x)\in[0,T]\times R^n, \ \xi\in R^n, \ |\xi|=1}} |\lambda_j(t,x,\xi)|$$

and define a *hyperbolic cone of symbol* P_m by

$$\Gamma := \{ (\lambda,\xi) \in R^{n+1} ; \lambda > \lambda_{max}|\xi| \} \qquad (0.4)$$

while Γ^* is dual cone of Γ that is

$$\Gamma^* := \{ (t,x) \in R^{n+1} ; t\lambda + x \cdot \xi \geq 0 \text{ for all } (\lambda,\xi) \in \Gamma \} \qquad (0.5)$$

and is called a *propagation cone of symbol* P_m.

For the Cauchy problem (0.1) we can prove the following theorem.

Theorem 0. 1 *Assume* (0.2) *and* (0.3). *Then for any* $u_j \in H^\infty(\mathbf{R}^n)$ *and for any* $f \in C^\infty([0,T]; H^\infty(\mathbf{R}^n))$ *there exists a unique solution* $u \in C^\infty([0,T]; H^\infty(\mathbf{R}^n))$ *of the Cauchy problem* (0.1). *For the support of the solution the following formula holds*

$$\text{supp} \, u \subset \left\{ \bigcup_{y\in D_0} K^+(0,y) \right\} \bigcup \left\{ \bigcup_{(\tau,y)\in\Omega_0} K^+(\tau,y) \right\}, \qquad (0.6)$$

where

$$D_0 = \bigcup_{i=0}^{m-1} \text{supp} \, u_i, \qquad \Omega_0 = \text{supp} \, f, \qquad K^+(\tau,y) = (\tau,y) + \Gamma^*. \qquad (0.7)$$

Thus according to this theorem the solution propagates along the propagation cone.

Example 0. 2 *The second-order operator*

$$P = D_t^2 - (\phi_1(t,x)^{2d} + \phi_2(t,x)^{2d})D_x^2 + (\phi_1(t,x)^{d-1} + \phi_2(t,x)^{d-1})D_x,$$

satisfies the conditions (0.2), (0.3), *if we take* $\phi_1 = \chi(\frac{t-x}{\sqrt{t^2+x^2+1}})$ *and* $\phi_2 \in B^(R^2)$, *where* $\chi \in C^\infty(\mathbf{R})$ *satisfing* $\chi(t) = t, |t| \leq 1$ *and* $\chi(t) = \pm 2, \pm t \geq 2$,*and* $\varphi = (\phi_1^2 + \phi_2^2)^{\frac{1}{2}}$ *does not vanish outside a compact set and* $\varphi_0 = \phi_1$. *See* [3] *for more general hyperbplic operators with double characteristics.*

Example 0.3 *For the operator* $P = P_3 + P_2 + P_1$, $P_s = \sum_{j+|\alpha|=s} a_{j,\alpha}(t)\lambda^j \xi^\alpha$, $s = 1, 2, 3$, *with the principal symbol*

$$P_3 = \lambda(\lambda^2 - (\phi_1(t,x)^{2d} + \phi_2(t,x)^{2d})|\xi|^2)$$

the condition (0.3) will be satisfied if

$$a_{0,\alpha}(t,x) = \sum_{j=1}^{2d} a_j \phi_1^{2d-j} \phi_2^{j-1}, \quad |\alpha| = 2,$$

$$a_{1,\alpha}(t,x) = \sum_{j=1}^{d} b_j \phi_1^{d-j} \phi_2^{j-1}, \quad a_{0,\alpha}(t,x) = \sum_{j=2}^{d} c_j \phi_1^{d-j} \phi_2^{j-2}, \quad |\alpha| = 1,$$

where ϕ_1, ϕ_2 *are the same as in Example 0.2 and* $a_j, b_j c_j$ *are in* $B^\infty(R^2)$.

The microlocal version of Theorem 0.1 will be given in [7]. For the operators with the multiple characteristics a microlocal energy method developed in [4], [5]; allows to prove the well-posedness and gives a complete picture of the propagation of singularities. When we can take $\varphi_0 = t$, the case that d in (0.2) depends on j is treated in [8]. When $d = 1$, Theorem is already proved in [6]. Therefore we consider here the case of $d \geq 2$. We shall describe the outline of the proof of Theorem 0.1 in the next section.

1 The outline of the proof

In this section we shall give the outline of proof of an a priori estimate for the Cauchy problem (0.1) under the assumptions (0.2) and (0.3). The proof will be done in the following seven steps.

Step 1: Reduction to the problem in usual Sobolev spaces

The equation $P(t, x, D_t, D_x)u(t, x) = f(t, x)$ obtained from (0.1) one can replace by the equation

$$P_\gamma(t, x, D_t, D_x)u_\gamma(t, x) = f_\gamma(t, x), \qquad f_\gamma(t, x) := e^{-\gamma t} f(t, x), \tag{1.1}$$

for new unknown function $u_\gamma(t, x) = e^{-\gamma t} u(t, x)$, where $f_\gamma \in H^\infty(R^{n+1})$, while

$$P_\gamma(t, x, D_t, D_x)v = P(t, D_t - i\gamma, D_x)v = e^{-\gamma t} P(t, D_t, D_x)(e^{\gamma t} v). \tag{1.2}$$

If we define v by

$$u_\gamma(t, x) := e^\Lambda(t, x, D_x)v(t, x) = \int e^{ix\xi + \Lambda(t,x,\xi)} \hat{v}(t, \xi) d\xi,$$

with weight function $\Lambda(t, x, \xi) \in C^\infty(R^{2n+1})$ defined by (1.8) below then the function $v(t, x)$ satisfies

$$P_{\gamma,\Lambda}(t, x, D_t, D_x)v(t, x) = f_{\gamma,\Lambda}(t, \xi), \qquad f_{\gamma,\Lambda}(t, x) := e^\Lambda(t, x, D_x)^{-1} f_\gamma(t, x), \tag{1.3}$$

where

$$P_{\gamma,\Lambda}(t,x,D_t,D_x) := e^{\Lambda}(t,x,D_x)^{-1}P(t,x,D_t - i\gamma, D_x)e^{\Lambda}(t,x,D) \qquad (1.4)$$

and the inverse $e^{\Lambda}(t,x,D_x)^{-1}$ is assured by the ellipticity of the symbol $e^{\Lambda(t,x,\xi)}$. Our aim is to obtain *a priori* estimate for the solutions of the differential equation (1.3):

$$\|v\|_{H^s(\mathbf{R}^{n+1})} \le \frac{C}{\sqrt{\gamma}}\|P_{\gamma,\Lambda}(t,x,D_t,D_x)v(\cdot)\|_{H^s(\mathbf{R}^{n+1})} \quad \text{for all} \quad v \in H^{\infty}(\mathbf{R}^{n+1}). \qquad (1.5)$$

This estimate holds for all $\gamma \ge \gamma_0$, where γ_0 will be chosen and fixed, while C is independent of γ. Since $P(t,x,D_t,D_x) = e^{\gamma t}e^{\Lambda}(t,x,D_x)P_{\gamma,\Lambda}(t,x,D_t,D_x)e^{-\gamma t}e^{\Lambda}(t,x,D_x)^{-1}$, from the last inequality we obtain

$$\|e^{-\gamma t}e^{\Lambda}(t,x,D_x)^{-1}u\|_{H^s(\mathbf{R}^{n+1})} \le C\|e^{-\gamma t}e^{\Lambda}(t,x,D_x)^{-1}Pu\|_{H^s(\mathbf{R}^{n+1})} \qquad (1.6)$$

for $u \in e^{\gamma t}H^{\infty}(\mathbf{R}^{n+1})$.

Let $d \ge 2$ be an integer given in (0.2) and φ_l functions given in (0.9) and define

$$\omega(t,x,\xi) := \frac{1}{\sqrt{\varphi(t,x)^2 + \langle\xi\rangle_h^{-\delta}}}, \quad \delta = \frac{2}{d+1}, \quad \langle\xi\rangle_h = \sqrt{h^2 + |\xi|^2} \qquad (1.7)$$

$$\tau(t,x,\xi) := \langle\xi\rangle_h^{\frac{2}{d+1}}\omega(t,x,\xi)^{-1} \qquad (1.8)$$

$$\Lambda_0(t,x,\xi) := M\int_0^{\varphi(t,x)}\left(s^2 + \langle\xi\rangle_h^{-\delta}\right)^{-1/2}ds, \qquad (1.9)$$

$$\Lambda(t,x,\xi) := \int \chi(|t-s|\omega(s,y,\eta)\chi(|x-y|\omega(s,\eta.))\chi(|\xi-\eta|\tau(s,y,\eta)) \qquad (1.10)$$
$$\times \ \Lambda_0(s,y,\eta)\omega^n(s,y,\eta)\tau^{-(n-1)}(s,y,\eta)dyd\eta,$$

$$g_{t,x,\xi}(s,y,\eta,\zeta) := \omega(t,x,\xi)^2(s^2 + |y|^2) + \tau(t,x,\xi)^{-2}\eta^2 + \langle\xi\rangle_h^{-2}|\zeta|^2, \qquad (1.11)$$

$$g_{t,x,\xi}^{\sigma}(s,y,\eta,\zeta) := \tau(t,x,\xi)^2s^2 + \langle\xi\rangle_h^2|y|^2 + \omega(t,x,\xi)^{-2}(\eta^2 + |\zeta|^2), \qquad (1.12)$$

where $h,\gamma,M \in \mathbf{R}^+$ are parameters. The metric $g_{t,x,\xi}(s,y,\eta,\zeta)$ is a Riemann metric in $T^*(\mathbf{R}^{n+1})$ and slowly varying in the sense of Definition 18.4.1 [1] that follows from the fact that $\omega(t,x,\xi)$ and $\tau(t,x,\xi)$ are weights with respect to g. The constant M counts the loss of regularity, while γ controls the support of solution and reflect the finite propagation speed.

Step 2: Leray-Gårding's method

In this step we prove estimate (1.5). Denote $H_{\tau,\zeta} = \tau\partial_{\lambda} + \sum_{j=1}^n \zeta_j\partial_{\xi_j}$ and $(\hat{\tau},\hat{\zeta}) = \frac{\gamma+\Lambda_t,\Lambda_x}{\sqrt{\gamma^2+\Lambda_t^2+|\Lambda_x|^2}}$. Set $Q(t,x,\lambda,\xi) := (H_{\hat{\tau},\hat{\zeta}}P_m)(t,x,\lambda-i(\gamma+\Lambda_t),\xi-i\Lambda_x)$ for the symbol of the *separating operator*. To prove estimate (1.5) it is enough to show that there exists a constant $C_0 > 0$ such that an estimate

$$\text{Im}\,(P_{\gamma,\Lambda}v, Qv)_{H^s(\mathbf{R}^{n+1})} \ge C_0\|v\|_{H^s(\mathbf{R}^{n+1})}^2 \qquad (1.13)$$

holds for all $v \in H^{\infty}(\mathbf{R}^{n+1})$. We note

$$\text{Im}\,(P_{\gamma,\Lambda}v, Qv)_{H^s(\mathbf{R}^{n+1})} = \frac{1}{2i}\left\{(Q^*P_{\gamma,\Lambda}v, v)_{H^s(\mathbf{R}^{n+1})} - (P_{\gamma,\Lambda}^*Qv, v)_{H^s(\mathbf{R}^{n+1})}\right\}. \qquad (1,14)$$

Therefore we define an auxiliary operator

$$S(t, x, D_t, D_x) := \operatorname{Im} Q^*(t, x, D_t, D_x) P_{\gamma, \Lambda}(t, x, D_t, D_x) = \frac{1}{2i} \left\{ Q^* P_{\gamma, \Lambda} - P^*_{\gamma, \Lambda} Q \right\}, \quad (1.15)$$

and rewrite (1.14) as follows

$$\operatorname{Im} (P_{\gamma, \Lambda} v, Qv)_{H^s(\mathbf{R}^{n+1})} = (S(t, x, D_t, D_x)v, v)_{H^s(\mathbf{R}^{n+1})}. \quad (1.16)$$

For the following calculations will be very important the principal symbol

$$S_0(t, x, \lambda, \xi) := \operatorname{Im} \overline{Q(t, x, \lambda, \xi)} P_m(t, x, \lambda - i(\gamma + \Lambda_t), \xi - i\Lambda_x),$$

of the operator $S(t, x, D_t, D_x)$. The remaining lower order part will be denoted by

$$S'(t, x, D_t, D_x) := S(t, x, D_t, D_x) - S_0(t, x, D_t, D_x). \quad (1.17)$$

To continue we need some calculus of pseudo-differential operators.

Step 3: The calculus of pseudo-differential operators

Let $\varphi_j(x, \xi)$, $\psi_j(x, \xi)$ $(j = 1, ...n)$ be positive functions defined in $\mathbf{R}^{2n}$. We introduce at (x, ξ) an orthogonal basis with respect to the Riemannian metric

$$g_{x, \xi}(y, \eta) = \sum_{j=1}^{n} \{\varphi_j(x, \xi)^{-2} y_j^2 + \psi_j(x, \xi)^{-2} \eta_j^2\}.$$

We assume that this metric is slowly varying and σ temperate in the sense of Definition 18.4.1 [1] that is there are positive constants c and C such that

$$g_{x, \xi}(y, \eta) < c \Longrightarrow C^{-1} g_{x, \xi}(\tilde{y}, \tilde{\eta}) \leq g_{x+y, \xi+\eta}(\tilde{y}, \tilde{\eta}) \leq C g_{x, \xi}(\tilde{y}, \tilde{\eta})$$

for any $(\tilde{y}, \tilde{\eta}) \in \mathbf{R}^{2n}$ and

$$g_{x+y, \xi+\eta}(\tilde{y}, \tilde{\eta}) \leq C g_{x, \xi}(\tilde{y}, \tilde{\eta})(1 + g_{x, \xi}(y, \eta)),$$

for any $(x, \xi), (y, \eta) \in \mathbf{R}^{2n}$.

Definition 1. 1 *Let g be $\sigma-$ temperate. A positive function $m(x, \xi)$ is said to be g-continuous if there exist positive numbers c_0, C such that*

$$g_{x, \xi}(y, \eta) < c_0 \Longrightarrow C^{-1} \leq m(x + y, \xi + \eta)/m(x, \xi) \leq C.$$

For $(y, \eta) \in \mathbf{R}^{2n}$ we define a dual metric of $g_{x, \xi}$

$$g^{\sigma}_{x, \xi}(y, \eta) = \sum_{j=1}^{n} \{\psi_j(x, \xi)^2 y_j^2 + \varphi_j(x, \xi)^2 \eta_j^2\}.$$

Definition 1. 2 *A positive function $m(x, \xi)$ is said to be $g - \sigma$-temperate if there exist positive numbers C, N such that*

$$m(x + y, \xi + \eta) \leq C m(x, \xi)(1 + g^{\sigma}_{x, \xi}(y, \eta))^N$$

Example 1.3 *If*

$$g_{x,\xi}(y,\eta) = |y|^2 + \langle\xi\rangle_h^{-2}|\eta|^2, \qquad g_{x,\xi}^\sigma(y,\eta) = \langle\xi\rangle_h^2|y|^2 + |\eta|^2,$$

then $m = \langle\xi\rangle_h$ *is g-continuous and σ-temperate.*

Definition 1.4 *A positive function $m(x,\xi)$ is said to be a weight with respect to g if m is g continuous and $\sigma - g$-temperate.*

Definition 1.5 *(symbol of pseudo-differential operator) For given m and g a set $S(m,g)$ consists of all smooth functions $a(x,\xi) \in C^\infty(\mathbf{R}^{2n})$ satisfying for every $\alpha,\beta \in \mathbf{Z}_+^n$*

$$|a_{(\beta)}^{(\alpha)}(x,\xi)| \le C_{\alpha,\beta}m(x,\xi)\prod_{j=1}^n\{\psi_j(x,\xi)^{-\alpha_j}\varphi_j(x,\xi)^{-\beta_j}\} \quad \text{for all} \quad (x,\xi) \in \mathbf{R}^{2n}.$$

For a given symbol $a \in S(m,g)$ we define pseudo-differential operator $a(x,D_x)$ by

$$a(x,D_x)u = (2\pi)^{-n}\int e^{ix\cdot\xi}a(x,\xi)\hat{u}(\xi)\,d\xi, \qquad u \in S.$$

If $u \in S$ and $a_i \in S(m_i,g)$, $i = 1,2$, then $a_i(x,D_x)u \in S(i = 1,2)$. Hence one can define
$a(x,D_x)u := a_1(x,D_x)(a_2(x,D_x)u)$. According to the next proposition we have $a \in S(m_1m_2,g)$.

Proposition 1.6 *Let $a_i \in S(m_i,g)$, $i = 1,2$. Assume $g \le g^\sigma$. Then the product $a(x,D_x) = a_1(x,D_x)a_2(x,D_x)$ is also a pseudo-differential operator with a symbol $a \in S(m_1m_2,g)$. Moreover*

$$a(x,\xi) \sim \sum_\alpha a_1^{(\alpha)}(x,\xi)a_{2(\alpha)}(x,\xi)/\alpha!$$

in the sense that for every N one has

$$a(x,\xi) - \sum_{|\alpha|<N} a_1^{(\alpha)}(x,\xi)a_{2(\alpha)}(x,\xi)/\alpha! \in S(m_1m_2(\max_j\{\varphi_j\psi_j\}^{-1})^N,g).$$

Proposition 1.7 *Let $a \in S(1,g)$, then operator $a(x,D_x) : L_2 \longrightarrow L_2$ is bounded, so that*

$$\|a(x,D_x)u\|_{L_2} \le C\|u\|_{L_2}, \qquad u \in S.$$

Proposition 1.8 *Let $m(x,\xi) > 0$ be a weight and $m(x,\xi) \in S(m,g)$. If $(\max_j(\varphi_j\psi_j)^{-1}(x,\xi) \le C_0 \ll 1$ then there exists symbol $q(x,\xi) \in S(m^{-1},g)$ such that*

$$m(x,D_x)q(x,D_x) = q(x,D_x)m(x,D_x) = I \quad \text{(identity operator) in } L^2.$$

Step 4: Construction of the weight function and the metric

The metric $g_{t,x,\xi}(s,y,\eta,\zeta)$ defined by (1.11) is slowly varying in the sense of Definition 18.4.1 [1] that is there are positive constants c and C independent of γ (large enough), and M (large enough) such that if $g_{t,x,\xi}(s,y,\eta,\zeta) < c$

$$C^{-1}g_{t,x,\xi}(\tilde{t},\tilde{y},\tilde{\lambda},\tilde{\zeta}) \le g_{t+s,x+y,\xi+\zeta}(\tilde{t},\tilde{y},\tilde{\lambda},\tilde{\zeta}) \le Cg_{t,x,\xi}(\tilde{t},\tilde{y},\tilde{\lambda},\tilde{\zeta})$$

for all $(\tilde{t},\tilde{y},\tilde{\lambda},\tilde{\zeta}) \in \mathbf{R}^{2(n+1)}$.

we can see that $\omega(t,x,\xi)$, $\tau(t,x,\xi)$ given by (1.7),(1.9) are g-continuous and σ-temperate, too.

Let $h_j(t,x,\lambda,\xi)$ be the polynomial in λ and ξ defined as

$$|P_m(t,x,\lambda-i\gamma,\xi)|^2 = \sum_{j=0}^{m} h_j(t,x,\lambda,\xi)\gamma^{2(m-j)}.$$

Then for these symmetric polynomials we have

$$\begin{cases} h_0 = 1, \quad h_m(t,x,\lambda,\xi) = |P_m(t,x,\lambda,\xi)|^2, \\[2mm] h_j(t,x,\lambda,\xi) = \sum_{1\le l_1<l_2<\cdots<l_j\le m} |q_{l_1}(t,x,\lambda,\xi)|^2 \cdots |q_{l_j}(t,x,\lambda,\xi)|^2, \end{cases} \tag{1.18}$$

where we denote $q_j = \lambda - \lambda_j(t,x,\xi)$. We denote $m_0 = h_0$ and for $j = 1,\ldots,m-1$,

$$m_j(t,x,\lambda,\xi) := \sum_{1\le l_1<\cdots<l_j\le m} \left((q_{l_1}(t,x,\lambda,\xi))^2 + (\gamma+\Lambda_t)^2\right)\cdots\left((q_{l_j}(t,x,\lambda,\xi))^2 + (\gamma+\Lambda_t)^2\right).$$

It will be proved in Lemma 4.4. that

$$\frac{m_j(t,x,\lambda,\xi)}{m_{j+1}(t,\lambda)} \le C\tau(t,x,\xi)^{-2}, \quad j = 0,1,\ldots,m-2.$$

We obtain from the last inequality for all $j = 1,\ldots,m-1$, the estimates

$$m_j(t,x,\lambda,\xi) \le C\tau(t,x,\xi)^{-2(m-1-j)}m_{m-1}(t,x,\lambda,\xi) \tag{1.19}$$

which point out the importance of symbol $m_{m-1}(t,x,\lambda,\xi)$. Therefore we define

$$\begin{aligned} m(t,x,\lambda,\xi) &:= m_{m-1}(t,x,\lambda,\xi) \\[2mm] &= \sum_{1\le l_1<l_2<\cdots<l_{m-1}\le m} \prod_{i=1}^{m-1}\left(q_{l_i}(t,x,\lambda,\xi)^2 + (\gamma+\Lambda_t)^2\right) \end{aligned} \tag{1.20}$$

and can prove that under the conditions (0.2) and (0.3) the function $m(t,x,\lambda,\xi)$ is g - continuous and that it is also a σ - temperate i.e.

$$m(t+s,x+y\lambda+\eta,\xi+\zeta) \le Cm(t,x,\lambda,\xi)\left(1 + g^\sigma_{t,x,\xi}(s,y,\eta,\zeta)\right)^N.$$

Step 5: A calculus of the pseudo-differential operators based on m and $g_{t,x,\xi}$

The σg - temperate function m allows to use a calculus of the pseudo-differential operators based on functions $m(t,x,\lambda,\xi)$ and $g_{t,x,\xi}(s,y,\eta,\zeta)$. In particular, assuming (0.2) and (0.3) we can prove that $P_s(t,x,\lambda,\xi)$ are symbols of this calculus:

$$|P_{s(k,\beta)}^{(l,\alpha)}(t,x,\lambda,\xi)| \leq C_{lk\alpha\beta}\omega(t,x,\xi)^{m-s+k+|\beta|}\tau(t,x,\xi)^{-l-(m-s)+1}\langle\xi\rangle_h^{-|\alpha|}\sqrt{m(t,x,\lambda,\xi)},$$

for all $l,k \in \mathbf{N}^1\,\alpha,\beta \in \mathbf{N}^n$. Hence, $P_s(t,x,\lambda,\xi) \in S(\tau(t,x,\xi)^{1-(m-s)}\sqrt{m(t,x,\lambda,\xi)},g)$. Furthermore due to the above estimate for symbol $P_{\gamma,\Lambda}(t,x,\lambda,\xi)$ of the operator $P_{\gamma,\Lambda}(t,x,D_t,D_x)$ given by (1.4) the estimate

$$\left|P_{\gamma,\Lambda(k,\beta)}^{(l+\alpha)}(t,\lambda,\xi)\right| \leq C_{kl\alpha\beta}\omega^k(t,x,\xi)\tau^{-l+1}(t,x,\xi)\langle\xi\rangle_h^{-|\alpha|}\sqrt{m(t,x,\lambda,\xi)}, \quad (l+k+|\alpha+\beta| \neq 0)$$

holds for all $t \in \mathbf{R}$, $\lambda \in \mathbf{R}$, $x,\xi \in \mathbf{R}^n$, while for the "lower order terms" $P'(t,x,\lambda,\gamma,\Lambda,\xi) := P_{\gamma,\Lambda}(t,\lambda,\xi) - P_m(t,\lambda-i(\gamma+\Lambda_t),\xi=i\Lambda_x)$ we have

$$P'(t,x,\lambda,\gamma,\Lambda,\xi) \in S(\omega\sqrt{m(t,x,\lambda,\xi)},g)\,.$$

Moreover, for the symbol $Q(t,x,\lambda,\xi)$ of the operator $Q(t,x,D_t,D_x)$, and for the symbol $Q^*(t,x,\lambda,D_x)$ of its adjoint operator $Q^*(t,x,D_t,D_x)$, the estimates

$$\left|Q_{(k,\beta)}^{(l,\alpha)}(t,x,\lambda,\xi)\right|+\left|Q_{(k,\beta)}^{*(l,\alpha)}(t,x,\lambda,\xi)\right| \leq C_{kl\alpha\beta}\omega(t,x,\xi)^{k+|\beta|}\tau(t,x,\xi)^{-l}\langle\xi\rangle_h^{-|\alpha|}\sqrt{m(t,x,\lambda,\xi)},$$

hold for all $t \in \mathbf{R}$, $\lambda \in \mathbf{R}$, $x,\xi \in \mathbf{R}^n$. Thus, $Q, Q^* \in S(\sqrt{m(t,x,\lambda,\xi)},g)$.

Simple calculations show the equality $S_0(t,x,\lambda,\xi) = (\gamma+\Lambda_t)m(t,x,\lambda,\xi)$, which allows in the definition of the classes of symbols to replace weight m with $S_0(t,x,\lambda,\xi)$. This leads to the following inclusions

$$\begin{aligned}
S_0(t,x,\lambda,\xi) &\in S(S_0,g)\,, \quad \sqrt{S_0(t,x,\lambda,\xi)} \in S(\sqrt{S_0},g)\,, & (1.21)\\
S'(t,x,\lambda,\xi) &\in S(S_0/M,g)\,, & (1.22)\\
C_0 S_0(t,x,\lambda,\xi) &\leq |S(t,x,\lambda,\xi)| \leq C_1 S_0(t,x,\lambda,\xi)\,, & (1.23)\\
S(t,x,\lambda,\xi) &\in S(S_0,g) & (1.24)
\end{aligned}$$

Step 6: A priori estimate

If we take a symbol

$$S_1(t,x,\lambda,\xi) = \sqrt{S_0(t,x,\lambda,\xi)}$$

then an operator $S_1(t,x,D_t,D_x)$ has a parametrix $S_1^\#(t,x,D_t,D_x)$ with the symbol $S_1^\#(t,x,\lambda,D_x) \in S((\sqrt{S_0})^{-1},g)$, that is the symbols of the operators

$$\sigma(I - S_1(t,x,D_t,D_x)S_1^\#(t,x,D_t,D_x))(t,x,\lambda,\xi),$$

$$\sigma(I - S_1^\#(t,x,D_t,D_x)S_1(t,x,D_t,D_x))(t,x,\lambda,\xi)$$

belong to $S(M^{-1}, g)$. That means there is the inverse operator $S_1(t, x, D_t, D_x)^{-1}$ to $S_1(t, x, D_t, D_x)$ in the space $H^\infty(\mathbf{R}^{n+1})$ provided that M is large enough. Therefore for such M we can write

$$S(t, x, D_t, D_x) = S_1^*(t, x, D_t, D_x)\,(I + R(t, x, D_t, D_x))\,S_1(t, x, D_t, D_x),$$

with $R(t, x, D_t, D_x) \in S(M^{-1}, g)$.

For the estimates of the norms of functions this brings

$$
\begin{aligned}
\mathrm{Im}\,(P_{\gamma,\Lambda}v, Qv)_{H^s(\mathbf{R}^{n+1})} &= (Sv, v)_{H^s(\mathbf{R}^{n+1})} \\
&= (S_1^*\,(I + R)\,S_1 v, v)_{H^s(\mathbf{R}^{n+1})} \\
&= \|S_1 v\|_{H^s(\mathbf{R}^{n+1})}^2 - CM^{-1}\|S_1 v\|_{H^s(\mathbf{R}^{n+1})}^2 \geq \frac{1}{2}\|S_1 v\|_{H^s(\mathbf{R}^{n+1})}^2,
\end{aligned}
$$

provided that M is large enough. On the other hand $Q \in S(\sqrt{m}, g)$ and $S_1 \in S(\sqrt{\gamma + \Lambda_t} \cdot \sqrt{m}, g)$. Hence

$$
\begin{aligned}
\mathrm{Im}\,(P_{\gamma,\Lambda}v, Qv)_{H^s(\mathbf{R}^{n+1})} &= \mathrm{Im}\,(f_{\gamma,M}, Qv)_{H^s(\mathbf{R}^{n+1})} \\
&= \mathrm{Im}\left((\sqrt{\gamma + \Lambda_t})^{-1} f_{\gamma,M}, (\sqrt{\gamma + \Lambda_t}\,QS_1^{-1})S_1 v\right)_{H^s(\mathbf{R}^{n+1})} \\
&\leq C\|(\gamma + \Lambda_t)^{-\frac{1}{2}} f_{\gamma,M}\|_{H^s(\mathbf{R}^{n+1})}\|S_1 v\|_{H^s(\mathbf{R}^{n+1})}.
\end{aligned}
$$

Thus we get

$$\|S_1 v\|_{H^s(\mathbf{R}^{n+1})} \leq C\|(\gamma + \Lambda_t)^{-\frac{1}{2}} f_{\gamma,M}\|_{H^s(\mathbf{R}^{n+1})}. \tag{1.25}$$

Since $S_1^{-1} \in S((\sqrt{S_0})^{-1}, g) \subset S(1, g)$, (1.25) implies (1.5),

$$
\begin{aligned}
\|v\|_{H^s(\mathbf{R}^{n+1})} &\leq C\|(\gamma + \Lambda_t)^{-\frac{1}{2}} f_{\gamma,M}\|_{H^s(\mathbf{R}^{n+1})} \\
&\leq \frac{C}{\sqrt{\gamma}}\|f_{\gamma,M}\|_{H^s(\mathbf{R}^{n+1})} \\
&= \frac{C}{\sqrt{\gamma}}\|P_{\gamma,\Lambda}v\|_{H^s(\mathbf{R}^{n+1})}
\end{aligned}
$$

for all $v \in H^\infty(\mathbf{R}^{n+1})$. This proves a priori estimate (1.5).

Since $P_{\gamma,\Lambda}(t, x, D_t, D_x) = e^\Lambda(t, x, D)^{-1}e^{-\gamma t}P(t, x, D_t, D_x)e^{\gamma t}e^\Lambda(t, x, D)$, we obtain (1.6),

$$\|e^\Lambda(t, x, D)^{-1}e^{-\gamma t}u\|_{H^s(\mathbf{R}^{n+1})} \leq C\|e^\Lambda(t, x, D)e^{-\gamma t}Pu\|_{H^s(\mathbf{R}^{n+1})} \tag{1.26}$$

for $u \in e^{\gamma t}e^\Lambda(t, x, D)H^\infty(\mathbf{R}^{n+1})$.

It is not difficult to see that the above estimate (1.26) implies Theorem 0.1.

References

[1] L. Hörmander. *The analysis of linear partial differential operators, Vol. 3: Pseudo-differential operators.* Springer-Verlag, Berlin et al. 1985.

[2] V. Ivrii and V. Petkov, Necessary correctness conditions for the Cauchy problem for non-strictly hyperbolic equations, Uspekhi Mat. Nauk, 1974, v. 29, no. 5, 3-70 (in Russian). Engl. transl. see

[3] K. Kajitani and S. Wakabayashi, The Cauchy problem for a class of hyperbolic operators with double characteristics, Funkcial. Ekvac., v. 39, no. 2, 1996, 235-307.

[4] K. Kajitani and S. Wakabayashi, Microlocal a priori estimates and the Cauchy problem I, Japan. J. Math. (N.S.) 19 (1993), no. 2, 353-418.

[5] K. Kajitani and S. Wakabayashi, Microlocal a priori estimates and the Cauchy problem II, Japan. J. Math. (N.S.) 20 (1994), no.1, 1-71.

[6] K. Kajitani, S. Wakabayashi, T. Nishitani, The Cauchy problem for hyperbolic operators of strong type, Duke Math. J. 75 (1994), no. 2, 353-408.

[7] K. Kajitani, S. Wakabayashi, K. Yagdjian, The C^∞-well posed Cauchy problem for hyperbolic operators with multiple characteristics vanishing with the different speeds.

[8] K. Kajitani, S. Wakabayashi, K. Yagdjian, The C^∞-well posed Cauchy problem for hyperbolic operators dominated by the time function.

[9] K. Yagdjian: The Cauchy Problem for Hyperbolic Operators, Multiple Characteristics, Micro-Local Approach, Akademie Verlag, Berlin, 1997.

A remark on the Cauchy problem
for a model hyperbolic operator

ENRICO BERNARDI AND ANTONIO BOVE

DIPARTIMENTO DI MATEMATICA, UNIVERSITÀ DI BOLOGNA, PIAZZA DI PORTA
S. DONATO 5, 40127 BOLOGNA, ITALIA

ABSTRACT. In this paper we prove a well posedness result in the
Gevrey category for a model hyperbolic operator with double characteristics exhibiting a Jordan block of dimension four in the canonical form of its fundamental matrix.

Dedicated to Prof. Jean Vaillant

1. INTRODUCTION AND STATEMENTS

The purpose of this paper is to prove a well posedness result in the
Gevrey category for a model hyperbolic operator with double characteristics. The reason why we consider Gevrey well posedness rather than
the more general C^∞ well posedness is due to what we believe to be a
fascinating intercourse between some partial differential equations with
symbols vanishing of order 2 and the behaviour of their Hamiltonian
flows.

More precisely we know that if $p(x,\xi)$, $(x,\xi) \in T^*\mathbb{R}^{n+1}$, is a C^∞
symbol of order 2 vanishing on a given submanifold $\Sigma \subset T^*\mathbb{R}^{n+1}\backslash 0$, such
that for every $\rho \in \Sigma$ the localization p_ρ is a homogeneous hyperbolic
quadratic form and, denoting by $F_p(\rho)$ the fundamental matrix of p at
ρ, (see below for a definition) the following alternatives hold:

- either $\operatorname{Sp} F_p(\rho) \cap \mathbb{R} \setminus 0 \neq \emptyset$ or
- $\operatorname{Sp} F_p(\rho) \subset i\mathbb{R}$, i.e. it is purely imaginary or zero.

The first case, called effectively hyperbolic in the literature, (see [6] for
more details) has been thoroughly investigated by several authors.

In the second situation things happen to be more complicated, essentially because the zero eigenvalue can make the associated Hamilton
system unstable. It was noted in [3] that if the canonical form of $F_p(\rho)$

has a Jordan block of order 4 at zero and a suitable third order derivative, which can be given an invariantly defined geometrical meaning,
is non zero, then there exist null bicharacteristic curves issued from
simply characteristic points and landing tangentially onto the double
manifold of the operator.

On the other hand in [2] it was shown that, if the condition on the
third order derivative is not verified, then the Cauchy problem in the
C^∞ category, is well posed, provided the usual Levi conditions on the
lower order term are verified.

In this paper we study a class of models generalizing the third order
derivative condition stated in [3] and compute the Gevrey indexes up to
which it is possible to prove a regularity result; our technique consists
in proving an energy estimate.

In the transversally non degenerate case we find a Gevrey well posedness threshold equal to 5, which can be pushed up to 6 if non zero first
order terms or positive trace contributions are involved. Up to now
wether these numbers are sharp or not is an open problem.

Besides the transversally non degenerate case we also considered
some transversally degenerate models parametrized by an integer $k > 0$
(see (2.1) below): here we are able to show that the Gevrey threshold
for well posedness is $2k+3$ or $4k+2$ if non zero lower order contributions
are present.

2. GEOMETRICAL PROPERTIES

We consider the following differential operator:

$$(2.1) \qquad -D_0^2 + 2x_1^k D_0 D_n + D_1^2 + bx_1^{2k+1} D_n^2.$$

Here $b \neq 0$, $x \in \mathbb{R}^{n+1}$, $x = (x_0, x_1, \ldots, x_n) = (x_0, x_1, x', x_n)$, where x_0
denotes the time variable. The operator (2.1) is a (microlocal) model
for a weakly hyperbolic operator with double characteristics exhibiting
a Jordan block in the canonical form of its Hamiltonian matrix F. We
recall that $F(\rho) = d_{(x,\xi)} H_p(\rho)$, p denoting the symbol – principal in this

case – of (2.1) and ρ being a point in the double characteristic manifold $\Sigma = \{(x,\xi) \mid (x,\xi) \in \mathbb{R}^{n+1} \setminus 0,\ x_1 = 0, \xi_0 = \xi_1 = 0\ \}$.

It is well known that, in the case $k = 1$, the Hamiltonian system of (2.1) is stable, i.e. there are no points on the simple characteristic manifold of (2.1) such that the bicharacteristic curves issued from them have limit points on Σ, if and only if $b = 0$.

In [2] it was shown that condition $b = 0$ for the case $k = 1$ has an invariant meaning and actually amounts to asserting that $H_S^3 p(\rho) = 0$ for $\rho \in \Sigma$. Here $S(x,\xi)$ is a smooth real function vanishing on Σ verifying (2.8) in [2].

In this section we recall the argument for the operator (2.1). Let us consider, near $\rho_0 = (0, e_n)$, the symbol of (2.1) whose Hamiltonian system is

(2.2)
$$
\begin{aligned}
\dot{x}_0 &= -2\xi_0 + 2x_1^k \xi_n & \dot{\xi}_0 &= 0 \\
\dot{x}_1 &= 2\xi_1 & \dot{\xi}_1 &= -2kx_1^{k-1}\xi_0\xi_n - (2k+1)bx_1^{2k}\xi_n^2 \\
\dot{x}' &= 0 & \dot{\xi}' &= 0 \\
\dot{x}_n &= 2x_1^k\xi_0 + 2bx_1^{2k+1}\xi_n & \dot{\xi}_n &= 0.
\end{aligned}
$$

We want to show that there is a null bicharacteristic curve $\gamma(t)$ of finite length in the (x,ξ) space, that has a limit point on Σ and is issued from a point outside Σ.

Let us denote by t the curve parameter; we easily see that $\xi_0 \equiv 0$, $\xi_n \equiv 1$, $x' \equiv y'$ and $\xi' \equiv \eta'$, where $\gamma(0) = (y_0, y_1, y', y_n; 0, \eta_1, \eta', 1)$.

Because of this the only equation we have to solve is

$$
\ddot{x}_1 = -2(2k+1)bx_1^{2k}, \qquad x_1(0) = y_1, \quad \dot{x}_1(0) = 2\eta_1.
$$

This gives

$$
x_1(t) = \left(-(2k-1)\operatorname{sgn}(\eta_1)\sqrt{-b}\,t + y_1^{-k+1/2}\right)^{-\frac{2}{2k-1}},
$$

where we have chosen $b < 0$, $y_1 > 0$ and $\eta_1^2 + by_1^{2k+1} = 0$. In the same way one gets

$$
x_0(t) = y_0 + 2\int_0^t \left[-(2k-1)\operatorname{sgn}(\eta_1)\sqrt{-b}\,s + y_1^{-k+1/2}\right]^{-\frac{2k}{2k-1}} ds,
$$

$$x_n(t) = y_n + 2b \int_0^t \left[-(2k-1)\operatorname{sgn}(\eta_1)\sqrt{-b}s + y_1^{-\frac{2k-1}{2}} \right]^{-\frac{2(2k+1)}{2k-1}} ds.$$

As $t \to +\infty$ we easily see that

$$(2.3) \quad \lim_{t \to +\infty} \gamma(t) =$$

$$(y_0 + \frac{1}{\operatorname{sgn}(\eta_1)\sqrt{-b}}\sqrt{y_1}, 0, y', y_n + \frac{\sqrt{-b}}{\operatorname{sgn}(\eta_1)}\frac{1}{2k+3}y_1^{(2k+3)/2}; 0, 0, \eta', 1)$$

Actually from the above formulas one can deduce that the points on Σ which are limits of null bicharacteristics issued from simply character- istic points are actually reached by two different bicharacteristic curves landing on Σ along the direction of the propagation cone.

In order to deduce an a priori estimate for (2.1) we rewrite it in a way reminiscent of the Ivrii decomposition (see e.g. [4])

$$(2.4) \qquad P(x,\xi) = -\Lambda(x,\xi)M(x,\xi) + Q(x,\xi'),$$

where Λ, M and Q vanish on Σ, Λ, M are homogenous of degree one and Q is homogeneous of degree 2 w.r.t. ξ'; moreover $Q(x,\xi') \geq 0$, $H_\Lambda(\rho) \in \ker F(\rho) \cap \operatorname{Im} F(\rho)^3$ and

$$H_{\{\Lambda,Q\}}(\rho) = 0, \qquad \forall \rho \in \Sigma.$$

Set

$$(2.5) \qquad \Lambda(x,\xi) = \xi_0 - \gamma x_1^k |x_1|\xi_n$$

$$(2.6) \qquad M(x,\xi) = \xi_0 + 2x_1^k\xi_n + \gamma x_1^k |x_1|\xi_n,$$

$$(2.7) \qquad Q(x,\xi') = \xi_1^2 - 2\gamma|x_1|^{2k+1}\xi_n^2 + bx_1^{2k+1}\xi_n^2 - \gamma^2|x_1|^{2k+2}\xi_n^2.$$

Here γ denotes a negative constant such that

$$\gamma < 0, \qquad -2\gamma + b > 0.$$

We point out that the above symbols are not C^∞ although they retain a minimal C^k regularity allowing us to deduce an energy estimate.

3. The Gevrey estimate

3.1. **Lower bounds.** This section is devoted to the proof of the following estimate:

Lemma 1. *Let h be a positive integer, $v \in C_0^\infty(\mathbb{R}^n)$, $\chi(x, \xi)$ a pseudo differential operator of order 0 properly supported with support in a conic neighborhood of $(0, e_n)$; denote by $u(x) = \chi(x, D)v(x)$ and*

$$(3.1.1) \qquad L_h(x, D) = D_1^2 + a|x_1|^h D_n^2,$$

where a is a positive constant. Then we have the estimate

$$(3.1.2) \qquad \langle L_h u, u \rangle \geq C_h \|u\|_{2/(h+2)}^2,$$

where C_h is a suitable positive constant.

Proof. In order to prove (3.1.1) we decompose the operator $L_h(x, D)$ in (3.1.2) as follows: let θ_j, μ_j be positive real numbers, $\mu_j < 1$, $j = 1, \ldots N$, and $N \in \mathbb{N}$ a suitable integer dependent on h to be fixed later, such that $\sum_{j=1}^{N} \theta_j^2 = 1$. The following identity is straightforward:

$$L_h(x, D) = \sum_{j=2}^{N} \theta_j^2 D_1^2 + a\mu_1 |x_1|^h D_n^2 + \theta_1^2 D_1^2 + a(1 - \mu_1)|x_1|^h D_n^2.$$

For the last two terms in the above formula we have:

$$\theta_1^2 D_1^2 + a(1 - \mu_1)|x_1|^h D_n^2 = Y_{\theta_1}^* Y_{\theta_1} + \theta_1 h/2 \sqrt{a(1 - \mu_1)}|x_1|^{h/2-1} D_n.$$

where we have set $Y_{\theta_1} = \theta_1 D_1 - i\sqrt{a(1 - \mu_1)}x_1|x_1|^{h/2-1} D_n$. Let us iterate once more the same procedure:

$$(3.1.3) \quad L_h(x, D) =$$

$$\sum_{j=3}^{N} \theta_j^2 D_1^2 + Y_{\theta_1}^* Y_{\theta_1} + a\mu_1 |x_1|^h D_n^2 + \theta_1 \frac{h}{2} \mu_2 \sqrt{a(1 - \mu_1)}|x_1|^{h/2-1} D_n +$$

$$\theta_2^2 D_1^2 + \theta_1 \frac{h}{2}(1 - \mu_2)\sqrt{a(1 - \mu_1)}|x_1|^{h/2-1} D_n.$$

Again the sum of the last two terms in (3.1.3) can be rewritten as:

$$Y_{\theta_2}^* Y_{\theta_2} + \theta_2 \left(\frac{h}{4} - \frac{1}{2} \right) \sqrt{\frac{h}{2}(1 - \mu_2)\theta_1 \sqrt{a(1 - \mu_1)}}|x_1|^{h/4-1/2-1} D_n^{1/2}.$$

As before we defined

$$Y_{\theta_2} = \theta_2 D_1 - i\sqrt{\theta_1 \frac{h}{2}(1 - \mu_2)\sqrt{a(1 - \mu_1)}x_1|x_1|^{h/4 - 1/2 - 1} D_n^{1/2}}.$$

Iterating this procedure it is therefore possible to write (3.1.1) in the following form:

$$(3.1.4) \quad L_h(x, D) =$$

$$\sum_{j=1}^{N} Y_{\theta_j}^* Y_{\theta_j} + \sum_{j=1}^{N+1} c_j(a, \mu_1, \ldots, \mu_N, \theta_1, \ldots, \theta_N)|x_1|^{\frac{h}{2^{j-1}} - \sum_{\ell=0}^{j-2}\frac{1}{2^\ell}} D_n^{\frac{2}{2^{j-1}}},$$

where $c_j(a, \mu_1, \ldots, \mu_N, \theta_1, \ldots, \theta_N)$ are positive constants and N is taken as the integer part of $\log_2(h + 2)$.

This choice of N implies that

$$\frac{h}{2^N} - \sum_{\ell=0}^{N-1}\frac{1}{2^\ell} < 0$$

and

$$\frac{h}{2^{N-1}} - \sum_{\ell=0}^{N-2}\frac{1}{2^\ell} > 0.$$

From (3.1.4) and from the above two inequalities we have

$$\langle L_h(x, D)u, u \rangle = \sum_{j=1}^{N} \|Y_{\theta_j}u\|^2$$

$$+ \sum_{j=1}^{N+1} c_j(a, \mu_1, \ldots, \mu_N, \theta_1, \ldots, \theta_N)\langle |x_1|^{\frac{h}{2^{j-1}} - \sum_{\ell=0}^{j-2}\frac{1}{2^\ell}} D_n^{\frac{2}{2^{j-1}}}u, u \rangle.$$

Now each term of the second sum in the right hand side of the above equation has a positive power of $|x_1|$ but the last one. We remark explicitly that in spite of this there is no problem about the convergence of the integrals.

Moreover, introducing the adimensional variable $y_1 = x_1 \xi_n^{2/(h+2)}$, we can write

$$\sum_{j=1}^{N+1} c_j(a, \mu_1, \ldots, \mu_N, \theta_1, \ldots, \theta_N)|x_1|^{\frac{h}{2^j-1} - \sum_{\ell=0}^{j-2} \frac{1}{2^\ell}} \xi_n^{\frac{2}{2^j-1}}$$

$$= \xi_n^{4/(h+2)} \sum_{j=1}^{N+1} c_j(a, \mu_1, \ldots, \mu_N, \theta_1, \ldots, \theta_N)|y_1|^{\frac{h}{2^j-1} - \sum_{\ell=0}^{j-2} \frac{1}{2^\ell}}.$$

Now the above polynomial estimates from above

$$\xi_n^{4/(h+2)} \left[c_N(a, \mu_1, \ldots, \mu_N, \theta_1, \ldots, \theta_N)|y_1|^{\frac{h}{2^N-1} - \sum_{\ell=0}^{N-2} \frac{1}{2^\ell}} \right.$$

$$\left. + c_{N+1}(a, \mu_1, \ldots, \mu_N, \theta_1, \ldots, \theta_N)|y_1|^{\frac{h}{2^N} - \sum_{\ell=0}^{N-1} \frac{1}{2^\ell}} \right] \geq c > 0,$$

where c is a constant depending on (a, θ, μ, h).

Taking into account that, on the microsupport of χ, $\xi_n \sim |\xi|$ we complete the proof of the lemma. $\qquad\square$

3.2. **The energy estimate.** In this section we prove the

Theorem 1. *The Cauchy problem for the operator (2.1) is well posed in the Gevrey G^s category, for $1 \leq s \leq 2k+1$. Moreover the following energy estimate holds:*

(3.2.1)
$$\int_0^{+\infty} \|e^{\tau x_0 D_n^{1/s}} u(x_0, \cdot)\|^2_{\frac{2}{2k+3}} \, dx_0 \leq C \int_0^{+\infty} \|e^{\tau x_0 D_n^{1/s}} Pu(x_0, \cdot)\|^2 \, dx_0,$$

where $u \in C_0^\infty(\Omega)$, Ω an open subset of $\mathbb{R}^{n+1}$ containing the origin.

Proof. Let u be as in the statement. We compute

(3.2.2) $2i \operatorname{Im}\langle Pu, -\Lambda u\rangle$

$$= 2i \operatorname{Im}\langle M\Lambda u, \Lambda u\rangle - 2i \operatorname{Im}\langle Qu, \Lambda u\rangle$$

$$= D_0 \left[\|\Lambda u\|^2 + \langle Qu, u\rangle\right] + 2i\gamma \operatorname{Im}\langle D_1^2 u, x_1^k |x_1| \xi_n u\rangle,$$

where

$$\langle u, v\rangle = \int_{\mathbb{R}^{n-1}} \hat{u}(x_0, x_1, x', \xi_n) \bar{\hat{v}}(x_0, x_1, x', \xi_n) \, dx_1 dx'$$

$\hat{u}$ denoting the partial Fourier transform with respect to x_n; here Λ, M and Q are defined in (2.5), (2.6) and (2.7).

Define

(3.2.3) $E(u)(x_0) = \|\Lambda u\|^2 + \langle Qu, u \rangle;$

in order to deduce a suitable estimate yielding well posedness in the Gevrey category we multiply (3.2.2) by $\exp\left(2\tau x_0 \xi_n^{1/s}\right)$, where τ is a large positive parameter and s a real number, $s \geq 1$.

We therefore have

$$(3.2.4) \quad 2\int_0^{+\infty} e^{2\tau x_0 \xi_n^{1/s}} \, \mathrm{Im}\langle Pu, -\Lambda u \rangle \, dx_0$$

$$= -\int_0^{+\infty} e^{2\tau x_0 \xi_n^{1/s}} \partial_0 E(u)(x_0) \, dx_0$$

$$+ 2\gamma \int_0^{+\infty} e^{2\tau x_0 \xi_n^{1/s}} \langle x_1^{k-1} |x_1| D_1 u, \xi_n u \rangle$$

$$- i\gamma \int_0^{+\infty} e^{2\tau x_0 \xi_n^{1/s}} \langle k(k+1) x_1^{k-2} |x_1| u, \xi_n u \rangle \, dx_0.$$

Hence

$$(3.2.5) \quad 2\int_0^{+\infty} e^{2\tau x_0 \xi_n^{1/s}} \, \mathrm{Im}\langle Pu, -\Lambda u \rangle \, dx_0$$

$$= 2\tau \xi_n^{1/s} \int_0^{+\infty} e^{2\tau x_0 \xi_n^{1/s}} E(u)(x_0) \, dx_0 + E(u)(0) + 2\gamma R_1 - i\gamma R_2,$$

where R_1 and R_2 stand for the corresponding terms in identity (3.2.4).

Let us consider $E(u)$:

$$(3.2.6) \quad E(u) = \|\Lambda u\|^2 + \langle Qu, u \rangle$$

$$= \|\Lambda u\|^2 + \langle \left(D_1^2 + 2|\gamma| e(x_1) |x_1|^{2k+1} \xi_n^2\right) u, u \rangle$$

$$\geq C\left(\|\Lambda u\|^2 + \|D_1 u\|^2 + \||x_1|^{k+1/2} \xi_n u\|^2 + \|u\|_{\frac{2}{2k+3}}^2\right),$$

where we denoted

$$e(x_1) = 1 + \frac{b}{2|\gamma|} \, \mathrm{sgn}(x_1) - \frac{|\gamma|}{2}|x_1|$$

and the result of Lemma 1 has been applied with a equal to a lower bound for the function e.

Let us now estimate the error terms R_1 and R_2. Let us start with R_1:

$$\left| 2\gamma \int_0^{+\infty} e^{2\tau x_0 \xi_n^{1/s}} \langle x_1^{k-1} |x_1| D_1 u, \xi_n u \rangle \right|$$

$$\leq 2|\gamma| \int_0^{+\infty} e^{2\tau x_0 \xi_n^{1/s}} \left(\|D_1 u\|^2 + \||x_1|^k \xi_n u\|^2 \right) \, dx_0.$$

The first summand above can be easily estimated using the energy $E(u)$ by (3.2.6). Let us take a look at the second summand in the above inequality. We will actually estimate $\xi_n^{-2/s} \||x_1|^k \xi_n u\|^2$ with the quantity $\||x_1|^{k+1/2} \xi_n u\|^2 + \|u\|^2_{\frac{2}{2k+3}}$.

Since ξ_n is a positive large parameter the above inequality reduces to the following

$$\xi_n^{-2/s} |x_1|^{2k} \xi_n^2 \lesssim |x_1|^{2k+1} \xi_n^2 + \xi_n^{\frac{4}{2k+3}}.$$

We have

$$\xi_n^{-2/s+2} |x_1|^{2k} \lesssim \left(|x_1|^{2k} \xi_n^\theta \right)^p + \left(\xi_n^{2-2/s-\theta} \right)^q,$$

where $\frac{1}{p} + \frac{1}{q} = 1$ and θ is chosen to be $\frac{2}{p}$. We easily verify that the right hand side in the above inequality can be estimated by $|x_1|^{2k+1} \xi_n^2 + \xi_n^{\frac{4}{2k+3}}$ if $s \leq 2k+3$.

Let us now turn to R_2.

Arguing as above we are led to proving the estimate:

$$(3.2.7) \qquad |x_1|^{k-1} \xi_n^{1-1/s} \lesssim |x_1|^{2k+1} \xi_n^2 + \xi_n^{\frac{4}{2k+3}}.$$

Using as before the Hölder inequality we can write

$$|x_1|^{k-1} \xi_n^{1-1/s} = \left(|x_1|^{k-1} \xi_n^\theta \right)^p \left(\xi_n^{1-1/s-\theta} \right)^q,$$

where $p = \frac{2k+1}{k-1}$, $q = \frac{2k+1}{k+2}$, $\theta = \frac{2(k-1)}{2k+1}$ (we are clearly arguing in the case $k \geq 2$, the case $k = 1$ being much simpler).

The right hand side can be estimated by the right hand side of (3.2.7) if $s \leq 2k+3$.

Summing up we proved that

$$\left| 2 \int_0^{+\infty} e^{2\tau x_0 \xi_n^{1/s}} \operatorname{Im}\langle Pu, -\Lambda u \rangle \, dx_0 \right|$$

$$\geq C \tau \xi_n^{1/s} \int_0^{+\infty} e^{2\tau x_0 \xi_n^{1/s}} E(u)(x_0) \, dx_0$$

from which we deduce

$$\int_0^{+\infty} e^{2\tau x_0 \xi_n^{1/s}} \| Pu \|^2 \, dx_0 \geq \int_0^{+\infty} e^{2\tau x_0 \xi_n^{1/s}} \| u \|^2_{\frac{2}{2k+3}} \, dx_0.$$

A final Fourier transform in the x_n variable proves the theorem.   □

An analogous argument also allows us to prove the following

Theorem 2. *Consider the operator*

(3.2.8) $-D_0^2 + 2x_1^k D_0 D_n + D_1^2 + b x_1^{2k+1} D_n^2 + q_2(x', D'),$

where

$$q_2(x', \xi') = \sum_{j=2}^{n-1} \alpha_j (D_j^2 + x_j^2 D_n^2) + \beta D_n,$$

and $\alpha_j \geq 0$, $\pm |\beta| + \sum_{j=2}^{n-1} \alpha_j > 0$.

Then the Cauchy problem for (3.2.8) is well posed in G^s *for* $1 \leq s \leq 4k + 2$. *Furthermore the following energy estimate holds*

$$\int_0^{+\infty} \| e^{\tau x_0 D_n^{1/s}} u(x_0, \cdot) \|_{\frac{1}{2}}^2 \, dx_0 \leq C \int_0^{+\infty} \| e^{\tau x_0 D_n^{1/s}} Pu(x_0, \cdot) \|^2 \, dx_0,$$

for $1 \leq s \leq 4k + 2$ *and* $u \in C_0^\infty$.

References

[1] E. BERNARDI, A. BOVE, *Propagation of Gevrey singularities for hyperbolic operators with triple characteristics, I*, Duke Math. J. **60** (1990), pp. 187-205.

[2] E. BERNARDI, A. BOVE AND C. PARENTI, *Geometric Results for a Class of Hyperbolic Operators with Double Characteristics, II*, J. Functional Analysis **116** (1993), pp. 62-82.

[3] E. BERNARDI AND A. BOVE, *Geometric Results for a Class of Hyperbolic Operators with Double Characteristics*, Comm. Part. Diff. Eq. **13** (1988), pp. 61-86.

[4] V.YA.IVRII, *The Well-posedness of the Cauchy Problem for Nonstrictly Hyperbolic Operators. III. The Energy Integral*, Trans. Moscow math. Soc. **34** (1978), pp. 149-168.

[5] T. NISHITANI, *Note on Some Non Effectively Hyperbolic Operators*, Sci. Rep. **32** (1983), pp. 9-17.

[6] T. NISHITANI, *The Hyperbolic Cauchy Problem*, Lecture Notes in Mathematics **1505** (1991).

DIPARTIMENTO DI MATEMATICA, UNIVERSITÀ DI BOLOGNA, PIAZZA DI PORTA S. DONATO 5, 40127 BOLOGNA, ITALIA
E-mail address: `bernardi@dm.unibo.it`

DIPARTIMENTO DI MATEMATICA, UNIVERSITÀ DI BOLOGNA, PIAZZA DI PORTA S. DONATO 5, 40127 BOLOGNA, ITALIA
E-mail address: `bove@dm.unibo.it`

Gevrey well-posedness of the Cauchy problem for systems

HIDEO YAMAHARA

Faculty of Engineering, Osaka Electro-Communication University
Hatsu-cho, Neyagawa-city, Osaka, 572-8530, Japan

e-mail : yamahara@isc.osakac.ac.jp

Dedicated to Professor Jean Vaillant

1 INTRODUCTION

We are concerned with the classical Cauchy problem to the linear system of first order differential operators.

Let L be a $m \times m$ system of the first order partial differential operators, more precisely,

$$L \equiv L(t, x, D_t, D_x) = ID_t + \sum_{k=1}^{\ell} A_k(t, x)D_{x_k} + B(t, x)$$

defined in $[0, T] \times \mathbf{R}_x^\ell$, where I is the $m \times m$ unit matrix, $A_k(t, x)$ $(1 \le k \le n)$ $B(t, x)$ are $m \times m$ matrices and

$$D_t = \frac{1}{\sqrt{-1}}\frac{\partial}{\partial t}, \quad D_{x_k} = \frac{1}{\sqrt{-1}}\frac{\partial}{\partial x_j} \quad (1 \le k \le n).$$

For this system, we study the Cauchy problem:

$$\text{(C.P.)} \quad \begin{cases} L(t, x, D_t, D_x)u(t, x) = 0 \quad \text{in } [0, T] \times \mathbf{R}_x^{\ell}, \\ u(0, x) = u_0(x) \quad \text{on } \mathbf{R}_x^{\ell}. \end{cases}$$

We remark at first that we limit our concern to study (C.P.) in a class of Gevrey, therefore we shall always assume that the polynomial $\det(I\tau + \sum_{k=1}^{\ell} A_k(t, x)\xi_k)$ has only real roots with respect to the variable τ. Secondly as our present interest is the relation between the Gevrey index and the matrix structure of the principal symbol, we assume the regularity of the coefficients as we need. The contexture is divided into four parts as follows.

At first after some notations and definitions we will explain an historical background of our problem, especially the difference from the case of single equations of higher order, and the difference from the case of constant multiplicities of characteristic roots. Secondly we treat a certain system of differential operators, which includes, in some sense, a typical problem for systems. And with this system we will show the necessary and sufficient conditions for the strongly Gevrey wellposedness. In the last two sections we will give a rough sketch of the proofs for the sufficiency and the necessity of our theorem.

2   STATEMENT OF RESULTS

2.1   Let Ω be an open set in $\mathbf{R}_x^{\ell}$, and $f(x)$ be a C^{∞} function in Ω.

DEFINITION 1   *We say that $f(x)$ belongs to the class of $\gamma^{(s)}(\Omega)$ for $s > 1$ when for any compact set K in Ω, there exists constant C_K such that the following inequality holds.*

$$|D^{\alpha}f(x)| \leq C_K^{|\alpha|+1} \cdot |\alpha!|^s \quad (\forall \alpha \in N^{\ell}, \ \forall x \in K).$$

We are concerned that the (C.P.) has a unique solution in the class of $\gamma^{(s)}(\Omega)$. We say it $\gamma^{(s)}$-wellposedness, precisely we have the following

DEFINITION 2   *(C.P.) is $\gamma^{(s)}$-wellposed when for any initial data $u_0(x) \in \gamma^{(s)}(\Omega)$, there exist a positive $\tilde{T}$ and an unique solution $u(x, t)$ of (C.P.) such that $u(x, t) \in C^1([0, \tilde{T}]; \gamma^{(s)}(\Omega))$.*

Moreover our present concern is "$\gamma^{(s)}$-wellposedness for any lower order term".

DEFINITION 3   (C.P.) *is strongly $\gamma^{(s)}$-wellposed when for any lower order term $B(t,x)$, (C.P.) is $\gamma^{(s)}$-wellposed.*

Therefore our problem can be described as follows. For a given principal symbol

$$A(t,x,\xi) = \sum A_k(t,x)\xi_k,$$

find the index "s", with such a "s" (C.P.) is strongly $\gamma^{(s)}$-wellposed.

Here we remark that from the definition we know that when $s_1 \leq s_2$, $\gamma^{(s_2)}$-wellposedness leads us to $\gamma^{(s_1)}$-wellposedness.

So our final goal is to find out the greatest index s_0 :

$$s_0 = \max\{s\,;\,(C.P.)\text{ is }\gamma^{(s)}-\text{wellposed}\}.$$

In general this is a difficult problem. Even in the case of a second order equation, we have not complete results yet. Anyway concerning to this problem we must recall the famous theorem.

THEOREM 0 (Ohya [1], Leray-Ohya [2], Bronstein [3], Kajitani [4] and Nishitani [5])   (C.P.) *is strongly $\gamma^{(s)}$-wellposed when*

$$s \leq \frac{r}{r-1} \quad (r > 1),$$

where r is the maximal multiplicity of the eigenvalues of $A(t,x,\xi)$ ($\xi \neq 0$).

REMARK 1   This theorem is, at first, showed by Leray-Ohya in the case of constant multiplicities. In general case, Bronstein at first, and after a short time Kajitani and Nishitani proved this theorem with their own methods.

REMARK 2   When the multiplicities of the characteristic roots are constant, the number $\frac{r}{r-1}$ is optimal in a case of single equation of higher order. But of course, in general case this is not optimal. The following operator is a typical example in this matter.

EXAMPLE   *Let $P = (\partial_t - \lambda_1(t)\partial_x)(\partial_t - \lambda_2(t)\partial_x)$ be a second order operator, where λ_1 and λ_2 are real and smooth. When we assume that $\lambda_1(t) - \lambda_2(t) \sim t^p$ near $t = 0$, then The Cauchy problem for P is strongly $\gamma^{(s)}$-wellposed if and only if $s \leq \frac{2p}{p-1}$ ($p > 1$).*

In connections with this example we refer that Colombini and Orrù obtained similar result in the case of a third order operator.

Here we are concerned with this problem in the case of systems. Of course in both cases Theorem 0 holds, but our assertion is that the index $\dfrac{r}{r-1}$ can be replaced by a larger one. For example when $A = A(t,x,\xi)$ is diagonal matrix, the multiplicities have no meaning for the Gevrey wellposedness. Therefore we propose the following

THEOREM 1   *When the multiplicities of the characteristic roots are constant, (C.P.) is strongly $\gamma^{(s)}$-wellposed if and only if*

$$s \le \frac{r_0}{r_0 - 1} \quad (r_0 > 1),$$

where r_0 is the maximal multiplicity of zeros of the minimal polynomial of $A(t,x,\xi)$ ($\xi \ne 0$). (Or we can say that r_0 is the maximal size of the Jordan cells of the Jordan normal form similar to $A(t,x,\xi)$.)

REMARK 3   We know that in the case of constant multiplicities Vaillant has written many papers. And also Matsumoto [6] and Taglialatela [7] and others, they established so-called the Levi conditions using a non-commutative algebra. Concerning to the Gevrey wellposedness Vaillant [8], [9] gave more detail results, which means that he verified completely the relations between the Gevrey indices and the Levi conditions when the multiplicities are at most 5.

REMARK 4   Our proof ([10]) of Theorem 1 is very simple and is near to the method of Vaillant. But we imposed that the coefficients depend on only time variable. Though we believe our method will work well in general case we have not established the proof yet.

In the case of non-constant multiplicities, we studied a simple example which is a 4×4 system of partial differential operators ([11]). In that example we gave complete conditions for the strongly $\gamma^{(s)}$-wellposedness.

This time to make the situation clearer we will study a certain system, which is just a generalization of the last one.

2.2   Let
$$L \equiv L(t, D_t, D_x) = I_m D_t + A(t)D_x + B(t)$$
be a $m \times m$ system of differential operators with one spatial variable, where I_m means the unit matrix of order m, and
$$\det(\tau I - A(t)) = (\tau - \lambda_1(t))^{m_1}(\tau - \lambda_2(t))^{m_2},$$
more precisely, $A(t)$ is just a following matrix:

$$A(t) = \left(\begin{array}{cc|cc} \lambda_1(t) & 1 & & \\ & \ddots & \ddots & \\ & & \ddots & 1 \\ & & & \lambda_1(t) & a(t) \\ \hline & & & \lambda_2(t) & 1 \\ & & & & \ddots & \ddots \\ & & & & & \ddots & 1 \\ & & & & & & \lambda_2(t) \end{array} \right).$$

Moreover we impose the following hypothesis:

(H-1) $\lambda_1(t) - \lambda_2(t) \sim t^p$ (near $t = 0$),

(H-2) $a(t) \sim t^q$ (near $t = 0$),

where p and q are positive integers. Without loss of generality, we assume that $m_1 \geq m_2$. Therefore r in Theorem 0 is equal to $m_1 + m_2$ and r_0 in Theorem 1 is equal to m_1.

Our result is as follows.

THEOREM 2 *Assume that $m_1 \geq m_2$ and $m_1 \geq 2$, then (C.P.) is strongly $\gamma^{(s)}$-wellposed if and only if*

$$\begin{cases} s \leq \dfrac{m_1}{m_1 - 1} & \text{(when} \quad m_2 p \leq q), \\[3mm] s \leq \dfrac{(m_1 + m_2)p - q}{(m_1 + m_2 - 1)p - q} & \text{(when} \quad m_2 p \geq q). \end{cases}$$

REMARK 5 In the necessity, especially when $m_2 p > q$, we can find a lower order term $B(t)$ such that the Cauchy problem is not strongly $\gamma^{(s)}$-wellposed for such a s that

$$\forall s > \frac{(m_1 + m_2)p - q}{(m_1 + m_2 - 1)p - q}.$$

Of course we see that

$$\frac{m_1}{m_1 - 1} > \frac{(m_1 + m_2)p - q}{(m_1 + m_2 - 1)p - q},$$

which means that the assertion of Theorem 1 is not true in general.

REMARK 6   When $m_1 = 1$ (therefore $m_2 = 1$), essentially the operator L means in this case that

$$L_1 = I_2 D_t + \begin{pmatrix} t^p & t^q \\ 0 & t^p \end{pmatrix} D_x + B(t) \quad (p \geq 1, \ q \geq 1).$$

For this operator L_1, we can show the following

PROPOSITION 2.1

   (1)   (C.P.) *is strongly C^∞-wellposed* $\Leftrightarrow$ $p \leq q + 1$.

   (2)   (C.P.) *is strongly $\gamma^{(s)}$-wellposed* $\Leftrightarrow$ $s \leq \dfrac{2p - q}{p - q - 1}$ $(p > q + 1)$.

3   A SKETCH OF THE PROOF OF THEOREM 2 (SUFFICIENCY)

We start with $Lu = 0$, precisely

(3.1) $I_m D_t u + A(t) D_x u + B(t) u = 0.$

As usual by Fourier transform with respect to the spatial variable x, we study the ordinary differential equations:

(3.2) $I_m D_t v + nA(t)v + B(t)v = 0,$

where we denote the dual variable by n instead of the usual notation ξ, and we regard it as a large parameter. The unknown v means that $v = v(t,n) = \widehat{u}(t, \cdot)$ is the partial Fourier transform of $u(t,x)$ with respect to x.

   To get the energy inequality we separate the interval $[0,T]$ by $t = n^{-\sigma}$, where σ is a positive constant to be determined later. At first on the interval $[0, n^{-\sigma}]$ we obtain the following estimate.

PROPOSITION 3.1   *Assume that $m_1 \geq m_2$, then for $0 \leq t \leq n^{-\sigma}$, the solution of (3.2) satisfies the estimate:*

(3.3) $|v(t,n)| \leq C_1 n^{M_1} \exp\left(\delta_1 n^{1-\varepsilon} t\right) \cdot |v(0,n)|,$

where C_1, M_1 and δ_1 are positive constants independent of n. And ε is determined in such a way that

$$
\varepsilon = \begin{cases} \dfrac{1}{m_1} & \left(\text{when} \quad q\sigma \geq \dfrac{m_2}{m_1} \right), \\[4mm] \dfrac{1+q\sigma}{m_1+m_2} & \left(\text{when} \quad q\sigma \leq \dfrac{m_2}{m_1} \right). \end{cases}
$$

Next in the second interval we can show the following

PROPOSITION 3.2   *Assume that $m_1 \geq m_2$ and $m_1 \geq 2$, then for $n^{-\sigma} \leq t_0 \leq t$, the solution of (3.2) satisfies the estimate:*

$$
(3.4) \qquad |v(t,n)| \leq C_2\, n^{M_2} \exp\left(\delta_2 n^{1-\varepsilon}(t-t_0)\right) \cdot |v(t_0,n)|,
$$

where C_2, M_2 and δ_2 are positive constants independent of n. And ε is determined in such a way that

$$
\varepsilon = \begin{cases} \dfrac{1}{m_1} & (\text{when} \ \ m_2 p \leq q, \ \ p\sigma \geq \dfrac{1}{m_1}), \\[4mm] \dfrac{1}{m_1}(1-(m_2 p - q)\sigma) & (\text{when} \ \ m_2 p \geq q, \ \ ((m_1+m_2)p - q)\sigma \leq 1). \end{cases}
$$

To prove these propositions we introduce the "change of weight" $W(n) = W(n;\varepsilon_1,\varepsilon_2,\varepsilon')$, precisely we set the diagonal matrix:

$$
W_k(n) = \begin{pmatrix} 1 & & & \\ & n^{-\varepsilon_k} & & \\ & & \ddots & \\ & & & n^{-(m_k-1)\varepsilon_k} \end{pmatrix} \qquad (k=1,2).
$$

And define our change of weight such that

$$
W(n) = W_1(n) \oplus n^{-\varepsilon'} W_2(n).
$$

On the interval $[0, n^{-\sigma}]$, we directly estimate v_1, defined by $v = W(n)v_1$ to the solution v of (3.2).

We expect that all terms are estimated by $n^{1-\varepsilon_1}$. For this purpose we divide our procedure in two cases; One is the case when $(m_1 - 1)\varepsilon_1$ is greater than $(m_2 - 1)\varepsilon_2 + \varepsilon'$ and the other case. To obtain the good estimate we choose ε_1 as large as possible. Thus we can determine ε_1 as ε in proposition 3.1. Other ε_2 and ε' are chosen suitably.

When $n^{-\sigma} < t$, we tranform $A(t)$ to the strict Jordan normal form:

$$
N(t)^{-1}A(t)N(t) = \begin{pmatrix} \lambda_1(t) & 1 & & & & & & \\ & & \cdot & \cdot & & & & \\ & & & \cdot & \cdot & & & \\ & & & & \cdot & 1 & & \\ & & & & & \lambda_1(t) & & \\ \hline & & & & & & \lambda_2(t) & 1 \\ & & & & & & & \cdot & \cdot \\ & & & & & & & & \cdot & \cdot \\ & & & & & & & & & \cdot & 1 \\ & & & & & & & & & & \lambda_2(t) \end{pmatrix}.
$$

Next we apply the change of weight $\tilde{W}(n)$ which is the same type as of $W(n)$. After these transformations v_2, defined by $v = N(t)\tilde{W}(n)v_2$ to the solution v of (3.2), satisfies the equation:

$$
D_t v_2 + nA_{11}v_2 + n\tilde{W}^{-1}A_{12}\tilde{W}v_2 + \tilde{W}^{-1}\left[N^{-1}BN + N^{-1}(D_tN)\right]\tilde{W}v_2 = 0.
$$

Here we divide the Jordan form $N(t)^{-1}A(t)N(t)$ to the diagonal part:

$$
A_{11} = \lambda_1(t) \oplus \cdots \oplus \lambda_1(t) \oplus \lambda_2(t) \oplus \cdots \oplus \lambda_2(t),
$$

and A_{12} which is the nilpotent Jordan form.

To obtain the following inequality:

$$
\left| n\tilde{W}^{-1}A_{12}\tilde{W} + \tilde{W}^{-1}\left[N^{-1}BN + N^{-1}(D_tN)\right]\tilde{W} \right| \leq \text{const.}\, n^{1-\varepsilon_1} \quad (t \geq n^{-\sigma}),
$$

we impose several conditions to the parameters ε_1, ε_2, ε' and σ.

In conclusion we can choose ε_1 as ε in Proposition 3.2 according to the case when $m_2p \leq q$ or not.

Concerning to the determinations of ε in Proposition 3.1 and Proposition 3.2, it is easy to observe that

$$
\begin{cases} m_2p \leq q \quad \text{and} \quad p\sigma \geq \dfrac{1}{m_1} & \Rightarrow \quad q\sigma \geq \dfrac{m_2}{m_1}, \\[2mm] m_2p \geq q \quad \text{and} \quad ((m_1+m_2)p - q)\sigma \leq 1 & \Rightarrow \quad q\sigma \leq \dfrac{m_2}{m_1}. \end{cases}
$$

Therefore we can connect the inequality in Proposition 3.1 and the inequality in Proposition 3.2. Precisely we obtain following

PROPOSITION 3.3   *Assume that $m_1 \geq m_2$ and $m_1 \geq 2$, then for large n, the solution of (3.2) satisfies the estimate:*

(I) *When $m_2 p \leq q$ and $p\sigma \geq \dfrac{1}{m_1}$, then*

$$(3.5) \qquad |v(t,n)| \leq C\, n^M \exp\left(\delta n^{1-\frac{1}{m_1}} t\right) \cdot |v(t_0,n)|.$$

(II) *When $m_2 p \geq q$ and $((m_1 + m_2)p - q)\sigma \leq 1$, then*

$$(3.6) \qquad |v(t,n)| \leq C\, n^M \exp\left(\delta n^{1-\frac{1}{m_1}(1-(m_2 p - q)\sigma)} t \; + \right.$$
$$\left. + \; \delta' n^{1-\frac{1+q\sigma}{m_1 + m_2} - \sigma}\right) \cdot |v(t_0,n)|,$$

where C, M, δ and δ' are positive constants independent of n.

When the initial data satisfies the estimate:

$$|v(0,n)| \leq C_0 \exp\left(-c_0 n^{\frac{1}{s}}\right),$$

we expect that the solution also satisfies that

$$|v(t,n)| \leq C \exp\left(-c n^{\frac{1}{s}}\right).$$

Therefore it is clear that when $m_2 p \leq q$, (C.P.) is uniformly $\gamma^{(s)}$-wellposed for any $s \leq \dfrac{m_1}{m_1 - 1}$. On the other hand in the case when $m_2 p \geq q$ we must choose σ to satisfy two conditions: $((m_1+m_2)p-q)\sigma \leq 1$, $1-\dfrac{1+q\sigma}{m_1 + m_2} \leq \dfrac{1}{s}$. This leads us to the conclusion of the second part of Theorem 2.

4   A SKETCH OF THE PROOF OF THEOREM 2 (NECESSITY)

We prove the necessity of the Theorem 2 by contradiction. Namely we assume that (C.P.) with a certain lower order term $B(t)$ is $\gamma^{(s)}$-wellposed even though the condition to the Gevrey index s are violated. Because the principal part $A(t)$ is very concrete, we can easily see the crucial elements in the lower order term, precisely

$$B_1 = \left(\begin{array}{c|c} b & O \\ \hline O & O \end{array} \right) \quad \text{or} \quad B_2 = \left(\begin{array}{c|c} O & O \\ \hline b & O \end{array} \right).$$

B_1 gives us that the (C.P.) is not $\gamma^{(s)}$-wellposed when $s > \dfrac{m_1}{m_1 - 1}$ ($m_1 \geq 2$). To prove this it is enough that we essentially study only $m_1 \times m_1$ system:

$$ID_t + \begin{pmatrix} \lambda_1 & 1 & & \\ & \lambda_1 & \ddots & \\ & & \ddots & 1 \\ & & & \lambda_1 \end{pmatrix} D_x + \begin{pmatrix} & O & \\ b & & \end{pmatrix}.$$

Therefore we omit the proof in this case. To prove that (C.P.) is not $\gamma^{(s)}$-wellposed even though $s < \dfrac{m_1}{m_1 - 1}$ ($m_1 \geq 2$), we employ B_2 as the lower order term. With this B_2 we will show that (C.P.) is not $\gamma^{(s)}$-wellposed when $m_2 p \geq q$ and when $s > \dfrac{(m_1 + m_2)p - q}{(m_1 + m_2 - 1)p - q}$ ($m_1 \geq 2$).

Similar to the last section, after the partial Fourier transformation with respect to x (we denote the dual variable by n), let us start with the following equations:

(4.1) $I_m D_t v + n(\lambda_1 I_{m_1} \oplus \lambda_2 I_{m_2}) v +$

$$+ n \left(\begin{array}{c|c} J_1 & O \\ \hline & a(t) \\ O & J_2 \end{array} \right) v + \begin{pmatrix} & O \\ b & \end{pmatrix} v = 0,$$

where

$$J_k = \begin{pmatrix} 0 & 1 & & \\ & 0 & \ddots & \\ & & \ddots & 1 \\ & & & 0 \end{pmatrix} \qquad (m_k \times m_k \text{ matrix}, \ k = 1, 2).$$

Our purpose is to derive the inequality for large n which leads us a contradiction to the $\gamma^{(s)}$-wellposedness.

At first we take an asymptotic transformation:

$$t = n^{-\sigma} \tau.$$

Then (4.1) turns to

$$(4.2) \qquad n^\sigma I_m D_\tau \tilde{v} + n A_1^{(0)}(\tau) \tilde{v} +$$

$$+ n \left(\begin{array}{c|c} J_1 & O \\ \hline & \tilde{a}(\tau) \\ \hline O & J_2 \end{array} \right) \tilde{v} + \left(\begin{array}{cc} & O \\ b & \end{array} \right) \tilde{v} = 0,$$

where $\tilde{v} = \tilde{v}(\tau) = v(n^{-\sigma}t), \cdots$, etc, and $A_1^{(0)}(\tau) = \tilde{\lambda}_1 I_{m_1} \oplus \tilde{\lambda}_2 I_{m_2}$.

Next we take, so-called, a change of weight that we used in the last section as follows:

$$W = W(n) = W(n; \varepsilon, \varepsilon') = W_1(n; \varepsilon) \oplus n^{-\varepsilon'} W_2(n; \varepsilon),$$

where

$$W_k = W_k(n; \varepsilon) = \left(\begin{array}{cccc} 1 & & & \\ & n^{-\varepsilon} & & \\ & & \ddots & \\ & & & n^{-(m_k-1)\varepsilon} \end{array} \right) \qquad (k = 1, 2).$$

For a given $m_k \times m_k$ matrix $A = (a_{ij})$, we can see that

$$W_k^{-1} A W_k = \left(n^{(i-j)\varepsilon} a_{ij} \right),$$

and for a given $m \times m$ matrix A, the effect of $\tilde{W} = I_{m_1} \oplus n^{-\varepsilon'} I_{m_2}$ is as follows:

$$A = \left(\begin{array}{c|c} A_{11} & A_{12} \\ \hline A_{21} & A_{22} \end{array} \right) \Rightarrow \tilde{W}^{-1} A \tilde{W} = \left(\begin{array}{c|c} A_{11} & n^{-\varepsilon'} A_{12} \\ \hline n^{\varepsilon'} A_{21} & A_{22} \end{array} \right).$$

Therefore in the last two terms in (4.2), the possibilities of the maximal order with respect to n are

$$n^{1-\varepsilon}, \quad n^{(m_1-1)\varepsilon+\varepsilon'} \text{ and } n^{1+(m_1-1)\varepsilon-q\sigma-\varepsilon'}.$$

We fix our arguments that these three orders are same. Thus ε and ε' are determined as follows:

$$\begin{cases} \varepsilon = \dfrac{1 + q\sigma}{m_1 + m_2}, \\[3mm] \varepsilon' = \dfrac{m_1 - m_2 q\sigma}{m_1 + m_2}. \end{cases}$$

With this W we set $\tilde{v} = W\tilde{v}_1$, then

(4.3) $$I_m D_\tau \tilde{v}_1 + n^{1-\sigma} A_1^{(0)}(\tau)\tilde{v}_1 + n^{1-\varepsilon-\sigma} A_1^{(1)} \tilde{v}_1 = 0,$$

where we denote that $a(n^{-\sigma}\tau) = \tilde{a}(\tau) = n^{-q\sigma}\tilde{\tilde{a}}(\tau)$ and

$$A_1^{(1)} = \left(\begin{array}{c|c} \begin{matrix} J_1 & \\ & \tilde{\tilde{a}}(\tau) \end{matrix} & O \\ \hline O & \begin{matrix} J_2 \end{matrix} \\ b & \end{array} \right).$$

From the structure of $A_1^{(1)}$, its eigenvalues $\{\mu_j(\tau)\}_{j=1}^m$ are distinct when $\tau \neq 0$, moreover we can assume that there exist a constant $m_\mu \geq 1$ and a constant $\delta_1 > 0$ such that

(4.4) $$\begin{cases} \operatorname{Im} \mu_j(\tau) \geq \delta_1 & (j = 1, \cdots, m_\mu), \\ \operatorname{Im} \mu_j(\tau) \leq -\delta_1 & (j = m_\mu + 1, \cdots, m), \end{cases}$$

for any interval $\tau \in [T_1, T_2]$ $(0 < T_1 < T_2)$.

We denote the diagonalizer matrix of $A_1^{(1)}(\tau)$ by $N(\tau)$ and set $\tilde{v}_1 = N(\tau)\tilde{v}_2$, then (4.3) turns to

(4.5) $$I_m D_\tau \tilde{v}_2 + n^{1-\sigma} N(\tau)^{-1} A_1^{(0)}(\tau) N(\tau)\tilde{v}_2 +$$

$$+ n^{1-\varepsilon-\sigma} \begin{pmatrix} \mu_1 & & \\ & \ddots & \\ & & \mu_m \end{pmatrix} \tilde{v}_2 + N(\tau)^{-1}(D_\tau N(\tau))\tilde{v}_2 = 0.$$

Denoting the elements of $\tilde{v}_2$ by $\tilde{v}_2 = {}^t(\tilde{v}_2^{(1)}, \cdots, \tilde{v}_2^{(m)})$, we define the energy form:

$$E_n(\tau) = \sum_{j=1}^{m_\mu} \left| \tilde{v}_2^{(j)}(\tau) \right|^2 - \sum_{j=m_\mu}^{m} \left| \tilde{v}_2^{(j)}(\tau) \right|^2.$$

From (4.5) and (4.4) we get the following inequality:

$$(4.6) \qquad \frac{d}{d\tau} E_n(\tau) \geq \delta_1 n^{1-\varepsilon-\sigma} E_n(\tau) - \text{const.} \, n^{-(1-\varepsilon-\sigma)} |g_n(\tau)|^2,$$

where

$$g_n(\tau) = n^{1-\sigma}(\tilde{\lambda}_1 - \tilde{\lambda}_2)N(\tau)^{-1}(I_{m_1} \oplus O_{m_2})N(\tau)\tilde{N}_2 + N(\tau)^{-1}(D_\tau N(\tau))\tilde{v}_2.$$

Here we denote the zero matrix of order m_2 by O_{m_2}.

For the remainder term we expect that $g_n(\tau) = o(n^{1-\varepsilon-\sigma})$ when n tends to infinity. For this purpose we impose the following inequalities:

$$(4.7) \qquad\qquad 1 - \sigma - p\sigma < 1 - \varepsilon - \sigma,$$
$$(4.8) \qquad\qquad 0 < 1 - \varepsilon - \sigma.$$

Thus we obtain

$$(4.9) \qquad\qquad E_n(\tau) \geq \exp\left(\delta_1' n^{1-\varepsilon-\sigma}(\tau - \tau_0)\right) E_n(\tau_0),$$

for $0 < \tau_0 < \tau$ and for large n, where δ_1' is a positive constant independent of n.

Next we choose the initial data such that $E_n(\tau_0) = 1$, precisely such that $\tilde{v}_2^{(1)}(\tau_0) = 1$ and $\tilde{v}_2^{(j)}(\tau_0) = 0$ $(2 \leq j \leq m)$. Then under the assumption that the (C.P.) is $\gamma^{(s)}$-wellposed, we obtain the inequality:

$$(4.10) \qquad E_n(\tau) \leq \text{const.} \, n^M \exp(-\delta_2 n^{\frac{1}{s}-\sigma} + \delta_2' n^{1-\varepsilon-\sigma}),$$

where M, δ_2 and δ_2' are positive constants independent of n.

Observing two inequalities (4.9) and (4.10), we see that when we impose the inequality:

$$(4.11) \qquad\qquad 1 - \varepsilon - \sigma < \frac{1}{s} - \sigma,$$

these lead us to a contradiction.

Taking account that

$$\frac{(m_1 + m_2)p - q}{(m_1 + m_2 - 1)p - q} < \frac{m_1}{m_1 - 1} \quad \Leftrightarrow \quad m_2 p > q,$$

the existence of σ which satisfies (4.7), (4.8) and (4.11) leads us to the proof of the necessity of Theorem 2.

References

[1] Y. Ohya, *Le problème de Cauchy pour les équations hyperboliques à caractéristiques multiples*, J. Math. Soc. Japan, **16** (1964), pp.268–286.

[2] J. Leray et Y. Ohya, *Systèmes linéaires, hyperboliques nonstricts*, Deuxième Colloque sur l'Analyse fonctionelle à Liège, 1964.

[3] M. D. Bronstein, *The parametrix of the Cauchy problem for hyperbolic operator with characteristics of variabe multiplicity*, Trudy Moscow Math., **41** (1980), pp.83–90.

[4] K. Kajitani, *Cauchy problem for nonstrictly hyperbolic systems in Gevrey classes*, J. Math. Kyoto Univ., **23** (1983), pp.599–616.

[5] T. Nishitani, *Cauchy problem for nonstrictly hyperbolic systems in Gevrey classes*, J. Math. Kyoto Univ., **23** (1983), pp.739–773.

[6] W. Matsumoto, *Normal form of systems of partial differential and pseudo-differential operators in formal symbol classes and applications*, J. Math. Kyoto Univ., **34** (1994), pp.15–40.

[7] G. Taglialatela, *Conditions d'hyperbolicité pour un opérateur matriciel à caractéristiques de multiplicité constant*, Bollettino U.M.I. (7) **11-B** (1997), pp.917–959.

[8] J. Vaillant, *Systèmes d'équations aux dérivées partielles et classes de Gevrey*, C. R. Acad. Sci. Paris, **320** (1995), pp.1469–1474.

[9] J. Vaillant, *Invariants des systèmes d'opérateurs différentiels et sommes formelles asymptotiques*, Japanese J. Math. 25 (1999), 1-154.

[10] H. Yamahara, *A remark of Cauchy problem for hyperbolic systems with constant multiplicities*, in preparation.

[11] H. Yamahara, *Cauchy problem for hyperbolic systems in Gevrey classes*, Ann. Fac. Sci. Toulouse, **IX** (2000), pp.147–160.

Gevrey well-posedness for pseudosymmetric systems with lower order terms

LORENZO MENCHERINI - Università di Firenze, Dipartimento di Matematica, viale Morgagni 67-A, I-50134 Firenze, mencheri@mail.dm.unipi.it

SERGIO SPAGNOLO - Università di Pisa, Dipartimento di Matematica, via Buonarroti 2, I-56127 Pisa, spagnolo@dm.unipi.it [1]

Dedicated to Jean Vaillant

1   INTRODUCTION

1.1   Pseudo-symmetric matrices

Let $A(t)$ be an $N \times N$ matrix whose entries $a_{rs}(t)$ are analytic functions on $[0, T]$, and let

$$I_N = \{1, 2, \ldots, N\}.$$

Following [4], we say that $A(t)$ is *pseudosymmetric*, and write $A(t) \in (ps)$, if for all $r, s, h_1, \ldots, h_\nu \in I_N$, we have

$$a_{rs}(t) \cdot a_{sr}(t) \geq 0 \tag{1}$$

and [2]

$$a_{h_1 h_2} \cdot a_{h_2 h_3} \cdots a_{h_\nu h_1} \equiv \bar{a}_{h_2 h_1} \cdot \bar{a}_{h_3 h_2} \cdots \bar{a}_{h_1 h_\nu}, \tag{2}$$

or, equivalently, (taking (1) into account)

$$|a_{h_1 h_2}| \cdot |a_{h_2 h_3}| \cdots |a_{h_\nu h_1}| \equiv |a_{h_2 h_1}| \cdot |a_{h_3 h_2}| \cdots |a_{h_1 h_\nu}|. \tag{2'}$$

---

[1] Research partially supported by MURST Programme "Teoria e applicazioni delle equazioni iperboliche lineari e non lineari"

[2] If $\varphi(t)$ is a function on $I = [0, T]$, $\varphi \equiv 0$ means that $\varphi(t)$ is identically zero on I, and $\varphi \not\equiv 0$ means that $\varphi(t) \neq 0$ for some $t \in I$. If φ, ψ are analytic on I, then $\varphi \cdot \psi \not\equiv 0$ iff $\varphi \not\equiv 0$ and $\psi \not\equiv 0$.

We want to solve the Cauchy Problem

$$\partial_t u = A(t)\,\partial_x u + B(t)\,u \qquad \text{on } [0,T] \times \mathbf{R}_x \tag{3}$$
$$u(0,x) = u_0(x), \tag{4}$$

where $B(t)$ is an arbitrary matrix with entries belonging to $L^\infty(0,T)$.
When $B(t) \equiv 0$, i.e., if there are no lower order terms, the well-posedness in $\mathcal{C}^\infty$ for $\{(3)\text{-}(4)\}$ was proved in [4], as an extension of the result of [2] for the scalar equations of the second order. Here we are interested in general case $B \not\equiv 0$. In such a case, we cannot hope to prove, in general, the well-posedness in $\mathcal{C}^\infty$, but only in Gevrey classes $\gamma^s \equiv \gamma^s(\mathbf{R}_x)$ for a suitable range of the Gevrey exponent, say

$$1 \leq s < 1 + \frac{1}{\kappa}.$$

The classical Bronšteĭn's theorem ([1], cf. also [6]) says that we can take, in the greatest generality,

$$\kappa = r - 1 \tag{5}$$

where r is the maximum multiplicity of the eigenvalues of $A(t)$, $1 \leq r \leq N$. Here, assuming that $A(t)$ is a *pseudosymmetric matrix*, we prove a different and more precise (even if not always stronger) expression of κ, which displays the dependence of κ on some relevant analytic and algebric characteristics of the matrix $A(t)$. A part of the paper is devoted to a reading of the notion of pseudosymmetry in the language of the graph theory.

1.2   Notations and statement of the Theorem

• Let $r,s \in I_N$, we say that $r \sim s$ when, either $r = s$, or $a_{rs} \cdot a_{sr} \not\equiv 0$, or else there are some indices $h_1, \ldots, h_\nu \in I_N$ for which

$$a_{rh_1} \cdot a_{h_1 r} \cdot a_{h_1 h_2} \cdot a_{h_2 h_1} \cdots a_{h_\nu s} \cdot a_{sh_\nu} \not\equiv 0. \tag{6}$$

This condition implies, by (2), that, either $a_{rs} \cdot a_{sr} \not\equiv 0$, or $a_{rs} \equiv a_{sr} \equiv 0$.
• We split I_N into equivalence classes say

$$I_N = \alpha_1 \cup \ldots \cup \alpha_m \,,$$

and consider the quotient set

$$\mathcal{X} = \{\alpha_1, \ldots, \alpha_m\} \equiv I_N / \sim \,.$$

• We define on $\mathcal{X}$ the relation

$$\alpha \leq \alpha' \overset{def}{\Longleftrightarrow} \exists\, r \in \alpha,\ s \in \alpha',\ h_1, \ldots, h_\nu \in I_N \quad \text{s.t. } a_{rh_1} \cdot a_{h_1 h_2} \cdots a_{h_\nu s} \not\equiv 0,$$

which results in being a (partial) order, by (2). In particular, we have $\alpha < \alpha'$ whenever there is some $r \in \alpha$ and some $s \in \alpha'$, for which $a_{rs} \not\equiv 0$ and $a_{sr} \equiv 0$

- For each pair $r \neq s$, for which $a_{rs} \not\equiv 0$, we define the non-negative integer

$$k_{rs}(t_0) = \text{order of vanishing of } a_{rs}(t) \text{ at } t = t_0, \tag{7}$$

and we put, for all $r \sim s$,

$$\Delta_{rs}(t) = [(k_{rh_1}(t) - k_{h_1 r}(t)] + [(k_{h_1 h_2}(t) - k_{h_2 h_1}(t)] + \cdots + [k_{h_\nu s}(t) - k_{sh_\nu}(t)] \tag{8}$$

where $(h_1, \ldots, h_\nu)$, is any chain satisfying (6). The definition is not ambiguous, by (2).

- For each class $\alpha \in \mathcal{X}$, we define

$$\Delta(\alpha, t) = \max_{r,s \in \alpha} \Delta_{rs}(t), \qquad \Delta(\alpha) = \max_{t \in [0,T]} \Delta(\alpha, t), \tag{9}$$

and we put

$$\tilde{\Delta}(\alpha) = \max \left\{ \Delta(\alpha) - 2, 0 \right\}. \tag{10}$$

With these notations we'll prove:

THEOREM 1 : Let $A(t)$ be a pseudosymmetric matrix with entries analytic on $[0, T]$, and let $B(t)$ be bounded and measurable. Therefore, the Cauchy Problem $\{(3)-(4)\}$ is well-posed in the Gevrey class $\gamma^s(\mathbf{R})$ for $1 \leq s < 1 + 1/\bar{\kappa}$, where

$$\bar{\kappa} = \max_{\alpha_{i_1} < \cdots < \alpha_{i_\nu}} \left\{ \frac{\tilde{\Delta}(\alpha_{i_1}) + \ldots + \tilde{\Delta}(\alpha_{i_\nu})}{2} + \nu - 1 \right\}, \tag{11}$$

the maximum being taken among all the chains $\{\alpha_{i_1} < \cdots < \alpha_{i_\nu}\} \subseteq \mathcal{X}$. If $\bar{\kappa} = 0$, the Problem is well-posed in C^∞.

When $B(t) \equiv 0$, we have always the C^∞ well-posedness.

1.3 Some special cases

1.   For each chain $\mathcal{C} = \{\alpha_1 < \alpha_2 < \cdots < \alpha_\nu\}$ of $\mathcal{X}$, let us put $|\mathcal{C}| = \nu$ (this is also called the *length* of $\mathcal{C}$), and define the *index of concatenation* of $A(t)$ as

$$\bar{\nu} = \max \left\{ |\mathcal{C}| : \mathcal{C} \text{ chain} \subseteq \mathcal{X} \right\}. \tag{12}$$

Then, from (11) we get, in particular, the estimate:

$$\bar{\kappa} \leq \frac{\bar{\nu}}{2} \cdot \max_{\alpha \in \mathcal{X}} \tilde{\Delta}(\alpha) + \bar{\nu} - 1.$$

2.   The matrix

$$A(t) = [t^{k_{rs}}], \qquad k_{rs} \text{ integers} \geq 0,$$

is pseudosymmetric on $[0, T]$ iff, for all triplet (r, s, p), we have

$$(k_{rs} - k_{sr}) + (k_{sp} - k_{ps}) + (k_{pr} - k_{rp}) = 0. \tag{13}$$

In this case there is only one equivalence class; thus, taking (13) into account, we find

$$\bar{\kappa} = \frac{1}{2} \max \left\{ \Delta - 2, 0 \right\}, \quad \text{where} \quad \Delta = \max_{1 \leq r,s \leq N} |k_{rs} - k_{sr}|.$$

In particular, $\{(3)-(4)\}$ is well posed in C^∞ whenever, for all r, s, one has

$$|k_{rs} - k_{sr}| \leq 2.$$

3.   The case $\bar{\kappa} = 0$ could occur only when $\bar{\nu} = 1$ (that is, the equivalence classes are mutually non-comparable) and moreover $\Delta(\alpha, t) \leq 2$ for all α, i.e., $\Delta_{rs}(t) \leq 2$ for all r, s. A special case is that in which $\bar{\nu} = 1$ and $\Delta(\alpha, t) = 0$ for all α, which means $k_{rs}(t) = k_{sr}(t)$ for all $r \sim s$. We can easily see that this is equivalent to say that

$$a_{rs}(t) = a_{rs}^{\circ}(t) \cdot b_{rs}(t), \qquad r, s = 1, \ldots N,$$

where $A^{\circ}(t) = [a_{rs}^{\circ}(t)]$ is a Hermitian matrix, and $B(t) = [b_{rs}(t)]$ a (ps) matrix with $b_{rs}(t) \neq 0$ for all r, s, and all $t \in [0, T]$. Now, every matrix of this form results to be *smoothly symmetrizable*, thus the C^{∞} well-posedness is well known in this case.

4.   If $A(t) \equiv A$ is constant in time, we have $\Delta(\alpha, t) = 0$ for all α, hence $\kappa = \bar{\nu} - 1$.

2   GRAPH THEORY

We recall here some basic definitions or notations which will be used in the following sections. A *graph* is a pair of sets

$$\mathcal{G} = (\mathcal{V}, \mathcal{E})$$

where $\mathcal{E} \subseteq \mathcal{V} \times \mathcal{V}$; hence $\mathcal{G}$ is a set endowed with a binary relation. The elements of $\mathcal{V}$ are the *vertices* of the graph, those of $\mathcal{E}$ the *edges*.

If $x, y \in \mathcal{V}$, we call *path* from x to y any ordered sequence $z_1, z_1, \ldots z_\nu$ of (possibly non distinct) vertices of $\mathcal{G}$ with $z_1 = x$ and $z_\nu = y$, such that $(z_i, z_{i+1}) \in \mathcal{E}$ for all $i = 1, \ldots, \nu - 1$. We shall refer to ν as the *length* of the path. A *cycle* of $\mathcal{G}$ is a path connecting a vertex x to itself.

When there exists some path from x to y, or when $x = y$, we say that x *precedes* y, and we write

$$[x] \leq [y] .$$

Clearly, this is a transitive relation, hence a $pre - order$ on $\mathcal{V}$. In order to get a (partial) order, it is sufficient to pass to the quotient with respect to the equivalence relation:

$$x \sim y \qquad \overset{def}{\Longleftrightarrow} \qquad [x] \leq [y] \quad \text{and} \quad [y] \leq [x].$$

If $x \sim y$ we say that x *is connected* to y, and also write $[x] = [y]$. The relation $[x] \leq [y]$ is a partial order on the quotient set

$$\mathcal{X} = \mathcal{V}/\sim .$$

Finally, in the case that $\mathcal{V}$ is a finite set, we define the *index of concatenation* of $\mathcal{G}$ as

$$\bar{\nu}(\mathcal{G}) := \max \left\{ \nu : \exists \, x_1, \ldots, x_\nu \in \mathcal{V} \quad \text{s.t.} \quad [x_1] < [x_2] < \cdots [x_\nu] \right\}, \qquad (14)$$

i.e., the length of the longest *chain* of the ordered set $(\mathcal{X}, \leq)$.

3   THE GRAPH OF A PSEUDOSYMMETRIC MATRIX

Let us fix a real interval $[0, T]$ and an integer N, and let

$$A(t) = [a_{ij}(t)]_{i,j=1,\cdots,N}, \qquad a_{ij} \in \mathcal{A}([0, T]), \tag{15}$$

be an $N \times N$ matrix with (complex valued) analytic entries in a neighbourhood of $[0, T]$. We associate to such a matrix the graph

$$\mathcal{G}_A = (I_N, \mathcal{E}_A)$$

where $I_N = \{1, \ldots, N\}$ and

$$\mathcal{E}_A = \{(r, s) \in I_N \times I_N : \quad \text{either} \quad a_{rs} \not\equiv 0, \quad \text{or} \quad r = s\}.$$

In this context, the notions of §2 read (for all $r \neq s$):

$$[r] \leq [s] \iff \exists\, h_1, \ldots, h_\nu \in I_N \quad \text{s.t.} \quad a_{rh_1} \cdot a_{h_1 h_2} \cdots a_{h_\nu s} \not\equiv 0 \tag{16}$$

$$r \sim s \iff \exists\, h_i, k_j \in I_N \quad \text{s.t.} \quad a_{rh_1} \cdot a_{h_1 h_2} \cdots a_{h_\nu s} \cdot a_{sk_1} \cdot a_{k_1 k_2} \cdots a_{k_\mu s} \not\equiv 0. \tag{17}$$

Assume now that the matrix $A(t)$ is pseudosymmetric, i.e., satisfies the conditions (1) and (2). In the language of graphs, (2) can be expressed by saying that, for any cycle $(h_1 \rightarrow h_2 \rightarrow \ldots \rightarrow h_\nu \rightarrow h_1)$ of $\mathcal{G}(A)$, the opposite $(h_1 \rightarrow h_\nu \rightarrow \ldots \rightarrow h_2 \rightarrow h_1)$ is also a cycle of $\mathcal{G}(A)$, and, moreover, the functions

$$\beta_{rs}(t) = \frac{a_{rs}(t)}{\bar{a}_{sr}(t)} \equiv \frac{|a_{rs}(t)|}{|a_{sr}(t)|} \tag{18}$$

satisfy the cyclic condition

$$\beta_{h_1 h_2} \cdot \beta_{h_2 h_3} \cdots \beta_{h_\nu h_1} \equiv 1. \tag{19}$$

As to condition (1), this means that in the matrix $A(t)$, the entries which are symmetric with respect to the diagonal, must have opposite phases.

Finally, we observe that for a (ps) matrix, (17) takes the simpler form (6).

4   FILLING OF A PSEUDOSYMMETRIC MATRIX

In this section, we assume that A is a *constant* (ps) matrix. Then, we prove that it is always possible to "fill" the null entries of A in such a way that the resulting matrix is still pseudosymmetric, and, moreover, is arbitrarily "close" to A. More precisely:

THEOREM 2: Let $A = [a_{rs}]_{r,s=1,\ldots,N}$ be a pseudosymmetric, constant matrix. Then:

i) It is possible to construct another (ps) matrix $\widetilde{A}$ such that

$$\widetilde{a}_{rs} \neq 0 \quad \forall r, s \in I_N, \quad \text{and} \quad \widetilde{a}_{rs} \equiv a_{rs} \quad \text{if} \quad a_{rs} \not\equiv 0. \tag{20}$$

ii) For all $\epsilon > 0$, it is possible to find a (ps) matrix A_ϵ which satisfies (20), and also

$$\|A_\epsilon - A\| \leq C\epsilon.$$

LEMMA 1 : Let $D \subseteq I_N \times I_N$ be a symmetric set, i.e., such that $(r,r) \in D$ for all r, and $(s,r) \in D$ whenever $(r,s) \in D$. Let $\beta : D \to]0,+\infty[$ be a function satisfying

$$\beta_{h_1 h_2} \cdot \beta_{h_2 h_3} \cdots \beta_{h_\nu h_1} = 1 \qquad \forall\, h_1, h_2, \ldots h_\nu \in I_N \ . \tag{21}$$

Therefore, it is possible to find an extension

$$\widetilde{\beta} : I_N \times I_N \to]0,+\infty[$$

of β, which satisfies the identities

$$\widetilde{\beta}_{h_1 h_2} \cdot \widetilde{\beta}_{h_2 h_3} \cdots \widetilde{\beta}_{h_\nu h_1} = 1, \qquad \forall\, h_1, h_2, \ldots h_\nu \in I_N \ . \tag{22}$$

Proof of Lemma 1: Let us define on I_N the equivalence relation

$$r \overset{D}{\sim} s \quad \Longleftrightarrow \quad \exists\, h_1, \ldots h_\nu \in I_N \quad s.t. \quad (r,h_1), (h_1, h_2), \ldots, (h_\nu, s) \in D, \tag{23}$$

setting $r \overset{D}{\sim} r$ for all r, and let us consider the set

$$\widetilde{D} = \left\{ (r,s) \in I_N \times I_N \ : \ r \overset{D}{\sim} s \right\}.$$

We first extend β from D to $\widetilde{D}$, and then from $\widetilde{D}$ to the whole $I_N \times I_N$.

First extension. For all $(r,s) \in \widetilde{D}$, we choose an arbitrary path $(h_1, \ldots, h_\nu)$ connecting r with s in D, i.e., satisfying (23), and define

$$\widetilde{\beta}_{hk} = \beta_{r h_1} \cdots \beta_{h_\nu s}.$$

By (21), we see that $\widetilde{\beta}_{rs}$ does not depend on the path and that (22) holds on $\widetilde{D}$.

Second extension. Assume that $D \equiv \widetilde{D}$, that is, $r \overset{D}{\sim} s$ in the sense of (23) if and only if $(r,s) \in D$. This means that the cartesian projections of D on I_N coincide with the union $\alpha_1 \cup \ldots \cup \alpha_m$ of classes of D-equivalence. Therefore, we pick an arbitrary element $\bar{r}_j \in \alpha_j$ in each of these classes, and define, for all $r, s \in I_N$,

$$\widetilde{\beta}_{rs} = \beta_{r\bar{r}_i} \cdot \beta_{\bar{r}_j s}, \qquad \text{if} \qquad r \in \alpha_i \ , \ s \in \alpha_j.$$

In other words, we put $\widetilde{\beta}_{\bar{r}_i \bar{r}_j} = 1$, $i,j = 1, \ldots, m$.
In view of (21), let us consider an arbitrary cycle $(h_1, h_2, \ldots, h_\nu, h_1)$, with $h_j \in \alpha_{k_j}$. Therefore we have

$$\widetilde{\beta}_{h_1 h_2} \, \widetilde{\beta}_{h_2 h_3} \cdots \widetilde{\beta}_{h_\nu h_1} = (\beta_{h_1 \bar{r}_{k_1}} \beta_{\bar{r}_{k_2} h_2}) \cdot (\beta_{h_2 \bar{r}_{k_2}} \beta_{\bar{r}_{k_3} h_3}) \cdots (\beta_{h_\nu \bar{r}_{k_\nu}} \beta_{\bar{r}_{k_1} h_1})$$
$$= \beta_{h_1 \bar{r}_{k_1}} \cdot (\beta_{\bar{r}_{k_2} h_2} \beta_{h_2 \bar{r}_{k_2}}) \cdot (\beta_{\bar{r}_{k_3} h_3} \beta_{h_3 \bar{r}_{k_3}}) \cdots (\beta_{\bar{r}_{k_\nu} h_\nu} \beta_{h_\nu \bar{r}_{k_\nu}}) \cdot \beta_{\bar{r}_{k_1} h_1} = 1.$$

Proof of Theorem 2 : i) Let $D = \{(r,s) \in I_N \times I_N : a_{rs} \cdot a_{sr} \neq 0\}$, and $\beta : D \to]0,+\infty[$ be defined as

$$\beta_{rs} = \frac{a_{rs}}{\bar{a}_{sr}} , \qquad (r,s) \in D.$$

By (1), β is positive function which satisfies (21) on D: hence, by Lemma 1, it admits an extension $\widetilde{\beta} : I_N \times I_N \to]0,+\infty[$ still satisfying (21). Then we put

$$\widetilde{a}_{rs} = \begin{cases} a_{rs} & \text{if} \quad a_{rs} \neq 0, \\ \widetilde{\beta}_{rs} \bar{a}_{sr} & \text{if} \quad a_{rs} = 0, \ a_{sr} \neq 0, \\ 1 & \text{if} \quad a_{rs} = a_{sr} = 0, \ r \leq s, \\ \widetilde{\beta}_{rs} & \text{if} \quad a_{rs} = a_{sr} = 0, \ r > s. \end{cases} \tag{24}$$

Clearly, we have $\widetilde{a}_{rs} \neq 0$ for all r, s, and

$$\frac{\widetilde{a}_{rs}}{\widetilde{\overline{a}}_{sr}} = \widetilde{\beta}_{rs},$$

hence, the cyclic condition

$$\widetilde{a}_{h_1 h_2} \cdot \widetilde{a}_{h_2 h_3} \cdots \widetilde{a}_{h_\nu h_1} = \widetilde{\overline{a}}_{h_2 h_1} \cdot \widetilde{\overline{a}}_{h_3 h_2} \cdots \widetilde{\overline{a}}_{h_1 h_\nu}$$

is equivalent to the identity

$$\widetilde{\beta}_{h_1 h_2} \cdot \widetilde{\beta}_{h_2 h_3} \cdots \widetilde{\beta}_{h_\nu h_1} = 1.$$

ii) Let us take any function

$$\gamma : I_N \longrightarrow \{1, 2, 3, \cdots\}$$

which is *increasing* with respect to the relation (16), i.e., satisfies

$$[r] < [s] \implies \gamma(r) < \gamma(s).$$

A sample of such a function is the *index of minimality*

$$\gamma(r) = \max \left\{ k : \exists \, r_1, r_2, \ldots, r_k \in I_N \quad \text{with} \quad [r] < [r_1] \cdots < [r_k] \right\} + 1, \quad (25)$$

that is, the maximum length of the chains of $\mathcal{X}$ starting with r. Similarly to (24), we then define:

$$a_{rs}^\epsilon = \begin{cases} a_{rs} & \text{if} \quad a_{rs} \neq 0, \\ \epsilon^{\gamma(r)-\gamma(s)} \, \widetilde{\beta}_{rs} \, \bar{a}_{sr} & \text{if} \quad a_{rs} = 0, \ a_{sr} \neq 0, \\ \epsilon^{\gamma(r)} & \text{if} \quad a_{rs} = a_{sr} = 0, \ r \leq s, \\ \epsilon^{\gamma(r)} \, \widetilde{\beta}_{rs} & \text{if} \quad a_{rs} = a_{sr} = 0, \ r > s, \end{cases}$$

where $\widetilde{\beta}$ is as above. We have in particular

$$\frac{a_{rs}^\epsilon}{a_{sr}^\epsilon} = \widetilde{\beta}_{rs} \, \epsilon^{\gamma(r)-\gamma(s)}, \qquad r, s = 1, \ldots, N,$$

so that the cyclic condition for the matrix $[a_{rs}]$ follows from (22). Moreover, we have $|a_{rs} - a_{rs}^\epsilon| \leq C\epsilon$. Indeed, if $a_{rs} = 0$ and $a_{sr} \neq 0$, we have by definition $[s] < [r]$, hence $\gamma(r) - \gamma(s) \geq 1$. This completes the proof of Theorem 2.

5 QUASI-SYMMETRIZERS AND GEVREY WELL-POSEDNESS

Let us go back to a matrix $A(t) = [a_{rs}(t)]_{r,s=1,\ldots,N}$ with variable coefficients on $[0, T]$. In order to solve the Cauchy problem $\{(3)$-$(4)\}$ it is useful to look for a smooth *quasi-symmetrizer* for $A(t)$.

We call quasi-symmetrizer any family $\{Q_\epsilon(t)\}_{\epsilon>0}$ of Hermitian matrices, sufficiently smooth in t, which fulfils the conditions

$$C_1\, \epsilon^{2\delta}\, |v|^2 \;\leq\; (Q_\epsilon v, v) \;\leq\; C_2\, |v|^2 \tag{26}$$

$$\left|((Q_\epsilon A - A^* Q_\epsilon)\, v, v)\right| \;\leq\; C_3\, \epsilon^\theta\, (Q_\epsilon v, v) \tag{27}$$

$$\left|(Q_\epsilon'\, v, v)\right| \;\leq\; C_4\, \epsilon^{-\mu}\, (Q_\epsilon v, v), \tag{28}$$

for some non-negative constants δ, θ, μ and C_j. We'll always assume $\theta > 0$ and $C_1 > 0$. The existence of such a $\{Q_\epsilon(t)\}$ implies that the matrix $A(t)$ is *hyperbolic*, i.e., has real eigenvalues for all t.

Sometimes, it is possible to find a quasi-symmetrizer which satisfies, in place of (28), a condition like

$$|(Q_\epsilon''(t)\, v, v)| \;\leq\; C\, |v|^2\,.$$

In such a case by (26), using the Glaeser inequality, we get (28) with $\mu = \delta$.

In [3], a quasi-symmetrizer which satisfies (26), (27), (28), with $\delta = \mu = N - 1$ and $\theta = 1$, was constructed for a wide class of hyperbolic matrices including those of Sylvester type (see also [5]). We now recall the basic result of Gevrey well-posedness for a quasi-symmetrizable system.

PROPOSITION 1: If $A(t)$ has a quasi-symmetrizer satisfying the conditions (26), (27), (28), the Cauchy Problem $\{(3)\text{-}(4)\}$ is well-posed in γ^s for $1 \leq s < 1 + 1/\kappa$ with

$$\kappa \;=\; \frac{\max\{\delta, \mu\}}{\theta}, \tag{29}$$

and $\kappa = \mu/\theta$ if $B(t) \equiv 0$. If $\kappa = 0$, the Problem is well-posed in C^∞.

Proof of Proposition 1 : We prove (29) only for $s > 1$, since the well-posedness in the class of analytic functions is well-known. Each (ps) matrix is hyperbolic, hence for the equation (3) we have the *finite speed of propagation* property. Thus, it is not restrictive to assume that the initial datum $u_0(x)$ is a Gevrey function *with compact support* in $\mathbf{R}_x$ and look for an apriori estimate for a solution $u(t, x)$ with compact support in $\mathbf{R}_x$ for all $t \in [0, T]$.

Performing the Fourier transform with respect to x

$$v(t, \xi) \;=\; \int_{-\infty}^{+\infty} e^{-i\xi x}\, u(t, x)dx,$$

we convert (3) into the ordinary differential system

$$v' \;=\; i\xi\, A(t)\, v + B(t)\, v\,. \tag{30}$$

Our goal is to prove an apriori estimate like

$$|v(t, \xi)| \;\leq\; |v(0, \xi)| \cdot |\xi|^\nu \exp\left(C\, |\xi|^{\frac{\kappa}{\kappa+1}}\right), \qquad 0 \leq t \leq T, \tag{31}$$

with κ given from (29), for some ν, C, depending on T, C_1, C_2, C_3, C_4, and on

$$\beta \;=\; \sup_{0 \leq t \leq T} \|B(t)\|\,.$$

For this end, we define the *approximate energy* of the solution $v(t,\xi)$, as

$$E_\epsilon(t,\xi) = (Q_\epsilon(t)v, v). \tag{32}$$

Differentiating in t and recalling (30), we find

$$\begin{aligned}
E'_\epsilon(t,\xi) &= (Q_\epsilon v', v) + (Q_\epsilon v, v') + (Q'_\epsilon v, v) \\
&= i\,\xi\,[(Q_\epsilon Av, v) - (Q_\epsilon v, Av)] + 2\Re(Q_\epsilon v, Bv) + (Q'_\epsilon v, v).
\end{aligned}$$

Now we have, by (26),

$$|(Q_\epsilon v, Bv)| \leq (Q_\epsilon v, v)^{1/2}(Q_\epsilon Bv, Bv)^{1/2} \leq \beta\sqrt{C_2/C_1}\,\epsilon^{-\delta}(Q_\epsilon v, v), \tag{33}$$

hence, by (27) and (28), we find

$$E'_\epsilon(t,\xi) \leq K(\epsilon,\xi,t)\,E_\epsilon(t,\xi) \tag{34}$$

where

$$K(\epsilon,\xi,t) = C_3\,|\xi|\epsilon^\theta + 2\beta\sqrt{C_2/C_1}\,\epsilon^{-\delta} + C_4\,\epsilon^{-\mu}. \tag{35}$$

In particular, we have

$$K(\epsilon,\xi,t) \leq C\,(|\xi|\,\epsilon^\theta + \epsilon^{-\sigma}) \quad \text{with} \quad \sigma = \max\{\delta, \mu\}, \tag{36}$$

while $\sigma = \mu$ when $\beta = 0$.
For each given ξ, we choose

$$\epsilon = |\xi|^{-1/(\theta+\sigma)}, \tag{37}$$

so that it results $|\xi|\,\epsilon^\theta = \epsilon^{-\sigma}$. Hence, integrating in t, (34) gives

$$E_\epsilon(t,\xi) \leq E_\epsilon(0,\xi) \int_0^T K(\epsilon,\xi,t)\,dt \leq E_\epsilon(0,\xi) \cdot \exp\left(2CT\,|\xi|^{\sigma/(\theta+\sigma)}\right). \tag{38}$$

By (29) we have $\kappa = \sigma/\theta$, thus $\sigma/(\theta + \sigma) = \kappa/(\kappa+1)$. Moreover,

$$\sqrt{C_1}\,\epsilon^\delta\,|v| \leq \sqrt{E_\epsilon} \leq \sqrt{C_2}\,|v|,$$

thus using (37), we get the wished estimate

$$|v(t,\xi)| \leq |v(0,\xi)| \cdot \sqrt{C_2/C_1}\,|\xi|^\nu \exp\left(CT\,|\xi|^{\kappa/(\kappa+1)}\right), \qquad 0 \leq t \leq T, \tag{39}$$

for $\nu = \delta/(\theta + \sigma) = \delta/[\theta(\kappa + 1)]$.

To conclude the proof of the Gevrey well-posedness, we recall that a compact supported function $u(x)$ belongs to $\gamma^s(\mathbf{R}_x)$ if and only if its Fourier transform satisfies an estimate

$$|\widehat{u}(\xi)| \leq C\,e^{-\rho|\xi|^{1/s}}, \qquad \rho > 0. \tag{40}$$

Now, let $u(t,x)$ be a solution of $\{(3)\text{-}(4)\}$, with compact support in x for all t, and assume that $u(0,x) \in \gamma^s$ for some $s < (\kappa + 1)/\kappa$. Therefore, $\widehat{u}(0,\xi)$ satisfies (40), and then, by (31), also $\widehat{u}(t,\xi)$ satisfies a similar estimate, with the same s, for all $t \geq 0$. In other words, we have proved an apriori estimate in γ^s for each solution $u(t,x)$, which leads to the well-posedness by standard approximation arguments.

In the case when we have (26) and (28) with $\delta = \mu = 0$, i.e., $\kappa = 0$, we get the C^∞ well-posedness. Indeed, in (31) the exponential factor is now missing, and we have

$$|v(t, \xi)| \leq C \, |v(0, \xi)| \cdot |\xi|^\nu. \tag{41}$$

This ensures the well-posedness in C^∞ with loss of ν derivatives.

We can improve Proposition 1 by replacing (26), (27) and (28) by the following weaker assumptions of integral type, for $0 < \epsilon \leq \bar{\epsilon} < 1$,

$$\lambda_\epsilon(t) \, |v|^2 \leq (Q_\epsilon v, v) \leq \Lambda_\epsilon(t) \, |v|^2 \quad \text{with} \quad \int_0^T \sqrt{\frac{\Lambda_\epsilon}{\lambda_\epsilon}} \, dt \leq C_\circ \, \epsilon^{-\delta} \log \frac{1}{\epsilon} \tag{42}$$

$$\left| ((Q_\epsilon A - A^* Q_\epsilon) v, v) \right| \leq \omega_\epsilon(t) \, (Q_\epsilon v, v) \quad \text{with} \quad \int_0^T \omega_\epsilon \, dt \leq C_3 \, \epsilon^\theta \log \frac{1}{\epsilon} \tag{43}$$

$$\left| (Q'_\epsilon v, v) \right| \leq M_\epsilon(t) \, (Q_\epsilon v, v) \quad \text{with} \quad \int_0^T M_\epsilon \, dt \leq C_4 \, \epsilon^{-\mu} \log \frac{1}{\epsilon}. \tag{44}$$

In facts, proceeding as in the proof of Prop.1, but replacing now (35) by

$$K(\epsilon, \xi, t) = \omega_\epsilon(t) |\xi| \epsilon^\theta + 2\beta \sqrt{\Lambda_\epsilon / \lambda_\epsilon} + M_\epsilon(t),$$

we get:

PROPOSITION 2 : If $A(t)$ admits a quasi-symmetrizer satisfying the conditions (42)–(44), then the Cauchy Problem $\{(3)\text{-}(4)\}$ is well-posed in γ^s for $s < 1 + 1/\kappa$, with

$$\kappa = \frac{\max\{\delta, \mu\}}{\theta}, \tag{45}$$

and $\kappa = \mu/\theta$ if $B(t) \equiv 0$. If $\kappa = 0$, the Problem is well-posed in C^∞.

6   QUASI-SYMMETRIZERS OF DIAGONAL TYPE

We recall (see [4]) that the pseudosymmetric matrices $A(t)$ have the following important feature: they admit a quasi-symmetrizer of *diagonal type*, i.e.

$$Q_\epsilon(t) = \begin{pmatrix} \lambda_1^\epsilon(t) & 0 & \cdots & 0 \\ 0 & \lambda_2^\epsilon(t) & \cdots & 0 \\ \cdots & \cdots & \cdots & \cdots \\ 0 & 0 & \cdots & \lambda_N^\epsilon(t) \end{pmatrix}. \tag{46}$$

For such a quasi-symmetrizer, the conditions (42), (43) and (44) take, respectively, the following simpler form (where $r, s = 1, \ldots, N$)

$$\int_0^T \sqrt{\frac{\lambda_r^\epsilon}{\lambda_s^\epsilon}} \, dt \leq C_\circ \, \epsilon^{-\delta} \log \frac{1}{\epsilon} \tag{47}$$

$$\int_0^T \left| \sqrt{\frac{\lambda_r^\epsilon}{\lambda_s^\epsilon}} \, a_{rs} - \bar{a}_{sr} \sqrt{\frac{\lambda_s^\epsilon}{\lambda_r^\epsilon}} \right| dt \leq C_3 \, \epsilon^\theta \log \frac{1}{\epsilon} \tag{48}$$

$$\int_0^T \frac{|\partial_t \lambda_r^\epsilon|}{\lambda_r^\epsilon} \, dt \leq C_4 \, \epsilon^{-\mu} \log \frac{1}{\epsilon}. \tag{49}$$

In [4], for every (ps) matrix with *analytic entries* it was constructed a diagonal quasi-symmetrizer of type (46) which fulfils (49) with $\mu = 0$, thus obtaining the well-posedness in C^∞ in the case when $B(t) \equiv 0$.

7   PROOF OF THEOREM 1

Given the (ps) matrix $A(t)$, whose entries $a_{rs}(t)$, $r, s = 1, \ldots, N$, are analytic functions on $[0, T]$, we construct a diagonal quasi-symmetrizer of type (46) which satisfies the conditions (47), (48) and (49). For this end, we take a suitable partition of the interval

$$[0, T] = \bigcup_{j=0}^{k} [\tau_j, \tau_{j+1}] \qquad (0 = \tau_0 < \ldots < \tau_k = T),$$

and construct the wished quasi-symmetrizer $Q_\epsilon(t)$ on each one of the intervals $[\tau_j, \tau_{j+1}]$.

We choose the partition in such a way that, for each $j = 1, \ldots, k$, it results

$$\text{either} \qquad a_{rs}(t) \neq 0 \quad \text{on} \quad]\tau_j, \tau_{j+1}] \quad \forall\, r, s \quad \text{with} \quad a_{rs} \not\equiv 0, \qquad (50)$$

$$\text{or} \qquad a_{rs}(t) \neq 0 \quad \text{on} \quad [\tau_j, \tau_{j+1}[\quad \forall\, r, s \quad \text{with.} \quad a_{rs} \not\equiv 0. \qquad (51)$$

This is always possible thanks to the analyticity of the $a_{rs}(t)$'s.

Now we construct $Q_\epsilon(t)$ on each interval of the partition. We can consider only the intervals $[\tau_j, \tau_{j+1}]$ satisfying (50), the other case being quite similar; thus, after a linear change of variables, we are reduced to assume that the entries of $A(t)$ which are not identically zero, have the form

$$a_{rs}(t) = t^{k_{rs}} b_{rs}(t), \quad \text{with} \quad b_{rs}(t) \neq 0 \quad \forall t \in [0, T]. \qquad (52)$$

Note that the $b_{rs}(t)$'s are analytic functions on $[0, T]$. Moreover, from the pseudosymmetry of $A(t)$ it follows that, for all $r, s, h_1, \ldots, h_N \in I_N$,

$$b_{rs}(t)\, b_{sr}(t) \geq 0 \quad , k_{h_1 h_2} + k_{h_2 h_3} + \ldots + k_{h_{\nu-1} h_\nu} = k_{h_2 h_1} + k_{h_3 h_2} \ldots + k_{h_\nu h_{\nu-1}}. \quad (53)$$

Now (see the Notations of §1) we say that $r \sim s$ if it is possible to find some path $\{r \to h_1 \to \cdots \to h_\nu \to s\}$ connecting r with s, i.e., for which holds (6), and we denote by $\{\alpha_1, \alpha_2, \ldots, \alpha_m\}$ the equivalence classes in the set of indices I_N, putting $\mathcal{X} = I_N / \sim$. Next, for each $r \sim s$ we define the integer $\Delta(r, s)$ and the positive function $H_{rs}(t)$, as

$$\Delta(r, s) = (k_{rh_1} - k_{h_1 r}) + \cdots + (k_{h_\nu s} - k_{s h_\nu}) \qquad (54)$$

$$H_{rs}(t) = \frac{b_{rh_1}(t) \cdots b_{h_\nu s}(t)}{b_{h_1 r}(t) \cdots b_{s h_\nu}(t)} \equiv \left| \frac{b_{rh_1}(t)}{b_{h_1 r}(t)} \right| \cdots \left| \frac{b_{h_\nu s}(t)}{b_{s h_\nu}(t)} \right|, \qquad (55)$$

where $(h_1, \ldots h_\nu)$ is any path satisfying (6). By virtue of (53), $\Delta(r, s)$ and $H_{rs}(t)$ are independent on the path choosen. Moreover, for every triplet $r \sim s \sim p$, we have

$$\Delta(r, s) + \Delta(s, p) = \Delta(r, p), \qquad H_{rs}(t)\, H_{sp}(t) = H_{rp}(t). \qquad (56)$$

We also note that, for some constants $c_i > 0$, we have

$$c_1 \leq H_{rs}(t) \leq c_2 \quad \text{on } [0, T].$$

(57)

Now, for every equivalence class α, we fix an index $r_\alpha \in \alpha$ such that $\Delta(r_\alpha, r) \geq 0$ for all $r \in \alpha$. The existence of such an index is an easy consequence of the pseudosymmetry: indeed if, for all $r \in \alpha$, there was some $r' \in \alpha$ with $\Delta(r, r') < 0$, we could find a *cycle* $(r_1, r_2, \ldots, r_\nu) \subseteq \alpha$ for which

$$\Delta(r_1, r_2) < 0, \ \Delta(r_2, r_3) < 0, \ \ldots, \ \Delta(r_k, r_1) < 0,$$

in contradiction with (56). In conclusion, we have

$$0 \leq \Delta(r_\alpha, r) \leq \Delta(\alpha) = \max_{r, s \in \alpha} \Delta(r, s), \quad \forall \, r \in \alpha.$$

(58)

Finally, for each α, we define the non-negative integer $\sigma(\alpha)$, by putting $\sigma(\alpha) = 0$ if α is a *maximal* element of $\mathcal{X}$, otherwise

$$\sigma(\alpha) = \max_{\alpha < \alpha_{i_1} < \cdots < \alpha_{i_\nu}} \left(\tilde{\Delta}(\alpha_{i_1}) + \ldots + \tilde{\Delta}(\alpha_{i_\nu}) + 2(\nu - 1) \right),$$

(59)

the maximum being taken among all the chains (with not assigned length) of $\mathcal{X}$ which start with α. We recall that

$$\tilde{\Delta} = \max \left\{ \Delta - 2, 0 \right\},$$

(60)

and $\alpha < \alpha'$ means that $a_{rs} \not\equiv 0$ and $a_{sr} \equiv 0$, for some $r \in \alpha$ and $s \in \alpha'$. We now define the quasi-symmetrizer $Q_\epsilon(t) = \text{diag}\{\lambda_1^\epsilon(t), \ldots, \lambda_N^\epsilon(t)\}$ by setting

$$\lambda_r^\epsilon(t) = \epsilon^{\sigma(\alpha)} (t + \epsilon)^{\Delta(r_\alpha, r)} H_{r_\alpha r}(t) \quad \text{for} \quad r \in \alpha.$$

(61)

LEMMA 2 : The functions $\{\lambda_r^\epsilon(t)\}_{r=1,\ldots,N}$ defined by (61) fulfil the conditions (42), (43) and (44) with $\mu = 0$, $\theta = 1$, and

$$\delta = \frac{1}{2} \max_{\alpha \in \mathcal{X}} \left(\sigma(\alpha) + \tilde{\Delta}(\alpha) \right).$$

(62)

Proof of Lemma 2 : Let us preliminarily observe that

$$\int_0^T (t + \epsilon)^{-\Delta/2} \, dt \leq C \epsilon^{-\tilde{\Delta}/2} \log \frac{1}{\epsilon}, \quad \text{where} \quad \tilde{\Delta} = \max\{\Delta - 2, 0\}.$$

(63)

i) Proof of (44) : By (61) we have

$$\frac{\partial_t \lambda_r^\epsilon(t)}{\lambda_r^\epsilon(t)} = \frac{\partial_t H_{r_\alpha r}(t)}{H_{r_\alpha r}(t)} + \frac{\Delta(r_\alpha, r)}{t + \epsilon},$$

hence recalling (57), we find (44) with $\mu = 0$.

ii) Proof of (42) : Recalling (58), we have

$$C_1 e^{\sigma(\alpha)/2} (t + \epsilon)^{\Delta(\alpha)/2} \leq \sqrt{\lambda_s^\epsilon(t)} \leq C_2, \qquad \text{for} \ \ s \in \alpha$$

hence, using (63),

$$\int_0^T \sqrt{\lambda_r^\epsilon/\lambda_s^\epsilon} \, dt \ \leq \ C \, \epsilon^{-(\sigma(\alpha) + \tilde{\Delta}(\alpha))/2} \log \frac{1}{\epsilon}, \qquad \forall r, s \text{ with } \ s \in \alpha.$$

This gives (42) with δ as in (62).

iii) Proof of (43) : If $\alpha_{rs} \equiv \alpha_{sr} \equiv 0$, (43) is trivial. Thus, we consider only two cases:

• *Case I* : $a_{rs}(t) \not\equiv 0$, $a_{sr}(t) \not\equiv 0$. In this case, we have $r \sim s$, i.e., $r, s \in \alpha$ for some α. Therefore, by (61) and (56),

$$\frac{\lambda_r^\epsilon(t)}{\lambda_s^\epsilon(t)} \ = \ (t + \epsilon)^{\Delta(r_\alpha, r) - \Delta(r_\alpha, s)} \frac{H_{r_\alpha r}(t)}{H_{r_\alpha s}(t)} \ = \ (t + \epsilon)^{\Delta(s, r)} \, H_{sr}(t).$$

Hence, setting for sake of brevity

$$I_{rs}^\epsilon(t) \ := \ \left| \sqrt{\frac{\lambda_r^\epsilon(t)}{\lambda_s^\epsilon(t)}} \, a_{rs}(t) \ - \ \bar{a}_{sr}(t) \sqrt{\frac{\lambda_s^\epsilon(t)}{\lambda_r^\epsilon(t)}} \, \right|, \tag{64}$$

and recalling (52), we get

$$I_{rs}^\epsilon(t) \ = \ \left| (t + \epsilon)^{\Delta(s,r)/2} \sqrt{H_{sr}(t)} \, t^{k_{rs}} \, b_{rs}(t) \ - \ (t + \epsilon)^{\Delta(r,s)/2} \sqrt{H_{rs}(t)} \, t^{k_{sr}} \, \bar{b}_{sr}(t) \right|.$$

In our case, the functions a_{rs} and a_{sr} are not zero, hence we have, by (54) and (55), $\Delta(r, s) = k_{rs} - k_{sr}$, $H_{rs}(t) = b_{rs}(t)/\bar{b}_{sr}(t)$. Therefore, recalling (53), we see that

$$\bar{b}_{sr} \sqrt{H_{rs}} \ = \ b_{rs} \sqrt{H_{sr}} \ = \ (b_{rs}/\bar{b}_{rs}) \cdot \sqrt{b_{rs} b_{sr}}$$

$$\frac{\Delta(r, s)}{2} + k_{sr} \ = \ \frac{\Delta(s, r)}{2} + k_{rs} \ = \ \frac{k_{rs} + k_{sr}}{2}.$$

Hence

$$I_{rs}^\epsilon(t) \ = \ (t + \epsilon)^{(k_{rs} + k_{sr})/2} \sqrt{b_{rs}(t) b_{sr}(t)} \, \left| \left(\frac{t}{t + \epsilon} \right)^{k_{rs}} - \left(\frac{t}{t + \epsilon} \right)^{k_{sr}} \right|.$$

If $k_{rs} = k_{sr} = 0$, then $I_{sr}^\epsilon(t) \equiv 0$ and (43) is trivial. If $k_{rs} + k_{sr} \geq 1$, by the inequality $|x^h - x^k| \leq C(1 - x)$ with $x = t/(t + \epsilon)$, we get

$$I_{rs}^\epsilon(t) \ \leq \ C \, (t + \epsilon)^{1/2} \cdot \frac{\epsilon}{t + \epsilon} \ = \ \frac{C \epsilon}{\sqrt{t + \epsilon}} \ < \ \frac{C \epsilon}{\sqrt{t}}.$$

Thus,

$$\int_0^T I_{rs}^\epsilon(t) \, dt \ \leq \ C \epsilon. \tag{65}$$

- *Case II* : $a_{rs}(t) \not\equiv 0$, $a_{sr}(t) \equiv 0$. We have now $r \in \alpha$, $s \in \alpha'$, with $\alpha < \alpha'$. Therefore,

$$I_{rs}^\epsilon(t) = \sqrt{\frac{\lambda_r^\epsilon(t)}{\lambda_s^\epsilon(t)}} \, |a_{rs}(t)| \leq C \, \epsilon^{(\sigma(\alpha)-\sigma(\alpha'))/2} \, (t+\epsilon)^{-\Delta(\alpha')/2},$$

whence, taking (63) into account,

$$\int_0^T I_{rs}^\epsilon(t) \, dt \leq C \, \epsilon^{(\sigma(\alpha)-\sigma(\alpha')-\widetilde{\Delta}(\alpha'))/2}. \tag{66}$$

Thus, in order to get (43) with $\theta = 1$, we have only to prove that

$$\sigma(\alpha) - \sigma(\alpha') - \widetilde{\Delta}(\alpha') \geq 2, \quad \text{for} \quad \alpha < \alpha'.$$

This inequality follows directly from the definition (59). Indeed, let $\{\alpha_{i_1}, \ldots, \alpha_{i_\nu}\}$ be a chain, with $\alpha' < \alpha_{i_1} < \ldots < \alpha_{i_\nu}$, for which the maximum is taken, that is, such that

$$\sigma(\alpha') = \widetilde{\Delta}(\alpha_{i_1}) + \ldots + \widetilde{\Delta}(\alpha_{i_\nu}) + 2(\nu - 1).$$

Therefore $\alpha < \alpha' < \alpha_{i_1} < \ldots < \alpha_{i_\nu}$, so that, by (59), we find

$$\sigma(\alpha) \geq \widetilde{\Delta}(\alpha') + \widetilde{\Delta}(\alpha_{i_1}) + \ldots + \widetilde{\Delta}(\alpha_{i_\nu}) + 2\nu = \widetilde{\Delta}(\alpha') + \sigma(\alpha') + 2.$$

Thus, for each r, s, we have proved (65), that is (43) with $\theta = 1$. This completes the proof of Lemma 2.

Conclusion of the proof of Theorem 1

To conclude the proof of Theorem 1, we have only to observe that the number δ defined by (62) and (59), can be also written as

$$\delta = \frac{1}{2} \max_{\alpha_{i_1} < \cdots < \alpha_{i_\nu}} \left(\widetilde{\Delta}(\alpha_{i_1}) + \cdots + \widetilde{\Delta}(\alpha_{i_\nu}) + 2(\nu - 1) \right),$$

and then apply Proposition 2.

REFERENCES

[1] M.D. Bronšteĭn, *The Cauchy problem for hyperbolic operators with characteristics of variable multiplicity*, Trudy Moscow Math. Soc. 41 (1980), Trans. Moscow Math. Soc., Vol. 1 (1982), 87-103.

[2] F. Colombini, E. Jannelli and S. Spagnolo, *Well-posedness in the Gevrey classes of the Cauchy problem for a non-strictly hyperbolic equation with coefficients depending on time*, Ann. Scu. Norm. Sup. Pisa, Vol. 10 (1983), 291–312.

[3] P. D'Ancona and S. Spagnolo. *Quasi-symmetrization of hyperbolic systems and propagation of the analitic regularity*, Boll. Un. Mat. It., (8) Vol.1-B (1998), 169–185.

[4] P. D'Ancona and S. Spagnolo, On pseudosymmetric hyperbolic systems, Ann. Scu. Norm. Sup. Pisa, Vol. 25 (1997), 397–417.

[5] E. Jannelli, Linear Kovalewskian systems with time dependent coefficients, Comm. P.D.E., Vol. 9 (1984), 1373–1406.

[6] K. Kajitani, Local solution of the Cauchy problem for hyperbolic systems in Gervrey classes, Hokkaido Math. J., Vol. 12 (1983), 434–460.

[7] T. Nishitani and S. Spagnolo, On pseudosymmetric systems with one space variable, to appear on Ann. Scu. Norm. Sup. Pisa.

Le rôle des fonctions spéciales dans les problèmes de Goursat pour des équations aux dérivés partielles à coefficients constants

JAIME CARVALHO E SILVA, Departamento de Matemática, Universidade de Coimbra, Coimbra, Portugal

1 INTRODUCTION

Dans [1] et [2] Vaillant étudie des solutions asymptotiques pour des systèmes d'opérateurs différentiales linéaires à partie principale à coefficients constants et fortement hyperboliques quand le cône normal caractéristique admet un point double différent de zéro au voisinage duquel la multiplicité varie. Les solutions asymptotiques sont obtenues à partir de la solution d'un problème de Goursat à deux faces pour un certain système. Le problème de Goursat à deux faces pour une équation a été étudié par la première fois par Hasegawa [3]; la méthode repose sur la construction d'une solution particulière sous la forme de fonction de Bessel, et, avec une solution asymptotique de cette fonction, on prouve que la continuité de la solution par rapport aux conditions initiales n'est pas satisfaite si un certain nombre de coefficients est différent de zéro. Cette méthode a été généralisée pour certaines équations du troisième ordre dans [4] et [5]. La méthode est semblable mais on construit des solutions particulières à l'aide de fonctions hypergéométriques de plusieurs variables. La théorie des fonctions spéciales ne permet pas, pour le moment, d'aller plus loin dans cette méthode. Les fonctions spéciales ont ainsi un rôle important dans l'étude de certains problèmes de Goursat. On va essayer de prouver pourquoi il serait intéressant d'avoir dans l'étude des problèmes de Goursat plus de

---

Avec le support de CMUC, Coimbra.

résultats sur les fonctions spéciales.

2 SYSTÈMES AVEC MULTIPLICITÉ VARIABLE

Soit h un opérateur différentiel linéaire matriciel à coefficients constants

$$h(D) = \left(h_B^A(D) \right), 1 \le A \le m, 1 \le B \le m$$

où chaque h_B^A est un opérateur différentiel d'ordre au plus t. La partie principale de $h(D)$ est à coefficients constants réels et fortement hyperbolique. On suppose que le cône normal caractéristique admet un point double π différent de zéro au voisinage duquel la multiplicité varie.

Dans [1] et [2] Vaillant obtient des solutions asymptotiques de la forme

$$y^B(x,r) = e^{iw\pi\varphi^+(x,r)} \left(Y_t^{+B}(x,r) + \tfrac{1}{iw} Y_{t+1}^{+B}(x,r) + ... \right) +$$
$$+ e^{iw\varphi^-(x,r)} \left(Y_t^{-B}(x,r) + \tfrac{1}{iw} Y_{t+1}^{-B}(x,r) + ... \right) +$$
$$+ \int_{-x^0}^{x^0} e^{iw\Phi(x,\tau,r)} \left(Z_t^B(x,\tau,r) + \tfrac{1}{iw} Z_{t+1}^B(x,\tau,r) + ... \right) d\tau$$

où $Y_p^{\pm B}(x,r)$ et $Z_p^B(x,\tau,r)$, $p = t, t+1, ...$, sont des fonctions analytiques dans un voisinage de zéro. Les fonctions de phase φ^+, φ^- et Φ sont des fonctions voisines "par arcs" de la phase singulière φ telle que $grad\varphi = \pi$. Pour obtenir ces solutions il faut, par exemple, déterminer des fonctions $V_t^{\pm}$ telles que

$$Z_t^B(x,\tau,r) = V_t^+(x,r)d^{+B}(r) + V_t^-(x,r)d^{-B}(r)$$

où $d^{\pm B}(r)$ sont des fonctions analytiques convenables. Les fonctions $V_t^{\pm}$ doivent satisfaire un système de deux équations aux derivées partielles dont la matrice est fortement hyperbolique, avec données sur deux hyperplans caractéristiques (qui se résout par réduction à une certaine équation hyperbolique).

On peut aussi obtenir des solutions asymptotiques avec des coefficients $Y_p^{\pm B}(x,r)$ et $Z_p^B(x,\tau,r)$ qui soient C^∞ si r est assez petit.

Si on veut généraliser ces solutions asymptotiques à des systèmes avec un cône normal caractéristique admettant un point triple ou supérieur il va falloir résoudre des problèmes de Goursat pour certains systèmes avec des données sur trois ou plus hyperplans caractéristiques.

3 ÉQUATIONS DU SECOND ORDRE

Le problème de Goursat à deux faces pour une équation à coefficients constants réels a été étudié pour la première fois par Hasegawa [3]; Un problème pour l'équation suivante est étudié

$$\partial_t \partial_x u = \sum_{\substack{i+j+|\alpha| \le 2 \\ i+j \le 1}} a_{ij\alpha} \partial_t^i \partial_x^j \partial_y^\alpha u$$
$$t \ge 0, x \in R, y \in R^n, \alpha \in N_0^n, |\alpha| = \alpha_1 + ... + \alpha_n$$

Les données sont

$$u(0, x, y) = \varphi(x, y)$$
$$u(t, 0, y) = \psi(t, y)$$

avec les nécessaires conditions de compatibilité. D'abord l'opérateur est transformé en

(*) $\qquad [\partial_t - \sum_i a_i \partial_{y_i} + a_0][\partial_x - \sum_i b_i \partial_{y_i} + b_0] u = \sum_{|\alpha| \leq 2} C_\alpha \partial_y^\alpha u$

et le théorème suivante est démontré:

THÉORÈME 1. *Si le problème donné est C^∞-bien posé alors $C_\alpha = 0, |\alpha| \geq 1$ et vice versa.*

La preuve dépend de la construction d'une solution particulière correspondant aux données initiales (où η est un paramètre réel)

$$u(0, x, y) = e^{i\eta y}$$
$$u(t, 0, y) = e^{i\eta y}$$

Cette solution est une fonction de Bessel, et, si on suppose que $C_\alpha \neq 0$ alors, quand η tend vers ∞ la fonction $u(t, x, y)$ a une croissance exponentielle avec η ce qui prouve que la continuité de la solution par rapport aux conditions initiales n'est pas satisfaite si un certain nombre de coefficients est différent de zéro.

4 ÉQUATIONS DU TROISIÈME ORDRE

Le problème de Goursat à trois faces pour une équation à coefficients constants est étudié dans [4] et [5]. Le problème est une généralisation partielle du problème de Hasegawa (parce qu'ici on ne peut pas faire une décomposition simple comme (*)). Si on considère d'abord le cas simple suivant (semblable à la situation (*)):

$$\partial_t \partial_x \partial_y u = \sum_{|\alpha| \leq 3} C_\alpha \partial_z^\alpha u$$
$$t, x, y \in R, z \in R^n, \alpha \in N_0^n, |\alpha| = \alpha_1 + ... + \alpha_n$$

avec les données initiales

$$u(0, x, y, z) = g_1(x, y, z)$$
$$u(t, 0, y, z) = g_2(t, y, z)$$
$$u(t, x, 0, z) = g_3(t, x, z)$$

et les nécessaires conditions de compatibilité, alors on a un théorème égal à celui d'Hasegawa, mais dans la preuve les fonctions de Bessel sont remplacées par des fonctions hypergéométriques d'une variable $_0F_2(w)$. Une difficulté essentielle est de prouver que cette fonction hypergéométrique a une croissance exponentielle avec η.

Si on considère un cas un peu plus général avec l'équation suivante:

$$\partial_t \partial_x \partial_y u = \sum_{\substack{i+j+k+|\alpha| \leq 3 \\ i+j+k \leq 1}} a_{ijk\alpha} \partial_t^i \partial_x^j \partial_y^k \partial_z^\alpha u$$
$$t, x, y \in R, z \in R^n, \alpha \in N_0^n, |\alpha| = \alpha_1 + ... + \alpha_n$$

où les données initiales sont

$$u(0, x, y, z) = g_1(x, y, z)$$
$$u(t, 0, y, z) = g_2(t, y, z)$$
$$u(t, x, 0, z) = g_3(t, x, z)$$

avec les nécessaires conditions de compatibilité, alors on a un théorème semblable à celui d'Hasegawa mais cette fois la preuve exige des fonctions spéciales beaucoup plus compliquées: les fonctions hypergéométriques à quatre variables

$$_2F_5(w_1, w_2, w_2, w_4).$$

Ici il est très difficile de prouver que ces fonctions hypergéométriques ont une croissance exponentielle avec η (et que la somme a aussi une croissance exponentielle avec η). Il n'est pas possible d'aller plus loin maintenant parce que les fonctions hypergéométriques deviennent plus difficiles (et on ne peut pas obtenir des développements asymptotiques) et parce qu'une décomposition comme (*) devient impossible pour des cas plus généraux.

Dans [5] un problème de Goursat du même type est étudié mais pour l'équation

$$\partial_t \partial_x \partial_y u = \left(\bar{a}_1 \partial_t^2 \partial_z^\alpha + b_1 \partial_x^2 \partial_z^\alpha + c_1 \partial_y^2 \partial_z^\alpha + a_2 \partial_t^2 + b_2 \partial_x^2 + c_2 \partial_y^2 \right) u$$
$$t, x, y \in R, z \in R^n, \alpha \in N_0^n$$

Dans ce cas un théorème comme celui de Hasegawa est obtenu mais on construit une solution particulière à partir de données initiales différentes

$$u(0, x, y, z) = e^{i\eta z} \left(\tfrac{1}{4} x^2 y^2 + \tfrac{1}{4!} y^4 + \tfrac{1}{4!} z^4 \right)$$
$$u(t, 0, y, z) = e^{i\eta z} \left(\tfrac{1}{4} t^2 y^2 + \tfrac{1}{4!} t^4 + \tfrac{1}{4!} y^4 \right)$$
$$u(t, x, 0, z) = e^{i\eta z} \left(\tfrac{1}{4} t^2 x^2 + \tfrac{1}{4!} t^4 + \tfrac{1}{4!} x^4 \right)$$

et on obtient des sommes de plusieurs fonctions hypergéométriques $_pF_{p+1}$ de deux et trois variables.

5 ÉQUATIONS D'ORDRE QUELCONQUE

Dans [6] et [7] Nishitani et Hasegawa étudient un problème de Goursat pour l'équation suivante

$$\sum_{\substack{i+j+|\alpha|\leq m \\ i \leq m-s}} a_{ij\alpha} \partial_t^i \partial_x^j \partial_y^\alpha u = 0$$
$$t \geq 0, x \in R, y \in R^n, \alpha \in N_0^n, |\alpha| = \alpha_1 + ... + \alpha_n$$

avec des données initiales (plus conditions de compatibilité)

$$\partial_t^i u(0, x, y) = \varphi_i(x, y), 0 \leq i \leq m - s - 1$$
$$\partial_x^j u(t, 0, y) = \psi_j(t, y), 0 \leq j \leq s - 1$$

On suppose que l'hyperplan $\{t = 0\}$ est caractéristique d'ordre s; donc

$$a_{ij\alpha} = 0, i + j + |\alpha| = m, i > m - s$$

Hasegawa a prouvé qu'on doit avoir toujours $a_{m-s,s,0} \neq 0$. Dans le cas où

$$(**) \qquad a_{ij\alpha} = 0, i + j + |\alpha| < m, i > m - s$$

Nishitani montre que l'opérateur doit être hyperbolique:

THÉORÈME 2. *Le problème de Goursat est C^∞-bien posé si et seulement si il existe $\varepsilon > 0$ tel quer pour tout δ avec $0 < |\delta| < \varepsilon$ l'opérateur est hyperbolique par rapport à $(1, \delta, 0)$.*

Pour caractériser plus en détail l'opérateur, dans [7] Hasegawa prouve que si la condition (**) n'est pas vérifiée alors le problème n'est pas C^∞-bien posé. Pour arriver à cette conclusion il faut étudier les propriétés de la solution du problème de Goursat avec des données initiales (où η est un paramètre complexe et p, q, ϱ, α sont des constantes convenables)

$$\psi_j(t, y) = 0, 0 \leq j \leq \varrho - 1$$
$$\psi_j(t, y) = (\alpha \eta^q)^j e^{\eta^p t}, \varrho \leq j \leq s - 1$$
$$\varphi_i(x, y) = \eta^{pi} \sum_{k=\rho}^{s-1} \frac{\alpha^k \eta^{qk}}{k!} x^s, 0 \leq i \leq m - s - 1$$

Hasegawa montre que la solution croit exponentiellement avec $|\eta|$ mais les données initiales ont seulement une croissance polynomiale avec $|\eta|$. La preuve n'a pas besoin d'une forme explicite pour la solution mais est très compliquée et difficile à suivre.

6 CONCLUSION

On a vu comme les fonctions spéciales sont importantes dans un certain nombre de problèmes de Goursat quand il faut construire une solution explicite. Pour étudier complètement le cas d'une équation du troisième ordre (et d'ordres supérieures) et appliquer les conclusions aux problèmes des systèmes à caractéristiques de multiplicité variable supérieure à deux, il faudrait élargir ces recherches mais il y a plusieurs difficultés.

D'abord pour construire les solutions explicites on obtient des relations de récurrence très difficiles à étudier. Normalement, dans la théorie des fonctions spéciales on se donne une fonction et on cherche une relation de récurrence; mais si on a seulement la relation de récurrence comment obtenir la fonction correspondante? Il y a très peu de méthodes pour ça. Aussi les fonctions hypergéométriques qu'on rencontre sont de plusieurs variables dont on connaît peu de développements asymptotiques. Peut-être aussi difficile est de découvrir quelles conditions initiales sont adéquates pour obtenir une bonne fonction solution. La théorie des fonctions spéciales se développe beaucoup et peut-être la théorie des q-fonctions pourra fournir de bons outils.

Tous les théorèmes présentés ici montrent qu'un problème de Goursat à plusieurs faces n'est pas bien posé pour des opérateurs trop généraux. Il est donc normal d'espérer que les opérateurs deviennent assez simples et donc la construction de solutions particulières parait une bonne méthode pour ça. Il faut seulement espérer que les futurs développements de la théorie des fonctions hypergéométriques de plusieurs variables et leurs développements asymptotiques permettent d'aller plus loin.

REFERENCES

[1] J Vaillant, Systémes à caractéristiques de multiplicité variable et solutions asymptotiques, C.R.Acad.Sci.Paris, 276, 1973.

[2] J Vaillant, Solutions asymptotiques d'un système à caractéristiques de multiplicité variable, J. Math. Pures et Appl., 53, 1974.

[3] Y Hasegawa, On the C^∞-Goursat problem for 2nd order equations with real constant coefficients, Proc. Japan Academy, 51(7):516-519, 1975.

[4] J Carvalho e Silva, Problème de Goursat-Darboux généralisé pour un opérateur du troisième ordre, Série I, no. 6, C.R.Acad. Sci. Paris, 303:223-226, 1986.

[5] J Carvalho e Silva, Carlos Leal, The generalized Goursat-Darboux problem for a third order operator, Proc. A.M.S., 125(2): 471-475, 1997.

[6] T Nishitani, On the C^∞-well posedness for the Goursat problem with constant coefficients, J. Math. Kyoto Univ., 20, 1980.

[7] Y Hasegawa, A remark on the C^∞-Goursat problem II, Publ. RIMS Kyoto Univ., 28, 1992.

Influence of the hyperbolic part on decay rates in 1-d thermoelasticity

Ya-Guang Wang* and Michael Reissig**

*Department of Applied Mathematics, Shanghai Jiao Tong University, 200030 Shanghai, P.R.China, e-mail: ygwang@online.sh.cn
**   Faculty of Mathematics and Computer Science, TU Bergakademie Freiberg, Bernhard-von-Cotta-Str. 2, D-09596 Freiberg, Germany, e-mail:reissig@math.tu-freiberg.de

In honour of Prof. J.Vaillant

1   Introduction

To prove global existence results for solutions of the Cauchy problem for nonlinear wave equations so-called $L_p - L_q$ decay estimates for solutions of the linear wave equation play an essential role (see [2],[3] and references therein). That is the following estimate for the solution $u = u(t, x)$ of the Cauchy problem

$$u_{tt} - \triangle u = 0 \ , \ u(0, x) = 0 \ , \ u_t(0, x) = u_1(x) \ ,$$

where $u_1(x)$ belongs to $C_0^\infty(\mathbf{R}^n)$ (see [8]):

there exist constants C and M depending on p and n such that

$$\|u_t(t, \cdot)\|_{L_q(\mathbf{R}^n)} + \|\nabla u(t, \cdot)\|_{L_q(\mathbf{R}^n)} \leq C(1 + t)^{-\frac{n-1}{2}(\frac{1}{p} - \frac{1}{q})}\|u_1\|_{W_p^M(\mathbf{R}^n)} \ , \tag{1.1}$$

where $1 \leq p \leq 2$ and $1/p + 1/q = 1$.

The use of the powerful tool of $L_p - L_q$ decay estimates is not restricted to wave equations, it is also used to study nonlinear equations (systems) which are related to applied problems, e.g. thermoelasticity, elasticity, Klein-Gordon equation, Maxwell system ([3], [11]). For the one-dimensional model of thermoelasticity (α, β, γ_1, γ_2 are positive constants)

$$
\begin{aligned}
&u_{tt} - \alpha^2 u_{xx} + \gamma_1 \theta_x = 0 \ , \\
&\theta_t - \beta^2 \theta_{xx} + \gamma_2 u_{tx} = 0 \ , \\
&u(0, x) = u_0(x) \ , \ u_t(0, x) = u_1(x) \ , \ \theta(0, x) = \theta_0(x) \ ,
\end{aligned}
\tag{1.2}
$$

one gets the $L_p - L_q$ decay estimate ([11])

$$\left\|\left(u_t(t, \cdot), u_x(t, \cdot), \theta(t, \cdot)\right)\right\|_{L_q(\mathbf{R})} \leq C(1 + t)^{-\frac{1}{2}(\frac{1}{p} - \frac{1}{q})}\|(u_1, u_0', \theta_0)\|_{W_p^2(\mathbf{R})} \ . \tag{1.3}$$

89

In opposite to (1.1) for $n = 1$ the estimate (1.3) gives a decay rate for the model (1.2) which is typical for solutions of the Cauchy problem for one-dimensional heat or Schrödinger equations.

The results of papers [5], [6], [7] generalize the question for $L_p - L_q$ decay estimates to wave equations with time-dependent coefficients. Let us explain some of the results by the aid of the model problem

$$u_{tt} - \lambda^2(t)b^2(t) \triangle u = 0 \, , \, u(0, x) = u_0(x) \, , \, u_t(0, x) = u_1(x) \, .$$

The coefficient of $\triangle u$ consists of an increasing part $\lambda = \lambda(t)$ (improving influence on decay estimates) and an oscillating part $b = b(t)$ (deteriorating influence on decay estimates). There exists an interesting action and reaction between both influences, which can be described by the condition

$$|D_t b(t)| \leq C \frac{\lambda(t)}{\Lambda(t)} (\ln \Lambda(t))^\delta \text{ for large } t, \tag{1.4}$$

where $\Lambda(t) = \int_0^t \lambda(s)ds$. If

• $\delta < 1$, then the oscillations are called *slow* and one has $L_p - L_q$ decay estimates similar to (1.1) (the increasing part is dominating) [7];
• $\delta = 1$, then the oscillations are called *fast* and one has $L_p - L_q$ decay estimates for large dimension n (optimal action and reaction) [6];
• the condition (1.4) is not satisfied for $\delta = 1$, then the oscillations are called *very fast* and one has in general no $L_p - L_q$ decay estimates (the oscillating part is dominating) [5].

These observations motivate us to consider models with time-dependent coefficients containing the wave operator.

Here, we are interested in the Cauchy problem for the model of one-dimensional thermoelasticity with time-dependent coefficients of the form

$$\begin{aligned}
&u_{tt} - \lambda^2(t)b^2(t)u_{xx} + \gamma_1(t)\theta_x = 0 \, , \\
&\theta_t - \beta^2(t)\theta_{xx} + \gamma_2(t)u_{tx} = 0 \, , \\
&u(0, x) = u_0(x) \, , \, u_t(0, x) = u_1(x) \, , \, \theta(0, x) = \theta_0(x) \, .
\end{aligned} \tag{1.5}$$

This model has more freedoms than the constant coefficient model (1.2) because we are able to describe different growth conditions for coefficients. The case, where coefficients have the same growth is studied in [10]. In the present paper we want to study the case, where

• *the speed of growth of β^2 dominates those ones of the other coefficients.*

Thus, one should expect that in opposite to (1.2) the decay rate, which comes from the heat operator $\theta_t - \beta^2(t)\theta_{xx}$, has no essential influence on the model (1.5). Consequently,

• *the decay rate should be predicted by the wave operator.*

Can we get a decay rate for (1.5)? Pay attention that $n = 1$ gives no decay in (1.1), which means that $\gamma_1 = \gamma_1(t) \equiv 0$ allows no decay for u in (1.5). What is the influence of the coefficients γ_1, γ_2? These and related questions will be discussed in the present paper. Setting $\gamma_1 = \gamma_2 \equiv 0$ it follows that the classification of oscillations coming from (1.4) is the right one for our model (1.5). It is reasonable to suppose that

• *all coefficients of model (1.5) contain only slow oscillations.*

The considerations for (1.5) are basing on the following assumptions:
• A1: the functions $\lambda = \lambda(t)$, $b = b(t)$, belonging to $C^2[0, \infty)$ are bounded from below by a positive constant;
• A2: there exist positive constants C_k, $k = 1, 2$, such that for $t > 0$ it holds

$$\lambda'(t) > 0, \quad |D_t^k \lambda(t)| \leq C_k \left(\tfrac{\lambda(t)}{\Lambda(t)}\right)^k \lambda(t),$$

where $\Lambda(t) := \int\limits_0^t \lambda(\tau)d\tau$;
• A3: there exist positive constants C_k, $k = 0, 1, 2$, and a constant $\delta \in [0, 1)$ such that for large t it holds

$$|D_t^k b(t)| + \tfrac{1}{\lambda(t)}\left(|D_t^k \gamma_1(t)| + |D_t^k \gamma_2(t)|\right) \leq C_k \left(\tfrac{\lambda(t)}{\Lambda(t)}(\ln \Lambda(t))^\delta\right)^k;$$

• A4: there exists a positive constant C such that for large t it holds

$$|\gamma_1(t)| \int\limits_0^t |\gamma_2(s)|ds \leq C \frac{\Lambda(t)\lambda(t)}{(\ln \Lambda(t))^\delta};$$

• A5: there exist positive constants C and C_k, $k = 1, 2$, such that for large t it holds

$$\left|\tfrac{1}{\beta^2}(t)\right| \leq \frac{C}{\Lambda(t)\lambda(t)}, \quad \left|D_t^k \tfrac{1}{\beta^2}(t)\right| \leq \frac{C_k}{\Lambda(t)\lambda(t)} \left(\tfrac{\lambda(t)}{\Lambda(t)}(\ln \Lambda(t))^\delta\right)^{k+1}.$$

Remark 1.1 *The assumption A1 excludes zeros for λ and b and gives the necessary regularity. It is fixed by the diagonalization procedure modulo $S_{\delta,0}\{-1, -1, 2\}$ presented in Section 2.3.2. By assumption A2 the function λ is an increasing one with suitable properties for its derivatives. These inequalities which are not very restrictive allow logarithmic, potential, exponential or superexponential growth for λ. The condition A3 restricts the influence of $b = b(t)$ on the oscillating behaviour of the coefficient of $\triangle$, it allows only slow oscillations. The assumption A5 guarantees the dominating role of the coefficient $\lambda^2 b^2$. The form of the dominate speed of the growth of β^2 (see the first part of A5) is inspired by the papers [6], [7]. If $\lambda = \exp t$, then this condition is satisfied but does not express a dominating behaviour. Consequently, $L_p - L_q$ decay estimates can be only expected for logarithmic or potential growth (see the following examples). We want to underline that the assumptions describe a model which generalizes the constant coefficient case in a suitable way.*

Under the above assumptions we will prove $L_p - L_q$ estimates (see Theorem 3.6) which yield the following two $L_p - L_q$ decay estimates for models from one-dimensional thermoelasticity.

Example 1.2 *Let us consider the next model:*

$$u_{tt} - (\ln(e+t))^2 b^2(t)u_{xx} + \gamma_1 \ln(e+t)\theta_x = 0 ,$$
$$\theta_t - \ln(e+t)(1+t\ln(e+t))\theta_{xx} + \gamma_2 \ln(e+t)u_{tx} = 0 ,$$
$$u(0,x) = u_0(x) , \ u_t(0,x) = u_1(x) , \ \theta(0,x) = \theta_0(x) .$$

Here $b = b(t)$ is a positive smooth function belonging to $S_{\delta,2}\{0,0,0\}$ (see Definition 2.1), for example $b = b(t) := 2 + \sin((\ln(2+t))^{\delta+1})$, γ_1 and γ_2 are positive constants satisfying $\gamma_1\gamma_2 < 2$. Then we have the $L_p - L_q$ decay estimate

$$\|u_t(t,\cdot)\|_{L_q(\mathbf{R})} + \ln(e+t)\|u_x(t,\cdot)\|_{L_q(\mathbf{R})} + \|\theta(t,\cdot)\|_{L_q(\mathbf{R})}$$
$$\leq C \frac{(\ln(e+t))^{\frac{1}{2}}}{\left(1+t\ln(e+t)\right)^{\frac{\gamma_1\gamma_2}{2}}} \left(\|u_0\|_{W_p^3(\mathbf{R})} + \|u_1\|_{W_p^2(\mathbf{R})} + \|\theta_0\|_{W_p^2(\mathbf{R})}\right) ,$$

where $1/p + 1/q = 1$, $1/p - 1/q > \frac{\gamma_1\gamma_2}{2}$.

Example 1.3 *Let us consider with $\alpha > 0$ the next model:*

$$u_{tt} - (1+t)^{2\alpha}b^2(t)u_{xx} + \gamma_1(1+t)^{\alpha}\theta_x = 0 ,$$
$$\theta_t - (1+t)^{2\alpha+1}\theta_{xx} + \gamma_2(1+t)^{\alpha}u_{tx} = 0 ,$$
$$u(0,x) = u_0(x) , \ u_t(0,x) = u_1(x) , \ \theta(0,x) = \theta_0(x) .$$

Here $b = b(t)$ is a positive smooth function belonging to $S_{\delta,2}\{0,0,0\}$ (see Definition 2.1), for example $b = b(t) := 2 + \sin((\ln(2+t))^{\delta+1})$, γ_1 and γ_2 are positive constants satisfying $\alpha(1+2\kappa) < \gamma_1\gamma_2 < \alpha(1+2\kappa) + 2$ with a suitable $\kappa \in [0, 1/2]$. Then we have the $L_p - L_q$ decay estimate

$$\|u_t(t,\cdot)\|_{L_q(\mathbf{R})} + (1+t)^{\alpha}\|u_x(t,\cdot)\|_{L_q(\mathbf{R})} + \|\theta(t,\cdot)\|_{L_q(\mathbf{R})}$$
$$\leq C (1+t)^{-\frac{(\gamma_1\gamma_2-\alpha(1+2\kappa))}{2}} \left(\|u_0\|_{W_p^3(\mathbf{R})} + \|u_1\|_{W_p^2(\mathbf{R})} + \|\theta_0\|_{W_p^2(\mathbf{R})}\right) ,$$

where $1/p + 1/q = 1$, $1/p - 1/q > \frac{\gamma_1\gamma_2-(2\kappa-1)\alpha}{2(\alpha+1)}$.

The remainder of this paper is arranged as follows: In Section 2 we introduce tools of our approach and derive WKB-representations for solutions of auxiliary Cauchy problems by using decoupling and diagonalizing techniques. In Section 3 we formulate a result for the transformed problem which gives immediately the representation for the solution of the model of 1-d-thermoelasticity. Corresponding $L_p - L_q$ estimates will be proved which lead in special cases to decay estimates for the models from Examples 1.2 and 1.3 which are dominated by the hyperbolic part.

2   WKB representations

2.1   Tools of the approach

2.1.1   Zones

We split the set $\{(t,\xi) \in [0,\infty) \times (\mathbb{R} \setminus \{0\})\}$ into subdomains which will be called *zones*. With two large parameters N_1 and N_2 and the constant δ from assumption A3 we define the *pseudodifferential zone* $Z_{pd}(N_1, N_2)$ by

$$Z_{pd}(N_1, N_2) := \left\{ (t,\xi) : \Lambda(t)|\xi| \leq N_1(\ln(N_2 + \Lambda(t)))^\delta \right\},$$

and the *hyperbolic zone* $Z_{hyp}(N_1, N_2)$ by

$$Z_{hyp}(N_1, N_2) := \left\{ (t,\xi) : \Lambda(t)|\xi| \geq N_1(\ln(N_2 + \Lambda(t)))^\delta \right\}.$$

For $|\xi| \in (0,\infty)$ we define the function $t_\xi = t(|\xi|)$ as the unique solution of $\Lambda(t_\xi)|\xi| = N_1(\ln(N_2 + \Lambda(t_\xi)))^\delta$.

2.1.2   Classes of symbols

For the further considerations we need suitable classes of symbols of finite smoothness which are defined only in the hyperbolic zone $Z_{hyp}(N_1, N_2)$.

Definition 2.1 *For any given real numbers* m_1, m_2, m_3 , $\delta \in [0,1)$, *and* $k \in \mathbb{N}_0$, *we denote by* $S_{\delta,k}\{m_1, m_2, m_3\}$ *the set of all symbols* $a = a(t,\xi) \in C(Z_{hyp}(N_1, N_2))$ *of* k-th *order differentiable with respect to* t *satisfying*

$$|D_t^l a(t,\xi)| \leq C_l |\xi|^{m_1} \lambda(t)^{m_2} \left(\tfrac{\lambda(t)}{\Lambda(t)} \left(\ln(N_2 + \Lambda(t)) \right)^\delta \right)^{m_3+l}$$

for all $(t,\xi) \in Z_{hyp}(N_1, N_2)$, *all* $l \in \mathbb{N}_0$, $l \leq k$, *with constants* C_l *independent of* N_1 *and* N_2.

Let us list some simple rules of the symbolic calculus:

1. $S_{\delta,k}\{m_1, m_2, m_3\} \subset S_{\delta,k}\{m_1 + l, m_2 + l, m_3 - l\}$ for $l \geq 0$;

2. if $a \in S_{\delta,k}\{m_1, m_2, m_3\}$ and $b \in S_{\delta,k}\{n_1, n_2, n_3\}$, then $a\,b \in S_{\delta,k}\{m_1 + n_1, m_2 + n_2, m_3 + n_3\}$;

3. if $a \in S_{\delta,k}\{m_1, m_2, m_3\}$, then $D_t a \in S_{\delta,k-1}\{m_1, m_2, m_3 + 1\}$.

2.2   Consideration in the pseudodifferential zone $\mathbf{Z_{pd}(N_1, N_2)}$

Let us consider (1.5). After partial Fourier transformation with respect to the spatial variable we get the system

$$\hat{u}_{tt} + \lambda^2(t)b^2(t)\xi^2\hat{u} + i\gamma_1(t)\xi\hat{\theta} = 0 ,$$

$$\hat{\theta}_t + \beta^2(t)\xi^2\hat{\theta} + i\gamma_2(t)\xi\hat{u}_t = 0 ,$$

$$\hat{u}(0,\xi) = \hat{u}_0(\xi) , \quad \hat{u}_t(0,\xi) = \hat{u}_1(\xi) , \quad \hat{\theta}(0,\xi) = \hat{\theta}_0(\xi) .$$

By setting $u_1 := \hat{u}_t$, $u_2 := \lambda(t)\xi\,\hat{u}$, and $u_3 := \hat{\theta}$, this system can be transformed to the first order system

$$\partial_t u_1 + b^2(t)\lambda(t)\xi u_2 + i\gamma_1(t)\xi u_3 = 0 ,$$
$$\partial_t u_2 - \frac{\lambda'(t)}{\lambda(t)}u_2 - \lambda(t)\xi u_1 = 0 , \qquad (2.1)$$
$$\partial_t u_3 + \beta^2(t)\xi^2 u_3 + i\gamma_2(t)\xi u_1 = 0 ,$$

$$u_1(0,\xi) = \hat{u}_1(\xi) , \quad u_2(0,\xi) = \lambda(0)\xi\hat{u}_0(\xi) \quad , \quad u_3(0,\xi) = \hat{\theta}_0(\xi) .$$

Theorem 2.2 *To each positive ε and suitable chosen N_1, N_2 there exists a constant C_{ε,N_1} such that the energy estimate*

$$E(u_1,u_2,u_3)(t,\xi) \le C_{\varepsilon,N_1} E(u_1,u_2,u_3)(0,\xi)\lambda(t)(N_2 + \Lambda(t_\xi))^\varepsilon$$

holds in $Z_{pd}(N_1,N_2)$ for the solution of (2.1), where

$$E(u_1,u_2,u_3)(t,\xi) = \max_{s\in[0,t]}(|u_1(s,\xi)| + |u_2(s,\xi)| + |u_3(s,\xi)|).$$

Proof. Setting the transformation $u_3 = e^{-\int_0^t \beta^2(s)\xi^2 ds}\, v_3$ into the third equation of (2.1) gives for v_3 the relation

$$v_3(t,\xi) = u_3(0,\xi) - i\int_0^t \gamma_2(s)\xi e^{\int_0^s \beta^2(\sigma)\xi^2 d\sigma} u_1(s,\xi)ds . \qquad (2.2)$$

Using this transformation for u_3 in the first two equations of (2.1) gives immediately the system of integro-differential equations for (u_1,u_2) :

$$\partial_t u_1 + b^2(t)\lambda(t)\xi u_2 \;+\; i\gamma_1(t)\xi\Big(e^{-\int_0^t \beta^2(s)\xi^2 ds} u_3(0,\xi) - i\int_0^t \gamma_2(s)\xi e^{-\int_s^t \beta^2(\sigma)\xi^2 d\sigma} u_1(s,\xi)ds\Big) = 0 ,$$

$$\partial_t u_2 \;-\; \frac{\lambda'(t)}{\lambda(t)}u_2 - \lambda(t)\xi u_1 = 0 ,$$
$$u_1(0,\xi) = \hat{u}_1(\xi) , \quad u_2(0,\xi) = \lambda(0)\xi\hat{u}_0(\xi) .$$

By the aid of this transformation and the assumption A5 we understand that the growth of $\beta^2 = \beta^2(t)$ has no essential influence on the energy estimates for u_1 and u_2 in $Z_{pd}(N_1,N_2)$. Obviously, from the above system we obtain the next one of integral equations:

$$u_1(t,\xi) = u_1(0,\xi) - \int_0^t b^2(s)\lambda(s)\xi u_2(s,\xi)ds$$

$$- \int_0^t \gamma_1(s)\xi\Big(\int_0^s \gamma_2(\sigma)\xi e^{-\int_\sigma^s \beta^2(\tau)\xi^2 d\tau} u_1(\sigma,\xi)d\sigma\Big)ds$$

$$- i\int_0^t \gamma_1(s)\xi\, e^{-\int_0^s \beta^2(\sigma)\xi^2 d\sigma} u_3(0,\xi)ds; \qquad (2.3)$$

$$u_2(t,\xi) = u_2(0,\xi) + \int_0^t \frac{\lambda'(s)}{\lambda(s)}u_2(s,\xi)ds + \int_0^t \lambda(s)\xi u_1(s,\xi)ds .$$

To estimate (u_1, u_2) let us first list several simple results:

a) $\exp\left(\int_0^t \frac{\lambda'(s)}{\lambda(s)} ds\right) = \frac{\lambda(t)}{\lambda(0)}$;

b) $\exp\left(\int_0^t \lambda(s)|\xi| ds\right) \leq \exp\left(\Lambda(t)|\xi|\right) \leq \exp\left(N_1(\ln(N_2 + \Lambda(t)))^\delta\right) = (N_2 + \Lambda(t))^{\alpha(N_1, N_2, t)}$,

where $\alpha(N_1, N_2, t) = N_1/(\ln(N_2 + \Lambda(t)))^{1-\delta}$. Here we use the definition of $Z_{pd}(N_1, N_2)$. Thus the condition $\delta \in [0, 1)$ and a suitable choice of $N_2 \gg N_1$ guarantee that $\exp(\int_0^t \lambda(s)|\xi| ds) \leq (N_2 + \Lambda(t_\xi))^\varepsilon$ with an arbitrary small positive ε;

c) $\exp\left(\int_0^t |\gamma_1(s)| \int_0^s |\gamma_2(\sigma)| d\sigma \, ds \, \xi^2\right) \leq \exp\left(C \int_0^t \frac{\Lambda(s)\lambda(s)\xi^2}{(\ln(N_2 + \Lambda(s)))^\delta} ds\right)$

$\leq \exp(CN_1^2(\ln(N_2 + \Lambda(t)))^\delta) = (N_2 + \Lambda(t))^{\alpha(N_1, N_2, t)}$,

by using the assumption A4, where $\alpha(N_1, N_2, t) = CN_1^2/(\ln(N_2 + \Lambda(t)))^{1-\delta}$. As in b) we get

$\exp(\int_0^t |\gamma_1(s)| \int_0^s |\gamma_2(\sigma)| d\sigma \, ds\xi^2) \leq (N_2 + \Lambda(t_\xi))^\varepsilon$ with an arbitrary small $\varepsilon > 0$.

If we define $E(u_1, u_2)(t, \xi) = \max_{s \in [0, t]}(|u_1(s, \xi)| + |u_2(s, \xi)|)$ and $E(u_3)(t, \xi) = \max_{s \in [0, t]} |u_3(s, \xi)|$, then after application of Gronwall's inequality to (2.3) we get the energy estimate

$$E(u_1, u_2)(t, \xi) \leq E(u_1, u_2)(0, \xi) + \int_0^t |\gamma_1(s)||\xi| ds \, E(u_3)(0, \xi)$$

$$+ \int_0^t \left((\lambda(s)|\xi| + \frac{\lambda'(s)}{\lambda(s)} + |\xi|^2|\gamma_1(s)| \int_0^s |\gamma_2(\sigma)| d\sigma\right)$$

$$\times \left(E(u_1, u_2)(0, \xi) + \int_0^s |\gamma_1(\sigma)||\xi| d\sigma \, E(u_3(0, \xi))\right) \frac{\lambda(t)}{\lambda(s)} (N_2 + \Lambda(t_\xi))^\varepsilon\right) ds$$

by using the estimates from a) to c). Together with the assumptions $A3$ and $A4$ this implies

$$E(u_1, u_2)(t, \xi) \leq CE(u_1, u_2, u_3)(0, \xi)\lambda(t)(N_2 + \Lambda(t_\xi))^\varepsilon .$$

By the aid of (2.2) the energy of u_3 can be estimated by

$$E(u_3)(t, \xi) \leq CE(u_1, u_2, u_3)(0, \xi)\lambda(t)(N_2 + \Lambda(t_\xi))^\varepsilon .$$

This completes the proof. □

Corollary 2.3 *To each positive ε and suitable chosen N_1, N_2 there exists a constant C_{ε,N_1} such that the solution of (2.1) has in $Z_{pd}(N_1, N_2)$ the representation*

$$u_k(t,\xi) = a_{1,k}(t,\xi)u_1(0,\xi) + a_{2,k}(t,\xi)u_2(0,\xi) + a_{3,k}(t,\xi)u_3(0,\xi), \ k = 1,2,3,$$

where the functions $a_{k,l}$, $k,l = 1,2,3$, satisfy the estimates

$$|a_{k,l}(t,\xi)| \le C_{\varepsilon,N_1} \lambda(t)(N_2 + \Lambda(t_\xi))^\varepsilon .$$

Proof. It suffices to use the estimates from Theorem 2.2 for the data $(u_1, u_2, u_3)(0,\xi) = (1,0,0)$; $(0,1,0)$; $(0,0,1)$ separately, because a linear combination of these solutions gives a solution of the general Cauchy problem for (2.1).   □

2.3   Consideration in the hyperbolic zone $Z_{hyp}(N_1, N_2)$

2.3.1   Decoupling modulo $S_{\delta,1}\{-1,-1,2\}$

To decouple the system

$$\hat{u}_{tt} + \lambda^2(t)b^2(t)\xi^2\hat{u} + i\gamma_1(t)\xi\hat{\theta} = 0 ,$$
$$\hat{\theta}_t + \beta^2(t)\xi^2\hat{\theta} + i\gamma_2(t)\xi\hat{u}_t = 0 ,$$

we use a procedure proposed in [9]. If we define $u_1 := \left(\partial_t + i\lambda(t)b(t)\xi\right)\hat{u}$, $u_2 := \left(\partial_t - i\lambda(t)b(t)\xi\right)\hat{u}$ and $u_3 := \hat{\theta}$, then the above system is transformed to

$$\partial_t \begin{pmatrix} u_1 \\ u_2 \\ u_3 \end{pmatrix} + \begin{pmatrix} 0 & 0 & 0 \\ 0 & 0 & 0 \\ 0 & 0 & \beta^2(t)\xi^2 \end{pmatrix} \begin{pmatrix} u_1 \\ u_2 \\ u_3 \end{pmatrix} + \begin{pmatrix} -i\lambda(t)b(t)\xi & 0 & i\gamma_1(t)\xi \\ 0 & i\lambda(t)b(t)\xi & i\gamma_1(t)\xi \\ \frac{i\gamma_2(t)\xi}{2} & \frac{i\gamma_2(t)\xi}{2} & 0 \end{pmatrix} \begin{pmatrix} u_1 \\ u_2 \\ u_3 \end{pmatrix}$$

$$+ \frac{\partial_t(\lambda(t)b(t))}{2\lambda(t)b(t)} \begin{pmatrix} -1 & 1 & 0 \\ 1 & -1 & 0 \\ 0 & 0 & 0 \end{pmatrix} \begin{pmatrix} u_1 \\ u_2 \\ u_3 \end{pmatrix} = \begin{pmatrix} 0 \\ 0 \\ 0 \end{pmatrix} .$$

The matrix

$$\frac{\partial_t(\lambda(t)b(t))}{2\lambda(t)b(t)} \begin{pmatrix} -1 & 1 & 0 \\ 1 & -1 & 0 \\ 0 & 0 & 0 \end{pmatrix}$$

belongs to $S_{\delta,1}\{0,0,1\}$. Now let us decouple the above system modulo $S_{\delta,1}\{0,0,1\}$. After writing this system in the form

$$\partial_t U + A_{2,0}U + A_{1,0}U + A_{0,0}U = 0,$$

and setting the ansatz $U_1 = (I + K_1)U$ with $K_1 = K_1(\xi)$ being homogeneous of degree -1 in ξ we obtain

$$\begin{aligned}
\partial_t U_1 \ &+ \ A_{2,0}U_1 + (A_{1,0} + [K_1, A_{2,0}])U_1 + (A_{0,0} + [K_1, A_{1,0}] - [K_1, A_{2,0}]K_1)U_1 \\
&= \ -[K_1, A_{2,0}]K_1^2(I + K_1)^{-1}U_1 + [K_1, A_{1,0}]K_1(I + K_1)^{-1}U_1 \quad\quad (2.4) \\
&- \ [K_1, A_{0,0}]U_1 + [K_1, A_{0,0}]K_1(I + K_1)^{-1}U_1 + \partial_t K_1 U_1 - \partial_t K_1\, K_1(I + K_1)^{-1}U_1 ,
\end{aligned}$$

Now we choose K_1 in such a way that $A_{1,0} + [K_1, A_{2,0}]$ has block structure. This block structure will be produced by the aid of the decoupler

$$K_1 = \begin{pmatrix} 0 & 0 & \frac{-i\gamma_1(t)}{\beta^2(t)\xi} \\ 0 & 0 & \frac{-i\gamma_1(t)}{\beta^2(t)\xi} \\ \frac{i\gamma_2(t)}{2\beta^2(t)\xi} & \frac{i\gamma_2(t)}{2\beta^2(t)\xi} & 0 \end{pmatrix}.$$

By choosing the parameter N_1 sufficiently large we have $K_1 \in S_{\delta,2}\{-1,-1,1\}$, which gives the invertibility of $I + K_1$. Moreover, our assumptions $A4$ and $A5$ even yield the property $\left(\ln(N_2 + \Lambda(t))\right)^\delta K_1 \in S_{\delta,2}\{-1,-1,1\}$.

Thus the system (2.4) can be written in the form

$$\partial_t U_1 + A_{2,1}U_1 + A_{1,1}U_1 + A_{0,1}U_1 = R_1 U_1, \tag{2.5}$$

where $A_{2,1} = A_{2,0} = \text{diag}\,[0,0,\beta^2(t)\xi^2]$;

$$A_{1,1} = \begin{pmatrix} -i\lambda(t)b(t)\xi & 0 & 0 \\ 0 & i\lambda(t)b(t)\xi & 0 \\ 0 & 0 & 0 \end{pmatrix};$$

$$A_{0,1} = \frac{\partial_t(\lambda(t)b(t))}{2\lambda(t)b(t)} \begin{pmatrix} -1 & 1 & 0 \\ 1 & -1 & 0 \\ 0 & 0 & 0 \end{pmatrix} + \begin{pmatrix} \frac{\gamma_1(t)\gamma_2(t)}{2\beta^2(t)} & \frac{\gamma_1(t)\gamma_2(t)}{2\beta^2(t)} & \frac{\lambda(t)b(t)\gamma_1(t)}{\beta^2(t)} \\ \frac{\gamma_1(t)\gamma_2(t)}{2\beta^2(t)} & \frac{\gamma_1(t)\gamma_2(t)}{2\beta^2(t)} & \frac{\lambda(t)b(t)\gamma_1(t)}{\beta^2(t)} \\ \frac{\lambda(t)b(t)\gamma_2(t)}{2\beta^2(t)} & \frac{\lambda(t)b(t)\gamma_2(t)}{2\beta^2(t)} & \frac{-\gamma_1(t)\gamma_2(t)}{\beta^2(t)} \end{pmatrix};$$

and $R_1 U_1$ denotes the right-hand side of (2.4). The assumptions $A2$, $A3$ and $A5$ guarantee that $A_{1,1} \in S_{\delta,2}\{1,1,0\}$; $A_{0,1} \in S_{\delta,1}\{0,0,1\}$ and $\left(\ln(N_2 + \Lambda(t))\right)^\delta R_1 \in S_{\delta,1}\{-1,-1,2\}$. Here we use that if $A_{l,0}$ appears in one term of $R_1 U_1$, then K_1 appears at least $(l+1)$ times there.

The system (2.5) represents the decoupled system modulo $S_{\delta,1}\{0,0,1\}$. If we choose $K_2 = K_2(\xi)$ being homogeneous of degree -2 in ξ as

$$K_2 = \begin{pmatrix} 0 & 0 & \frac{-\gamma_1(t)\lambda(t)b(t)}{\beta^4(t)\xi^2} \\ 0 & 0 & \frac{-\gamma_1(t)\lambda(t)b(t)}{\beta^4(t)\xi^2} \\ \frac{\gamma_2(t)\lambda(t)b(t)}{2\beta^4(t)\xi^2} & \frac{\gamma_2(t)\lambda(t)b(t)}{2\beta^4(t)\xi^2} & 0 \end{pmatrix},$$

then $U_2 := (I + K_2)U_1$ satisfies

$$\partial_t U_2 + A_{2,2}U_2 + A_{1,2}U_2 + A_{0,2}U_2 = R_2 U_2, \tag{2.6}$$

where $A_{2,2} = A_{2,0}$, $A_{1,2} = A_{1,1}$, and

$$A_{0,2} = A_{0,1} - \begin{pmatrix} 0 & 0 & \frac{\lambda(t)b(t)\gamma_1(t)}{\beta^2(t)} \\ 0 & 0 & \frac{\lambda(t)b(t)\gamma_1(t)}{\beta^2(t)} \\ \frac{\lambda(t)b(t)\gamma_2(t)}{2\beta^2(t)} & \frac{\lambda(t)b(t)\gamma_2(t)}{2\beta^2(t)} & 0 \end{pmatrix}.$$

The assumptions $A1$ to $A3$, and $A5$ provide $\left(\ln(N_2 + \Lambda(t))\right)^{2\delta} K_2 \in S_{\delta,2}\{-2,-2,2\}$ which guarantees $\left(\ln(N_2 + \Lambda(t))\right)^\delta R_2 \in S_{\delta,1}\{-1,-1,2\}$. The system (2.6) represents the decoupled system modulo $S_{\delta,1}\{-1,-1,2\}$.

2.3.2   Diagonalization modulo $S_{\delta,0}\{-1,-1,2\}$

In order to diagonalize the system (2.6) for the first two components of U_2 let us consider the system

$$\partial_t V + A_1 V + A_0 V + HV = 0 \ , \tag{2.7}$$

where

$$A_1 = i\lambda(t)b(t)\xi \begin{pmatrix} -1 & 0 \\ 0 & 1 \end{pmatrix} \ , \quad A_0 = \frac{\partial_t(\lambda(t)b(t))}{2\lambda(t)b(t)} \begin{pmatrix} -1 & 1 \\ 1 & -1 \end{pmatrix} + \frac{\gamma_1(t)\gamma_2(t)}{2\beta^2(t)} \begin{pmatrix} 1 & 1 \\ 1 & 1 \end{pmatrix} \ ,$$

and $H \in S_{\delta,1}\{-1,-1,2\}$. Thus we have only a diagonalization modulo $S_{\delta,1}\{0,0,1\}$. Due to [7] there exists a diagonalizer N_1 such that

$$(\partial_t + A_1 + A_0 + H)N_1 = N_1(\partial_t + A_1 + B_1 + H_1) \ ,$$

where $B_1 = \text{diag } A_0$, $H_1 \in S_{\delta,0}\{-1,-1,2\}$ and $N_1 \in S_{\delta,1}\{0,0,0\}$.
The transformation $V = N_1 W$ reduces (2.7) to the system of first order

$$(\partial_t + A_1 + B_1 + H_1)W = 0.$$

Consequently, the transformation

$$U_2 = \begin{pmatrix} N_1 & 0 \\ 0 & 1 \end{pmatrix} W$$

reduces (2.6) to

$$\partial_t W + A_{2,2}W + A_{2,1}W + BW = RW \ , \tag{2.8}$$

where

$$B = \begin{pmatrix} -\frac{\partial_t(\lambda(t)b(t))}{2\lambda(t)b(t)} + \frac{\gamma_1(t)\gamma_2(t)}{2\beta^2(t)} & 0 & 0 \\ 0 & \frac{-\partial_t(\lambda(t)b(t))}{2\lambda(t)b(t)} + \frac{\gamma_1(t)\gamma_2(t)}{2\beta^2(t)} & 0 \\ 0 & 0 & \frac{-\gamma_1(t)\gamma_2(t)}{\beta^2(t)} \end{pmatrix} \ ,$$

and

$$R = \begin{pmatrix} -H_1 & 0 \\ 0 & 0 \end{pmatrix} + \begin{pmatrix} N_1 & 0 \\ 0 & 1 \end{pmatrix}^{-1} R_2 \begin{pmatrix} N_1 & 0 \\ 0 & 1 \end{pmatrix} \ .$$

It is easy to check that the first matrix of the remainder R belongs to $S_{\delta,0}\{-1,-1,2\}$ while the second one multiplied by $\left(\ln(N_2 + \Lambda(t))\right)^\delta$ belongs to $S_{\delta,1}\{-1,-1,2\}$. Thus the matrix $R = (r_{jk})$ has the remarkable property that $r_{jk}\left(\ln(N_2 + \Lambda(t))\right)^\delta$ belongs to $S_{\delta,1}\{-1,-1,2\}$ for $(j,k) \in \{(1,3);(2,3);(3,3);(3,2);(3,1)\}$. The above system (2.8) represents the diagonalization modulo $S_{\delta,0}\{-1,-1,2\}$.

2.3.3 WKB representations of solutions

If we denote by

$$\tau_1 = \tau_1(t,\xi) := -i\,\lambda(t)b(t)\xi - \frac{\partial_t(\lambda(t)b(t))}{2\lambda(t)b(t)} + \frac{\gamma_1(t)\gamma_2(t)}{2\beta^2(t)}\ ,$$

$$\tau_2 = \tau_2(t,\xi) := \ i\,\lambda(t)b(t)\xi - \frac{\partial_t(\lambda(t)b(t))}{2\lambda(t)b(t)} + \frac{\gamma_1(t)\gamma_2(t)}{2\beta^2(t)}\ ,$$

$$\tau_3 = \tau_3(t,\xi) := \ \beta^2(t)\xi^2 - \frac{\gamma_1(t)\gamma_2(t)}{\beta^2(t)} - r_{33}(t,\xi)\ ,$$

then the system (2.8) can be written in the following form:

$$\partial_t \begin{pmatrix} w_1 \\ w_2 \end{pmatrix} + \begin{pmatrix} \tau_1 & 0 \\ 0 & \tau_2 \end{pmatrix} \begin{pmatrix} w_1 \\ w_2 \end{pmatrix} = \begin{pmatrix} r_{11}w_1 + r_{12}w_2 + r_{13}w_3 \\ r_{21}w_1 + r_{22}w_2 + r_{23}w_3 \end{pmatrix}\ ; \tag{2.9}$$

$$\partial_t w_3 + \tau_3 w_3 = r_{31}w_1 + r_{32}w_2. \tag{2.10}$$

Moreover we formulate Cauchy conditions for (w_1, w_2, w_3) on $t = t_\xi$, where t_ξ solves $\Lambda(t_\xi)|\xi| = N_1\big(\ln(N_2 + \Lambda(t_\xi))\big)^\delta$. From the equation (2.10) we obtain

$$w_3(t,\xi) = e^{-\int_{t_\xi}^t \tau_3(s,\xi)ds}\, w_3(t_\xi,\xi) + \int_{t_\xi}^t e^{-\int_s^t \tau_3(\sigma,\xi)d\sigma}\big(r_{31}w_1 + r_{32}w_2\big)(s,\xi)ds\ .$$

Setting this relation into (2.9) gives the following system of integro-differential equations for (w_1, w_2):

$$\partial_t w_1 + \tau_1 w_1 \ = \ r_{11}w_1 + r_{12}w_2$$

$$+\ r_{13}\Big(e^{-\int_{t_\xi}^t \tau_3(s,\xi)ds}\, w_3(t_\xi,\xi) + \int_{t_\xi}^t e^{-\int_s^t \tau_3(\delta,\xi)d\delta}(r_{31}w_1 + r_{32}w_2)(s,\xi)ds\Big),$$

$$\tag{2.11}$$

$$\partial_t w_2 + \tau_2 w_2 \ = \ r_{21}w_1 + r_{22}w_2$$

$$+\ r_{23}\Big(e^{-\int_{t_\xi}^t \tau_3(s,\xi)ds}\, w_3(t_\xi,\xi) + \int_{t_\xi}^t e^{-\int_s^t \tau_3(\delta,\xi)d\delta}(r_{31}w_1 + r_{32}w_2)(s,\xi)ds\Big).$$

$$\tag{2.12}$$

Our goal is to show that the characteristic root τ_3 has no essential influence on WKB representations for the solution of the Cauchy problem for (2.9).

Lemma 2.4 *If $t_\xi \le s \le t$, then after a suitable choice of the parameters N_1 and N_2 we have the estimates*

$$e^{-\int_s^t \tau_3(\sigma,\xi)d\sigma} \le 1\ , \ \Big|e^{-\int_s^t (\tau_3-\tau_k)(\sigma,\xi)d\sigma}\Big| \le 1 \qquad \text{for}\ \ k = 1,2\ .$$

Proof. Due to the assumptions $A4, A5$ and the definition of $Z_{hyp}(N_1, N_2)$ we have

$$\beta^2(t)\xi^2 \geq \frac{C_1\lambda(t)}{\Lambda(t)}N_1^2\Big(\ln(N_2 + \Lambda(t))\Big)^{2\delta} \ ;$$

$$\frac{|\gamma_1(t)\gamma_2(t)|}{\beta^2(t)} \leq C_2\frac{\lambda(t)}{\Lambda(t)} \ , \ |r_{33}(t,\xi)| \leq C_3\frac{\lambda(t)}{\Lambda^2(t)|\xi|}\Big(\ln(N_2 + \Lambda(t))\Big)^{\delta} \leq C_3\frac{\lambda(t)}{\Lambda(t)N_1} \ ,$$

which implies $\tau_3(\sigma,\xi) \geq 0$ in $Z_{hyp}(N_1, N_2)$ for large N_1 and N_2.

To estimate $\left| e^{-\int\limits_s^t \mathrm{Re}(\tau_3-\tau_k)(\sigma,\xi)d\sigma} \right|, \quad k = 1,2$, we have only to estimate

$$-\frac{\partial_t(\lambda(t)b(t))}{2\lambda(t)b(t)} \ + \ \frac{\gamma_1(t)\gamma_2(t)}{2\beta^2(t)} \ .$$

The second term was already estimated before. For the first term we get due to the assumptions $A2$ and $A3$

$$\left| \frac{\partial_t(\lambda(t)b(t))}{2\lambda(t)b(t)} \right| \leq C_4 \frac{\lambda(t)}{\Lambda(t)}\Big(\ln(N_2 + \Lambda(t))\Big)^{\delta} \ ,$$

which implies in $Z_{hyp}(N_1, N_2)$

$$\mathrm{Re}(\tau_3 - \tau_k)(t,\xi) \geq \frac{\lambda(t)}{\Lambda(t)}\left(C_1N_1^2\Big(\ln(N_2 + \Lambda(t))\Big)^{2\delta} - C_2 - \frac{C_3}{N_1} - C_4\Big(\ln(N_2 + \Lambda(t))\Big)^{\delta}\right) \ .$$

A suitable choice of N_1, N_2 proves the second statement of the lemma . $\square$

Lemma 2.5 *There exist uniquely determined amplitude functions $a_1 = a_1(t,\xi)$ and $a_2 = a_2(t,\xi)$ satisfying in $Z_{hyp}(N_1, N_2)$ the estimates*

$$|a_k(t,\xi)| \leq CE(w_1, w_2, w_3)(t_\xi,\xi)\Big(1 + \tfrac{C_\varepsilon}{|\xi|^\varepsilon}\Big) \qquad \textit{for each positive } \varepsilon$$

such that $w_k = w_k(t,\xi) = a_k(t,\xi)\, e^{-\int\limits_{t_\xi}^t \tau_k(s,\xi)ds}$, $k = 1,2$, fulfill the system of integro-differential equations (2.11), (2.12) with the Cauchy conditions

$$w_k(t_\xi,\xi) = w_k(\xi) \quad \textit{on} \ \ t = t_\xi \ .$$

Proof. The existence of solutions to (2.11), (2.12) is obvious. For this reason it suffices for us to estimate $|a_k(t,\xi)|$.

Substituting the ansatz $w_k = a_k(t,\xi)e^{-\int\limits_{t_\xi}^t \tau_k(s,\xi)ds}$, $k = 1,2$, into the system of integro-differential equations (2.11), (2.12) it follows

$$a_1(t,\xi) \ = \ w_1(\xi) + \int\limits_{t_\xi}^t \left(r_{11}(s,\xi)a_1(s,\xi) + r_{12}(s,\xi)a_2(s,\xi)e^{\int\limits_{t_\xi}^s (\tau_1-\tau_2)(\sigma,\xi)d\sigma}\right)ds$$

$$+ \int_{t_\xi}^{t} r_{13}(s,\xi) e^{-\int_{t_\xi}^{s}(\tau_3-\tau_1)(\sigma,\xi)d\sigma} w_3(t_\xi,\xi)ds + \int_{t_\xi}^{t} r_{13}(s,\xi) \int_{t_\xi}^{s} e^{-\int_{\sigma}^{s}(\tau_3-\tau_1)(\eta,\xi)d\eta}$$

$$\times \left(r_{31}(\sigma,\xi)a_1(\sigma,\xi) + r_{32}(\sigma,\xi)a_2(\sigma,\xi) e^{-\int_{t_\xi}^{\sigma}(\tau_2-\tau_1)(\eta,\xi)d\eta} \right) d\sigma ds \ ;$$

$$a_2(t,\xi) = w_2(\xi) + \int_{t_\xi}^{t} \left(r_{21}(s,\xi)a_1(s,\xi)e^{\int_{t_\xi}^{s}(\tau_2-\tau_1)(\sigma,\xi)d\sigma} + r_{22}(s,\xi)a_2(s,\xi) \right) ds$$

$$+ \int_{t_\xi}^{t} r_{23}(s,\xi) e^{-\int_{t_\xi}^{s}(\tau_3-\tau_2)(\sigma,\xi)d\sigma} w_3(t_\xi,\xi)ds + \int_{t_\xi}^{t} r_{23}(s,\xi) \int_{t_\xi}^{s} e^{-\int_{\sigma}^{s}(\tau_3-\tau_2)(\eta,\xi)d\eta}$$

$$\times \left(r_{31}(\sigma,\xi)a_1(\sigma,\xi)e^{-\int_{t_\xi}^{\sigma}(\tau_1-\tau_2)(\eta,\xi)d\eta} + r_{32}(\sigma,\xi)a_2(\sigma,\xi) \right) d\sigma \ ds \ .$$

From the fact that $\tau_1 - \tau_2$ is pure imaginary we obtain with Lemma 2.4 the energy estimates

$$E(a_1)(t,\xi) \leq |w_1(\xi)| + \int_{t_\xi}^{t} \left(|r_{11}(s,\xi)|E(a_1)(s,\xi) + |r_{12}(s,\xi)|E(a_2)(s,\xi) \right) ds$$

$$+ \ C|w_3(t_\xi,\xi)| + C\int_{t_\xi}^{t} |r_{13}(s,\xi)|(E(a_1)(s,\xi) + E(a_2)(s,\xi))ds \ ,$$

$$E(a_2)(t,\xi) \leq |w_2(\xi)| + \int_{t_\xi}^{t} \left(|r_{21}(s,\xi)|E(a_1)(s,\xi) + |r_{22}(s,\xi)|E(a_2)(s,\xi) \right) ds$$

$$+ \ C|w_3(t_\xi,\xi)| + C\int_{t_\xi}^{t} |r_{23}(s,\xi)|(E(a_1)(s,\xi) + E(a_2)(s,\xi))ds \ ;$$

which imply

$$E(a_1,a_2)(t,\xi) \leq CE(w_1,w_2,w_3)(t_\xi,\xi) + C\int_{t_\xi}^{t} \frac{\lambda(s)}{\Lambda^2(s)|\xi|} \left(\ln(N_2 + \Lambda(s)) \right)^{2\delta} E(a_1,a_2)(s,\xi)ds,$$

by using $\left(\ln(N_2 + \Lambda(t)) \right)^{\delta} r_{jk} \in S_{\delta,1}\{-1,-1,2\}$ for $(j,k) \in \{(3,2),(3,1),(1,3),(2,3)\}$. The application of Gronwall's inequality gives immediately the energy estimate

$$E(a_1,a_2)(t,\xi) \leq E(w_1,w_2,w_3)(t_\xi,\xi)\left(1 + C\int_{t_\xi}^{t} \frac{\lambda(s)}{\Lambda^2(s)|\xi|} \left(\ln(N_2 + \Lambda(s)) \right)^{2\delta} \right.$$

$$\times \ e^{\int_{s}^{t} \frac{\lambda(\sigma)}{\Lambda^2(\sigma)|\xi|} \left(\ln(N_2+\Lambda(\sigma)) \right)^{2\delta} d\sigma} \ ds \bigg) \ ,$$

which implies

$$E(a_1, a_2)(t, \xi) \le C\, E(w_1, w_2, w_3)(t_\xi, \xi)\left(1 + \tfrac{C_\varepsilon}{|\xi|^\varepsilon}\right).$$

This proves the statement of the lemma.   □

Corollary 2.6 *There exist uniquely determined amplitude functions $a_3 = a_3(t, \xi)$ and $a_4 = a_4(t, \xi)$ satisfying in $Z_{hyp}(N_1, N_2)$ the estimates*

$$|a_k(t, \xi)| \le C E(w_1, w_2, w_3)(t_\xi, \xi)\left(1 + \tfrac{C_\varepsilon}{|\xi|^\varepsilon}\right) \qquad \text{for each positive } \varepsilon$$

such that $w_3 = w_3(t, \xi) = a_3(t, \xi)e^{-\int_{t_\xi}^t \tau_1(s, \xi)ds} + a_4(t, \xi)e^{-\int_{t_\xi}^t \tau_2(s, \xi)ds}$ fulfills together with w_1, w_2 from Lemma 2.5 the system of differential equations (2.9) and (2.10) with Cauchy conditions $w_k(t_\xi, \xi) = w_k(\xi)$ on $t = t_\xi$.

Proof. We set the representations from Lemma 2.5 into the differential equation (2.10) and get

$$w_3(t, \xi) = e^{-\int_{t_\xi}^t \tau_3(s,\xi)ds}\, w_3(\xi) \;+\; e^{-\int_{t_\xi}^t \tau_1(s,\xi)ds} \int_{t_\xi}^t e^{-\int_s^t (\tau_3 - \tau_1)(\sigma,\xi)d\sigma}\, r_{31}(s, \xi)a_1(s, \xi)ds$$

$$+\; e^{-\int_{t_\xi}^t \tau_2(s,\xi)ds} \int_{t_\xi}^t e^{-\int_s^t (\tau_3 - \tau_2)(\sigma,\xi)d\sigma}\, r_{32}(s, \xi)a_2(s, \xi)ds\,.$$

If we define

$$a_3(t, \xi) := \tfrac{1}{2}e^{-\int_{t_\xi}^t (\tau_3 - \tau_1)(s,\xi)ds}\, w_3(\xi) \;+\; \int_{t_\xi}^t e^{-\int_s^t (\tau_3 - \tau_1)(\sigma,\xi)d\sigma}\, r_{31}(s, \xi)a_1(s, \xi)ds\,,$$

$$a_4(t, \xi) := \tfrac{1}{2}e^{-\int_{t_\xi}^t (\tau_3 - \tau_2)(s,\xi)ds}\, w_3(\xi) \;+\; \int_{t_\xi}^t e^{-\int_s^t (\tau_3 - \tau_1)(\sigma,\xi)d\sigma}\, r_{32}(s, \xi)a_2(s, \xi)ds\,,$$

and use Lemmas 2.4, 2.5 and $\left(\ln(N_2 + \Lambda(t))\right)^\delta (r_{31}, r_{32}) \in S_{\delta,1}\{-1, -1, 2\}$, then the corollary is proved.   □

Corollary 2.7 *There exist uniquely determined amplitude functions $a_{k,l} = a_{k,l}(t, \xi)$, $k = 1, 2, 3$, $l = 1, 2$, satisfying in $Z_{hyp}(N_1, N_2)$ the estimates*

$$|a_{k,l}(t, \xi)| \le C\left(1 + \tfrac{C_\varepsilon}{|\xi|^\varepsilon}\right) \qquad \text{for each positive } \varepsilon$$

such that the functions

$$w_k = w_k(t, \xi) = e^{-\int_{t_\xi}^t \tau_1(s,\xi)ds}\, a_{k,1}(t, \xi)w_k(t_\xi, \xi) + e^{-\int_{t_\xi}^t \tau_2(s,\xi)ds}\, a_{k,2}(t, \xi)w_k(t_\xi, \xi)$$

fulfill the system of differential equations (2.9) and (2.10) with Cauchy conditions $w_k(t_\xi, \xi) = w_k(\xi)$ on $t = t_\xi$.

Proof. It suffices to use the estimates from Corollary 2.6 for the data $(w_1, w_2, w_3)(t_\xi, \xi) = (1, 0, 0)$; $(0, 1, 0)$; $(0, 0, 1)$ separately, because a linear combination of these solutions gives a solution of the general Cauchy problem for (2.9) and (2.10).   □

Corollary 2.8 *Let us consider in* $Z_{hyp}(N_1, N_2)$ *the Cauchy problem*

$$\partial_t u_1 + b^2(t)\lambda(t)\xi u_2 + i\gamma_1(t)\xi u_3 = 0,$$
$$\partial_t u_2 - \frac{\lambda'(t)}{\lambda(t)}u_2 - \lambda(t)\xi u_1 = 0,$$
$$\partial_t u_3 + \beta^2(t)\xi^2 u_3 + i\gamma_2(t)\xi u_1 = 0,$$
$$u_1(t_\xi, \xi) = u_1(\xi), \quad u_2(t_\xi, \xi) = u_2(\xi), \quad u_3(t_\xi, \xi) = u_3(\xi).$$

Then there exist amplitude functions $a_{k,l}^{(p)}$, $k = 1, 2, 3$, $l, p = 1, 2$, *satisfying*

$$|a_{k,l}^{(p)}(t, \xi)| \le C\left(1 + \frac{C_\varepsilon}{|\xi|^\varepsilon}\right) \quad \text{for each positive } \varepsilon$$

such that

$$u_k(t, \xi) = e^{-\int_{t_\xi}^t \tau_1(s, \xi)ds} \sum_{l=1}^3 a_{k,l}^{(1)}(t, \xi)u_l(\xi) + e^{-\int_{t_\xi}^t \tau_2(s, \xi)ds} \sum_{l=1}^3 a_{k,l}^{(2)}(t, \xi)u_l(\xi)$$

is the uniquely determined solution of this Cauchy problem.

Proof. If we set

$$V = (v_1, v_2, v_3) := (I + K_1)^{-1}(I + K_2)^{-1}\begin{pmatrix} N_1 & 0 \\ 0 & 1 \end{pmatrix}\begin{pmatrix} w_1 \\ w_2 \\ w_3 \end{pmatrix}$$

and take into consideration that $u_1 = \frac{v_1+v_2}{2}$, $u_2 = \frac{v_1-v_2}{2ib(t)}$, $u_3 = v_3$, then the statement follows directly from Corollary 2.7 and

$$K_1 \in S_{\delta,2}\{-1, -1, 1\}, \quad K_2 \in S_{\delta,2}\{-2, -2, 2\}, \quad N_1 \in S_{\delta,1}\{0, 0, 0\}.$$

□

3 $L_p - L_q$ decay estimates

3.1 Solutions to Cauchy problems

If we use the results from Corollary 2.3 and Corollary 2.8, then we are able to formulate final results for the transformed problem (2.1). But before we should take account of the next lemma, which can be obtained by direct computation.

Lemma 3.1 *Let us suppose that the condition*

$$C_1 \frac{\lambda(t)}{\Lambda(t)} \le \frac{\gamma_1(t)\gamma_2(t)}{2\beta^2(t)} \tag{3.1}$$

is valid for large t. Then

$$\left| e^{\int_{t_\xi}^t \left(\frac{\partial_s(\lambda(s)b(s))}{2\lambda(s)b(s)} - \frac{\gamma_1(s)\gamma_2(s)}{2\beta^2(s)}\right)ds} \right| \le \frac{C_\varepsilon}{\sqrt{\lambda(t_\xi)}} \frac{\sqrt{\lambda(t)}}{(1+\Lambda(t))^{C_1}}(1 + |\xi|^{-(\varepsilon+C_1)}).$$

Summarizing all results from the previous chapter leads to the following result.

Theorem 3.2 *Under the assumptions A1 to A5 and (3.1) the solution $\hat{u} = \hat{u}(t,\xi)$ to the Cauchy problem*

$$\hat{u}_{tt} + \lambda^2(t)b^2(t)\xi^2\hat{u} + i\gamma_1(t)\xi\hat{\theta} = 0 ,$$
$$\hat{\theta}_t + \beta^2(t)\xi^2\hat{\theta} + i\gamma_2(t)\xi\hat{u}_t = 0 ,$$
$$\hat{u}(0,\xi) = \hat{u}_0(\xi) , \quad \hat{u}_t(0,\xi) = \hat{u}_1(\xi) , \quad \hat{\theta}(0,\xi) = \hat{\theta}_0(\xi) ,$$

has the following representation:

$$
\lambda(t)\xi\hat{u}(t,\xi) = a_{0,1}(t,\xi)\xi\hat{u}_0(\xi)\exp(i\int_0^t \lambda(s)b(s)ds) + a_{0,2}(t,\xi)\xi\hat{u}_0(\xi)\exp(-i\int_0^t \lambda(s)b(s)ds)
$$

$$
+ \quad a_{1,1}(t,\xi)\hat{u}_1(\xi)\exp(i\int_0^t \lambda(s)b(s)ds) + a_{1,2}(t,\xi)\hat{u}_1(\xi)\exp(-i\int_0^t \lambda(s)b(s)ds)
$$

$$
+ \quad a_{2,1}(t,\xi)\hat{\theta}_0(\xi)\exp(i\int_0^t \lambda(s)b(s)ds) + a_{2,2}(t,\xi)\hat{\theta}_0(\xi)\exp(-i\int_0^t \lambda(s)b(s)ds),
$$

where the amplitudes satisfy for $l = 1, 2$, $k = 0, 1, 2$, and each $\kappa \in [0, 1/2]$ the estimates

$$|a_{k,l}(t,\xi)| \leq C_\varepsilon \frac{\lambda(t)^{1/2+\kappa}\lambda(t_\xi)^{1/2-\kappa}}{(1+\Lambda(t))^{C_1}}\left(1 + |\xi|^{-(\varepsilon+C_1)}\right) .$$

Moreover, we have similar representations for $\hat{u}_t = \hat{u}_t(t,\xi)$ with amplitude functions $b_{k,l} = b_{k,l}(t,\xi)$ and for $\hat{\theta} = \hat{\theta}(t,\xi)$ with amplitude functions $c_{k,l} = c_{k,l}(t,\xi)$ satisfying for $l = 1, 2$, $k = 0, 1, 2$ the same estimates as above.

Example 3.3   1) *If $\lambda(t) := \ln(e+t)$, then with $\kappa \in [0, 1/2]$ it holds*

$$|a_{k,l}| \leq C_\varepsilon \frac{\ln(e+t))^{1/2+\kappa}}{(1+t\ln(e+t))^{C_1}}(1 + |\xi|^{-(\varepsilon+C_1)}).$$

2) *If $\lambda(t) := (1+t)^\alpha$, $\alpha > 0$, then the amplitude functions can be estimated with $\kappa \in [0, 1/2]$ by*

$$|a_{k,l}| \leq C_\varepsilon \frac{(1+t)^{\alpha/2+\alpha\kappa}(1+t_\xi)^{\alpha/2-\alpha\kappa}}{(1+t)^{C_1(\alpha+1)}}(1 + |\xi|^{-(\varepsilon+C_1)}).$$

3) *If $\lambda(t) := \exp t$, then the amplitude functions can be estimated with $\kappa \in [0, 1/2]$ by*

$$|a_{k,l}| \leq C_\varepsilon \frac{\exp((1/2+\kappa)t)\exp((1/2-\kappa)t_\xi)}{(1+\exp t)^{C_1}}(1 + |\xi|^{-(\varepsilon+C_1)}).$$

3.2 Fourier multipliers

Let us study the model Fourier multiplier

$$F^{-1}\left(e^{i\int_0^t \lambda(s)b(s)\xi ds} a(t,\xi)F(u_0)(\xi)\right),$$ (3.2)

where

$$|a(t,\xi)| \le C_\varepsilon\left(1 + |\xi|^{-(\varepsilon+C_1)}\right)$$

in $\mathbf{R}_+^2$ with $C_1 < 1$. We begin to consider

$$F^{-1}\left(e^{i\int_0^t \lambda(s)b(s)\xi ds} a(t,\xi)\psi(\xi)F(u_0)(\xi)\right),$$

where ψ is a cut-off function localizing the amplitude near the origin.
The condition for $|a(t,\xi)|$ and Theorem 1.11 from [1] imply that

$$\left\|F^{-1}\left(e^{i\int_0^t \lambda(s)b(s)\xi ds} a(t,\xi)\psi(\xi)F(u_0)(\xi)\right)\right\|_{L_q(\mathbf{R})} \le C_\varepsilon\|u_0\|_{L_p(\mathbf{R})}$$

for all $\frac{1}{p} - \frac{1}{q} \ge \varepsilon + C_1$. Now let us devote to the Fourier multiplier

$$F^{-1}\left(e^{i\int_0^t \lambda(s)b(s)\xi ds} a(t,\xi)(1-\psi)(\xi)F(u_0)(\xi)\right).$$ (3.3)

To get $L_1 - L_\infty$ estimates it is enough to add the factor $|\xi|^r/|\xi|^r$ in the amplitude and to use the estimate

$$\left\|\frac{a(t,\xi)(1-\psi)(\xi)}{|\xi|^r}\right\|_{L_1(\mathbf{R})} \le C_\varepsilon(r)$$

for $r > 1$. Plancherel's theorem gives obviously an $L_2 - L_2$ estimate for (3.3). Thus, an interpolation argument yields

$$\left\|F^{-1}\left(e^{i\int_0^t \lambda(s)b(s)\xi ds} a(t,\xi)(1-\psi)(\xi)F(u_0)(\xi)\right)\right\|_{L_q(\mathbf{R})} \le C\|u_0\|_{W_p^2(\mathbf{R})}.$$

All together we have shown the next result.

Lemma 3.4 *Let us suppose that the amplitude function $a = a(t,\xi)$ satisfies for each small and positive ε the condition*

$$|a(t,\xi)| \le C_\varepsilon\left(1 + |\xi|^{-(\varepsilon+C_1)}\right)$$

with a positive constant $C_1 < 1$. Then the model Fourier multiplier (3.2) satisfies the $L_p - L_q$ estimate

$$\left\|F^{-1}\left(e^{i\int_0^t \lambda(s)b(s)\xi ds} a(t,\xi)F(u_0)(\xi)\right)\right\|_{L_q(\mathbf{R})} \le C_{p,q}\|u_0\|_{W_p^2(\mathbf{R})}$$

for all $1/p + 1/q = 1$, $1/p - 1/q > C_1$.

Remark 3.5 *We understand by the aid of Lemma 3.4 that the influence of the hyperbolic zone $Z_{hyp}(N_1, N_2)$, especially the definition of the function t_ξ leads to the condition $C_1 < 1$ which guarantees that the singularity at $|\xi| = 0$ in the kernel of the Fourier multiplier is weak.*

3.3   How to get decay estimates?

The last lemma is an important tool for getting our main result.

Theorem 3.6 *Assume that the conditions A1 to A5 are satisfied for the Cauchy problem*

$$u_{tt} - \lambda^2(t)b^2(t)u_{xx} + \gamma_1(t)\theta_x = 0 \ ,$$
$$\theta_t - \beta^2(t)\theta_{xx} + \gamma_2(t)u_{tx} = 0 \ ,$$
$$u(0,x) = u_0(x) \ , \ u_t(0,x) = u_1(x) \ , \ \theta(0,x) = \theta_0(x) \ .$$

We suppose that u_0, u_1 and θ_0 are compact supported smooth data. Moreover, we assume that for large t it holds the condition

$$C_1 \frac{\lambda(t)}{\Lambda(t)} \leq \frac{\gamma_1(t)\gamma_2(t)}{2\beta^2(t)}, \tag{3.4}$$

where the constant C_1 can be chosen in such a way, that for each fixed $\kappa \in [0, 1/2]$, the asymptotic behaviour of $\lambda(t_\xi)^{1/2-\kappa}|\xi|^{-C_1}$ near $|\xi| = 0$ can be controlled by $|\xi|^{-C_2}$ with $C_2 < 1$. Then we get the $L_p - L_q$ estimate

$$\|u_t(t, \cdot)\|_{L_q(\mathbf{R})} + \lambda(t)\|u_x(t, \cdot)\|_{L_q(\mathbf{R})} + \|\theta(t, \cdot)\|_{L_q(\mathbf{R})}$$
$$\leq C_{p,q} \frac{\lambda(t)^{1/2+\kappa}}{(1 + \Lambda(t))^{C_1}} \left(\|u_0\|_{W_p^3(\mathbf{R})} + \|u_1\|_{W_p^2(\mathbf{R})} + \|\theta_0\|_{W_p^2(\mathbf{R})} \right) \ ,$$

where $1/p + 1/q = 1$, $1/p - 1/q > C_2$.

Proof. This result follows immediately from Theorem 3.2 and Lemma 3.4.   □

In special cases the statement of the last theorem leads to $L_p - L_q$ decay estimates even for 1-d-thermoelasticity models with dominating hyperbolic influence. Let us explain how to get the Examples 1.2 and 1.3.

to Example 1.2: The coefficients satisfy the conditions A1 to A5. Due to 1) from Example 3.3 we need $C_1 < 1$ which follows from $\gamma_1\gamma_2 < 2$. The constant C_1 appearing in (3.4) is equal to $\gamma_1\gamma_2/2$. Consequently, the estimate from Example 1.2 is obtained by using Theorem 3.6.

to Example 1.3: The coefficients satisfy the conditions A1 to A5. Due to 2) from Example 3.3 we need

$$C_2 := C_1 + \tfrac{\alpha}{\alpha+1}(\tfrac{1}{2} - \kappa) < 1, \ C_1 = \tfrac{\gamma_1\gamma_2}{2(\alpha+1)}.$$

Consequently, $\gamma_1\gamma_2 < \alpha(1+2\kappa)+2$ with a suitable $\kappa \in [0,1/2]$ guarantees $C_2 < 1$. Thus the condition $\alpha(1+2\kappa) < \gamma_1\gamma_2 < \alpha(1+2\kappa)+2$ gives really a decay estimate. The estimate from Example 1.3 is obtained by using Theorem 3.6.

Remark 3.7 *If we try to apply 3) from Example 3.3 to a corresponding thermoelastic model with $\gamma_k(t) = \gamma_k \exp t$ $(k = 1, 2)$ and $\beta(t) = \exp t$, then we have $C_2 = C_1 + 1/2 - \kappa$ with $C_1 = \gamma_1\gamma_2/2$. The requirement $C_2 < 1$ gives $\gamma_1\gamma_2 < 1 + 2\kappa$. On the other hand, from Theorem 3.6, we know that the coefficient function on the right hand side of the decay estimate is $e^{(1/2+\kappa)t}(1 + e^t)^{-C_1}$, which implies $\gamma_1\gamma_2 > 1 + 2\kappa$ gives really a decay. Consequently, Theorem 3.6 gives for $\lambda = \exp t$ no $L_p - L_q$ decay estimates, which is the same phenomenon as we observed in [10]. As we mentioned in Remark 1.1, in this case the assumption A5 implies not necessary a dominating influence of the hyperbolic roots. Thus one can expect in opposite to the Examples 1.2 and 1.3 an essential influence of the parabolic root on decay estimates.*

Acknowledgement: The first author would like to express his gratitude to the Faculty of Mathematics and Computer Sciences of the TU Bergakademie Freiberg, Germany for the hospitality. His work is partially supported by the K. C. Wong Fellowship of DAAD, the Education Ministry of China, the NSFC and the Shanghai Qimingxing Foundation. The second author wants to thank Prof. Gaveau and his collaborators for the organization of the colloquium in honour of the retirement of Prof. J.Vaillant held at the University Paris VI Pierre and Marie Curie 26.06 - 30.06.2000.

References

[1] L. Hörmander, *Translation invariant operators in L^p spaces*, Acta Math. 104(1960), 93-140.

[2] Li Ta-tsien, *Global classical solutions for quasilinear hyperbolic systems*, John Wiley & Sons, 1994.

[3] R. Racke, *Lectures on nonlinear evolution equations*, Aspects of Mathematics, Vieweg, Braunschweig/Wiesbaden, 1992.

[4] R. Racke, *On the Cauchy problem in nonlinear 3-d-thermoelasticity*, Math. Z. 203(1990), 649-682.

[5] M. Reissig and K. Yagdjian, *One application of Floquet's theory to $L_p - L_q$ estimates*, Math. Meth. Appl. Sci. 22(1999), 937-951.

[6] M. Reissig and K. Yagdjian , *$L_p - L_q$ decay estimates for hyperbolic equations with oscillations in the coefficients*, Chin. Ann. of Math., Ser. B, 21(2000)2, 153-164.

[7] M. Reissig and K. Yagdjian, $L_p - L_q$ *estimates for the solutions of hyperbolic equations of second order with time dependent coefficients –Oscillations via growth–*, Preprint 98-5, Fakultät für Mathematik und Informatik, TU Bergakademie Freiberg.

[8] R. Strichartz, *A priori estimates for the wave-equation and some applications*, J. Funct. Anal. 5(1970), 218-235.

[9] Y.G. Wang, *Microlocal analysis in nonlinear thermoelasticity*, to appear.

[10] Y.G. Wang and M. Reissig, *Parabolic type decay rates for 1-d-thermoelastic systems with time-dependent coefficients*, in: Decay rates for 1-d thermoelastic systems with time-dependent coefficients, Preprint 2000-08, Fakultät für Mathematik und Informatik, TU Bergakademie Freiberg.

[11] S. Zheng, *Nonlinear parabolic equations and hyperbolic-parabolic coupled systems*, Pitman Monographs and Surveys in Pure and Applied Mathematics, vol.76, Longman Harlow, 1995.

Integration and singularities of solutions for nonlinear second order hyperbolic equation

Mikio TSUJI [†]
Department of Mathematics, Kyoto Sangyo University
Kita-Ku, Kyoto 603-8555, Japan

Abstract

We will consider the non-characteristic Cauchy problem for nonlinear hyperbolic equations with smooth data. Though the Cauchy problem has a classical solution in a neighbourhood of the initial curve, it does not generally have a global classical solution. This means that singularities appear in finite time. To develop the global theory, we have to extend the solution beyond the singularities. For this purpose, we will first lift the solution surface into cotangent space so that the singularities would disappear and extend the lifted solution in the cotangent space. Next, by projecting it to the base space, we will get a multi-valued solution. Our problem is how to construct a reasonable single-valued solution with singularities. This is a survey note of our recent researches on this subject.

1   Introduction

We will consider the non-characteristic Cauchy problem for nonlinear partial differential equations of hyperbolic type. It is well known that the Cauchy problem with smooth data has a smooth solution in a neighbourhood of the initial curve, and that singularities appear generally in finite time. But, even if singularities may appear in solutions, physical phenomena can exist with the singularities. Moreover it seems to us that the singularities might cause various kinds of interesting phenomena. We are interested in the global theory for the above Cauchy problem. Therefore our purpose is to extend the solutions beyond their singularities.

Here we recall a little the methods used to solve the problems of singularities which have often been a starting point of many subjects of mathematics. The most traditional and typical method has been the "resolution of singularities", that is to say, to lift the surfaces with singularities into higher dimensional space so that the singularities would

---

[†]Partially supported by Grant-in-Aid for Scientific Research (C), No. 13640226.

disappear. As the singularities have disappeared, we can extend the lifted surfaces there. Next we will project it to the base space. Then we will get a surface which is extended beyond the singularities. In this talk we would like to apply this method to the above problem of partial differential equations. But we are not the first who takes this kind of approach. For example, R. Thom [19] originated "Catastrophe theory", and he has applied his theory to understand various kinds of phenomena caused by "singularities". We have been stimulated by his work. We would like to understand the natural phenomena from this point of view. If we might take this approach for hyperbolic systems of conservation laws, we would be led to a definition of "weak solution" which is a little different from the well-known one introduced by P. D. Lax [14].

The classical method to solve first order partial differential equations is so-called the "characteristic method". Then the characteristic strips are obtained as "solutions of canonical differential equations" defined in cotangent space. Therefore the solutions of first order partial differential equations are naturally lifted into the cotangent space. For partial differential equations of second order also, we would like to develop similar discussions. To do so, we must rewrite partial differential equations in the cotangent space in another form. In §2 we will consider the method of integration of second order partial differential equations from this point of view and introduce a notion of "geometric solution" defined in the cotangent space. In §3 we will apply the results obtained in §2 to certain nonlinear hyperbolic equation. Then we shall be able to get a geometric solution globally. Next we will project it to the base space. Then our problem is what kinds of weak solutions we could get. In §4 we will consider the same problem for a certain system of conservation laws.

2   Integration of Monge-Ampère equations

In this section we will study the method of integration of second order nonlinear partial differential equations, especially of Monge-Ampère type as follows:

$$F(x, y, z, p, q, r, s, t) = Ar + Bs + Ct + D(rt - s^2) - E = 0 \qquad (2.1)$$

where $p = \partial z/\partial x, q = \partial z/\partial y, r = \partial^2 z/\partial x^2, s = \partial^2 z/\partial x \partial y$, and $t = \partial^2 z/\partial y^2$. Here we assume that A, B, C, D and E are real smooth functions of (x, y, z, p, q). Partial differential equations of second order which appear in physics and geometry are often written in the above form.

Before beginning our discussion, we will briefly explain some classical notions and notations from our point of view. We would like to cut off this part. But we had better repeat theses things again, because it seems to us that they are not our common knowledge today. Equation (2.1) is regarded as a smooth surface defined in eight dimensional space $\mathbf{R}^8 = \{(x, y, z, p, q, r, s, t)\}$. As p and q are first order derivatives of $z = z(x, y)$, we put the relation $dz = pdx + qdy$. Moreover, as r, s and t are second order derivatives of $z = z(x, y)$, we introduce the relations $dp = rdx + sdy$ and $dq = sdx + tdy$. Let us call $\{ dz = pdx + qdy, dp = rdx + sdy, dq = sdx + tdy \}$ the "contact structure of second order"

We define a solution of (2.1) as a maximal integral submanifold of the contact structure of second order in the surface $\{(x,y,z,p,q,r,s,t) \in \mathbf{R}^8; f(x,y,z,p,q,r,s,t) = 0\}$. We will use this geometric formulation to solve equation (2.1) in exact form. Let

$$\Gamma : (x,y,z,p,q) = (x(\xi), y(\xi), z(\xi), p(\xi), q(\xi)), \quad \xi \in \mathbf{R}^1,$$

be a smooth curve in $\mathbf{R}^5$. It is called a "strip" if it satisfies the following:

$$\frac{dz}{d\xi}(\xi) = p(\xi)\frac{dx}{d\xi}(\xi) + q(\xi)\frac{dy}{d\xi}(\xi). \tag{2.2}$$

Let Γ be any strip in $\mathbf{R}^5$, and consider equation (2.1) in an open neighbourhood of Γ. As a "characteristic" strip means that one can not determine the values of the second order derivatives of solution along the strip, we have

$$\det \begin{bmatrix} F_r & F_s & F_t \\ \dot{x} & \dot{y} & 0 \\ 0 & \dot{x} & \dot{y} \end{bmatrix} = F_t\dot{x}^2 - F_s\dot{x}\dot{y} + F_r\dot{y}^2 = 0 \tag{2.3}$$

where $F_t = \partial F/\partial t$, $F_s = \partial F/\partial s$, $F_r = \partial F/\partial r$, $\dot{x} = dx/d\xi$ and $\dot{y} = dy/d\xi$. Here we substitute the relations $\dot{p} = r\dot{x} + s\dot{y}$ and $\dot{q} = s\dot{x} + t\dot{y}$ into (2.3), then we get the following:

Definition 2.1 *A curve Γ in $\mathbf{R}^5 = \{(x,y,z,p,q)\}$ is a "characteristic strip" if it satisfies (2.2) and*

$$A\dot{y}^2 - B\dot{x}\dot{y} + C\dot{x}^2 + D(\dot{p}\dot{x} + \dot{q}\dot{y}) = 0. \tag{2.4}$$

A strip Γ is called "non-characteristic" when it does not satisfy (2.4). Next we will give the definition of "hyperbolicity". Denote the discriminant of (2.3) by Δ. Then it follows that

$$\Delta = F_s^2 - 4F_rF_t = B^2 - 4(AC + DE).$$

If $\Delta < 0$, equation (2.1) is called elliptic. If $\Delta > 0$, equation (2.1) is hyperbolic. In this note, we will treat the equations of hyperbolic type. Let λ_1 and λ_2 be the solutions of $\lambda^2 + B\lambda + (AC + DE) = 0$. Then, in the case where $D \neq 0$, the characteristic strip satisfies the following equations (see [3] and [6]):

$$\begin{cases} dz - pdx - qdy = 0, \\ Ddp + Cdx + \lambda_1 dy = 0, \\ Ddq + \lambda_2 dx + Ady = 0, \end{cases} \tag{2.5}$$

or

$$\begin{cases} dz - pdx - qdy = 0, \\ Ddp + Cdx + \lambda_2 dy = 0, \\ Ddq + \lambda_1 dx + Ady = 0. \end{cases} \tag{2.6}$$

Let us denote $\omega_0 = dz - pdx - qdy$, $\omega_1 = Ddp + Cdx + \lambda_1 dy$ and $\omega_2 = Ddq + \lambda_2 dx + Ady$

Exchanging λ_1 and λ_2 in ω_1 and ω_2, we define ϖ_1 and ϖ_2 by $\varpi_1 = Ddp + Cdx + \lambda_2 dy$ and $\varpi_2 = Ddq + \lambda_1 dx + Ady$. Take an exterior product of ω_1 and ω_2, and also of ϖ_1 and ϖ_2. Substitute into their product the relations of the contact structure $\omega_0 = 0$, $dp = rdx + sdy$ and $dq = sdx + tdy$. Then we get

$$\omega_1 \wedge \omega_2 = \varpi_1 \wedge \varpi_2 = D\left\{Ar + Bs + Ct + D(rt - s^2) - E\right\} dx \wedge dy . \qquad (2.7)$$

In the above we have assumed $D \neq 0$, though it is not essential for our discussion. The key point is to represent equation (2.1) as a product of one forms. For example, we will consider in §3 and §4 a certain case where $D \equiv 0$. It will be shown that, though the above decomposition might be a small idea, it would effectively work to solve equation (2.1) in exact form. In a space whose dimension is greater than two, it is generally impossible to do so. Here we recall briefly the characteristic method developed principally by G. Darboux [3] and E. Goursat [5, 6] "from our point of view", because this would help us to explain our problem. The idea of Darboux and Goursat is how to reduce the solvability of (2.1) to the integration of first order partial differential equations.

Definition 2.2 *A function $V = V(x, y, z, p, q)$ is called a "first integral" of $\{\omega_0, \omega_1, \omega_2\}$ if $dV \equiv 0 \mod\{\omega_0, \omega_1, \omega_2\}$.*

Proposition 2.3 *Assume that $\lambda_1 \neq \lambda_2$ and $D \neq 0$, and that $\{\omega_0, \omega_1, \omega_2\}$, or $\{\omega_0, \varpi_1, \varpi_2\}$, has two independent first integrals $\{u, v\}$. Then there exists a function $k = k(x, y, z, p, q) \neq 0$ satisfying*

$$du \wedge dv = k\,\omega_1 \wedge \omega_2 = kD\{Ar + Bs + Ct + D(rt - s^2) - E\}dx \wedge dy . \qquad (2.8)$$

If $\{\omega_0, \omega_1, \omega_2\}$, or $\{\omega_0, \varpi_1, \varpi_2\}$, has at least two independent first integrals, equation (2.1) is called *integrable in the sense of Monge*. But, if we may follow G. Darboux (p. 263 of [3]), it seems to us that we had better call it *integrable in the sense of Darboux*. Moreover, as E. Goursat had profoundly studied equations (2.1) satisfying the above condition, we would like to add the name of Goursat. By these reasons, we will call equations (2.1) with two independent first integrals *integrable in the sense of Darboux and Goursat*. Then the representation (2.8) gives the characterization of "Monge-Ampère equations which are integrable in the sense of Darboux and Goursat". This integrability condition is very strong. For example, see [9] where we have studied the integrability property of a certain nonlinear hyperbolic equation appearing in geometry.

Next we advance to the integration of the Cauchy problem for (2.1). Let $\{u, v\}$ be two independent first integrals of $\{\omega_0, \omega_1, \omega_2\}$. For any function g of two variables whose gradient does not vanish, $g(u, v) = 0$ is called an "intermediate integral" of (2.1). Let C_0 be an initial strip defined in $\mathbf{R}^5 = \{(x, y, z, p, q)\}$. If the strip C_0 is not characteristic, we can find an "intermediate integral" $g(u, v)$ which vanishes on C_0. Here we put $g(u, v) = f(x, y, z, p, q)$. The Cauchy problem for (2.1) satisfying the initial condition C_0 is to look for a solution $z = z(x, y)$ of (2.1) which contains the strip C_0, i.e., the two dimensional surface $\{(x, y, z(x, y), \partial z/\partial x(x, y), \partial z/\partial y(x, y))\}$ in $\mathbf{R}^5$ contains the strip C_0.

The representation (2.8) assures that a smooth solution of $f(x, y, z, \partial z/\partial x, \partial z/\partial y) = 0$ satisfies equation (2.1), because $du \wedge dv = 0$ on a surface $g(u, v) = 0$. Therefore we get the following:

Theorem 2.4 ([3], [5, 6]) *Assume that the initial strip C_0 is not characteristic. Then a function $z = z(x, y)$ is a solution of the Cauchy problem for (2.1) with the initial condition C_0 if and only if it is a solution of $f(x, y, z, \partial z/\partial x, \partial z/\partial y) = 0$ satisfying the same initial condition C_0.*

If we may assume the global existence of two independent first integrals, we can get the solution of (2.1) by solving first order partial differential equations. Therefore we can apply our results on the singularities of solutions of first order partial differential equations. Concerning this subject, see [22]. For first order nonlinear partial differential equations, refer to T. D. Van, M. Tsuji and N. D. Thai Son [26] where our methods and the detail refernces are given.

Next we will study the method of integration of (2.1) in the case where neither $\{\omega_0, \omega_1, \omega_2\}$ nor $\{\omega_0, \varpi_1, \varpi_2\}$ has not two independent first integrals. We start from the point at which equation (2.1) is represented as a product of one forms as (2.8). We suppose $D \neq 0$ for simplicity, though it is not indispensable for our study. The essential condition for our following discussion is $\Delta \neq 0$.

Let us pay attention to the property that the left hand side of (2.8) is a product of one forms defined in $\mathbf{R}^5 = \{(x, y, z, p, q)\}$. This suggests us to introduce a notion of "geometric solution" as follows:

Definition 2.5 *A regular geometric solution of (2.1) is a submanifold of dimension 2 defined in $\mathbf{R}^5 = \{(x, y, z, p, q)\}$ on which it holds that $dz = pdx + qdy$ and $\omega_1 \wedge \omega_2 = 0$.*

Remark In the above definition, we have added "regular" to "geometric solution". This means that we will soon introduce a "singular" geometric solution whose dimension may depend on each point. The problem to construct the geometric solution is similar to the Pfaffian problem. A difference between the classical Pfaffian problem and the above one is that we consider it in C^∞-space. Therefore we need some condition which is corresponding to "hyperbolicity".

Our problem is to find a "submanifold on which $\omega_0 = 0$ and $\omega_1 \wedge \omega_2 = 0$". First we will sum up the classsical method, though it is written in J. Hadamard [8], and also in R. Courant-D. Hilbert [2] a little. Let us consider the Cauchy problem for equation (2.1). The initial condition is given by a smooth strip C_0 which is defined in $\mathbf{R}^5 = \{(x, y, z, p, q)\}$ and written down as follows:

$$C_0 : (x, y, z, p, q) = (x_0(\xi), y_0(\xi), z_0(\xi), p_0(\xi), q_0(\xi)), \quad \xi \in \mathbf{R}^1.$$

The idea of the classical method is to represent the solution surface by a family of characteristic strips. Then they are determined as solutions of the following system of equations (see [15], [8], and [2]):

$$
\begin{cases}
\dfrac{\partial z}{\partial \alpha} - p\dfrac{\partial x}{\partial \alpha} - q\dfrac{\partial y}{\partial \alpha} = 0, \\[2mm]
D\dfrac{\partial p}{\partial \alpha} + C\dfrac{\partial x}{\partial \alpha} + \lambda_1\dfrac{\partial y}{\partial \alpha} = 0, \\[2mm]
D\dfrac{\partial q}{\partial \alpha} + \lambda_2\dfrac{\partial x}{\partial \alpha} + A\dfrac{\partial y}{\partial \alpha} = 0, \\[2mm]
D\dfrac{\partial p}{\partial \beta} + C\dfrac{\partial x}{\partial \beta} + \lambda_2\dfrac{\partial y}{\partial \beta} = 0, \\[2mm]
D\dfrac{\partial q}{\partial \beta} + \lambda_1\dfrac{\partial x}{\partial \beta} + A\dfrac{\partial y}{\partial \beta} = 0.
\end{cases}
\tag{2.9}
$$

The initial condition for system (2.9) is given by

$$
\begin{cases}
x(\xi,\xi) = x_0(\xi), \quad y(\xi,\xi) = y_0(\xi), \quad z(\xi,\xi) = z_0(\xi), \\[2mm]
p(\xi,\xi) = p_0(\xi), \quad q(\xi,\xi) = q_0(\xi), \quad \xi \in \mathbf{R}^1.
\end{cases}
\tag{2.10}
$$

The local solvability of the Cauchy problem (2.9)-(2.10) is already proved first by H. Lewy [16] and afterward by J. Hadamard [8]. Let $(x(\alpha,\beta), y(\alpha,\beta), z(\alpha,\beta), p(\alpha,\beta), q(\alpha,\beta))$ be a solution of (2.9)-(2.10). Then we can prove $\partial z/\partial \beta - p\partial x/\partial \beta - q\partial y/\partial \beta = 0$. Therefore we do not need to add this equation to system (2.9). This means that (2.9) is just a "determined" system. What we must do more is to represent $z = z(\alpha,\beta)$ as a function of (x,y). To do so, we calculate the Jacobian $D(x,y)/D(\alpha,\beta)$. As it holds along the initial strip C_0 that

$$
\begin{cases}
\dfrac{\partial x}{\partial \alpha}(\xi,\xi) = \dfrac{1}{\lambda_1 - \lambda_2}(D\dot{q}_0(\xi) + \lambda_1\dot{x}_0(\xi) + A\dot{y}_0(\xi)), \\[3mm]
\dfrac{\partial x}{\partial \beta}(\xi,\xi) = -\dfrac{1}{\lambda_1 - \lambda_2}(D\dot{q}_0(\xi) + \lambda_2\dot{x}_0(\xi) + A\dot{y}_0(\xi)), \\[3mm]
\dfrac{\partial y}{\partial \alpha}(\xi,\xi) = -\dfrac{1}{\lambda_1 - \lambda_2}(D\dot{p}_0(\xi) + C\dot{x}_0(\xi) + \lambda_2\dot{y}_0(\xi)), \\[3mm]
\dfrac{\partial y}{\partial \beta}(\xi,\xi) = \dfrac{1}{\lambda_1 - \lambda_2}(D\dot{p}_0(\xi) + C\dot{x}_0(\xi) + \lambda_1\dot{y}_0(\xi)),
\end{cases}
\tag{2.11}
$$

it follows immediately that

$$
\frac{D(x,y)}{D(\alpha,\beta)} = \frac{1}{\lambda_1 - \lambda_2}\{A\dot{y}_0{}^2 - B\dot{x}_0\dot{y}_0 + C\dot{x}_0{}^2 + D(\dot{p}_0\dot{x}_0 + \dot{q}_0\dot{y}_0)\}.
$$

As we have assumed that the initial strip C_0 is not characteristic, we see by (2.4) that the Jacobian $D(x,y)/D(\alpha,\beta)$ does not vanish in a neighbourhood of C_0. Therefore we can uniquely solve the system of equations $x = x(\alpha, \beta)$, $y = y(\alpha, \beta)$ with respect to (α, β) in a neighbourhood of each point of C_0 and denote them by $\alpha = \alpha(x, y)$ and $\beta = \beta(x, y)$. Then we get the solution of the Cauchy problem for (2.1) by $z(x, y) = z(\alpha(x, y), \beta(x, y))$. Summing up the above discussion, we obtain the following:

Theorem 2.6 ([15], [8]) *Assume that the initial strip C_0 is not characteristic. Then the Cauchy problem for (2.1) with the initial condition C_0 uniquely admits a smooth solution in a neighbourhood of each point of C_0.*

Remark If the equation and the solution are sufficiently differentiable, the solution is uniquely determined by the initial data. For example, the uniqueness of solution in C^∞-space is one of the classical known results (see H. Lewy [15] and J. Hadamard [8]). But the uniqueness of solution in C^2-space is a delicate problem. We will consider this subject in a forthcoming paper.

Before ending this section, we will give some remarks on the above characteristic method and the geometric solution in the sense of Definition 2.5. First we will give the meaning of (2.9) from our point of view. Suppose that a geometric solution is represented by two parameters as follows:

$$x = x(\alpha, \beta), y = y(\alpha, \beta), z = z(\alpha, \beta), p = p(\alpha, \beta), q = q(\alpha, \beta). \tag{2.12}$$

Then ω_i and ϖ_i $(i = 1, 2)$ are written as

$$\omega_i = c_{i1} d\alpha + c_{i2} d\beta, \quad \varpi_i = d_{i1} d\alpha + d_{i2} d\beta \quad (i = 1, 2).$$

Hence we have $\omega_1 \wedge \omega_2 = (c_{11} c_{22} - c_{12} c_{21}) d\alpha \wedge d\beta$ and $\varpi_1 \wedge \varpi_2 = (d_{11} d_{22} - d_{12} d_{21}) d\alpha \wedge d\beta$. As it holds that $\omega_1 \wedge \omega_2 = \varpi_1 \wedge \varpi_2 = 0$ on the solution surface, a sufficient condition so that (2.12) be a geometric solution of (2.1) is given by

$$c_{11} = c_{21} = d_{12} = d_{22} = 0. \tag{2.13}$$

This is also a necessary condition. In fact, if $\omega_1 \wedge \omega_2 = \varpi_1 \wedge \varpi_2 = 0$, then we can choose the parameters (α, β) so that (2.13) is satisfied. On the other hand, from the contact relation $dz = p dx + q dy$, we get the following two equations:

$$\frac{\partial z}{\partial \alpha} - p \frac{\partial x}{\partial \alpha} - q \frac{\partial y}{\partial \alpha} = 0, \quad \frac{\partial z}{\partial \beta} - p \frac{\partial x}{\partial \beta} - q \frac{\partial y}{\partial \beta} = 0. \tag{2.14}$$

Therefore we get totally six equations from (2.13) and (2.14). As we are looking for a "determined" system, we add only one equation of (2.14) to (2.13). Then we get system (2.9). This is the meaning of system (2.9) from our point of view. Next let us start from Definition 2.5. Then the contact relation $dz = p dx + q dy$ is much more fundamental than (2.13). Therefore the characteristic system should contain two equations (2.14). Next we can choose three equations from (2.13) so that a new system becomes equivalent to (2.9). Concerning this procedure, see [25].

3   Nonlinear hyperbolic equations

In this section we will consider the Cauchy problem for nonlinear hyperbolic equations as follows:

$$F(q, r, t) = \frac{\partial^2 z}{\partial x^2} - \frac{\partial}{\partial y} f\left(\frac{\partial z}{\partial y}\right) = r - f'(q)t = 0 \quad \text{in} \quad \{x > 0, y \in \mathbf{R}^1\} \equiv \mathbf{R}^2_+, \quad (3.1)$$

$$z(0, y) = z_0(y), \quad \frac{\partial z}{\partial x}(0, y) = z_1(y) \quad \text{on} \quad \{x = 0, y \in \mathbf{R}^1\} \quad (3.2)$$

where $f(q)$ is in $C^\infty(\mathbf{R}^1)$ and $f'(q) > 0$. Here $z = z(x, y)$ is an unknown function of $(x, y) \in \mathbf{R}^2$. We assume that the initial functions $z_i(y)$ $(i = 0, 1)$ are sufficiently smooth, and that $z_0'(y)$ is bounded. Equation (3.1) is of Monge-Ampère type which we have studied in §2. In fact, if we may put $A = 1, B = D = E = 0$, and $C = -f'(q)$ in (2.1), then we get (3.1).

It is well known that the Cauchy problem (3.1)-(3.2) does not have a classical solution in the large. For example, see N. J. Zabusky [28] and P. D. Lax [13]. After them, many people have considered the life-span of classical solutions. As the number of papers on this subject is too many, we do not mention here on that subject.

The above phenomenon means that singularities generally appear in finite time. Our main problem is how to extend the solutions of (3.1) beyond the singularities. We can see that the solutions take many values after the appearance of singularities. If we may consider this problem from the physical point of view, we would be obliged to construct single-valued solutions of (3.1). To solve the problem of this kind, we recall what we have done for nonlinear first order partial differential equations. First we have lifted the solution surface into cotangent space so that its singularities would disappear. Then we could extend the lifted solution so that it would be defined in the whole space. Next we have projected it to the base space and gotten a multi-valued solution. Our final problem has been how to choose a single value from many values of the projected solution so that the new single-valued solution should satisfy some additional conditions attached to some physical phenomena.

Now we will construct a geometric solution of (3.1)-(3.2) by the method introduced in §2. The first step is to represent equation (3.1) as a product of one forms. Let us denote $\omega_1 = dp \pm \lambda(q)dq$ and $\omega_2 = \pm \lambda(q)dx + dy$ where $\lambda(q) = \sqrt{f'(q)}$. Take an exterior product of ω_1 and ω_2, and substitute into their product the relations of the contact structure $\omega_0 = 0$, $dp = rdx + sdy$ and $dq = sdx + tdy$. Then we get

$$\omega_1 \wedge \omega_2 = \{r - f'(q)t\} \ dx \wedge dy . \quad (3.3)$$

Definition 2.5 and the decomposition (3.3) for equation (3.1) suggest us to consider the following system, which is similar to (2.9) for equation (2.1) in the general case:

$$\begin{cases} \dfrac{\partial p}{\partial \alpha} + \lambda(q)\dfrac{\partial q}{\partial \alpha} = 0, \\[2mm] \lambda(q)\dfrac{\partial x}{\partial \alpha} + \dfrac{\partial y}{\partial \alpha} = 0, \\[2mm] \dfrac{\partial p}{\partial \beta} - \lambda(q)\dfrac{\partial q}{\partial \beta} = 0, \\[2mm] -\lambda(q)\dfrac{\partial x}{\partial \beta} + \dfrac{\partial y}{\partial \beta} = 0. \end{cases} \qquad (3.4)$$

The initial condition corresponding to (3.2) is given by

$$x(\xi,\xi) = 0, \quad y(\xi,\xi) = \xi, \quad p(\xi,\xi) = z_1(\xi), \quad q(\xi,\xi) = z_0'(\xi), \quad \xi \in \mathbf{R}^1. \qquad (3.5)$$

Solving (3.4)-(3.5), we get the following:

Theorem 3.1 *The Cauchy problem (3.1)-(3.2) has globally a regular geometric solution.*

Before giving the proof of this theorem, we prepare the following two lemmata.

Lemma 3.2 *Assume that there exists α_0 satisfying $\psi_2'(\alpha_0) = 0$. Then $(\partial x/\partial \alpha)(\alpha_0,\beta) \neq 0$ for any $\beta \in \mathbf{R}^1$.*

Lemma 3.3 *Assume that there exists β_0 satisfying $\psi_1'(\beta_0) = 0$. Then $(\partial x/\partial \beta)(\alpha,\beta_0) \neq 0$ for any $\alpha \in \mathbf{R}^1$.*

Proof of Lemmas 3.2 and 3.3   Eliminating $y = y(\alpha,\beta)$ from the two equations of (3.7), we get

$$\frac{\partial}{\partial \beta}\left(\lambda(q)\frac{\partial x}{\partial \alpha}\right) + \frac{\partial}{\partial \alpha}\left(\lambda(q)\frac{\partial x}{\partial \beta}\right) = 0.$$

As $2\lambda(q)(\partial q/\partial \alpha) = -\psi_2'(\alpha)$, it follows that $(\partial q/\partial \alpha)(\alpha_0,\beta) = 0$. Hence we get

$$2\lambda(q)\frac{\partial^2 x}{\partial \alpha \partial \beta}(\alpha_0,\beta) + \lambda'(q)\frac{\partial q}{\partial \beta}(\alpha_0,\beta)\frac{\partial x}{\partial \alpha}(\alpha_0,\beta) = 0.$$

As this is just a first order linear ordinary differential equation with respect to $(\partial x/\partial \alpha)(\alpha_0,\beta)$, we can solve it exactly. As $(\partial x/\partial \alpha)(\alpha_0,\alpha_0) = -1/2\lambda(q(\alpha_0,\alpha_0)) \neq 0$, (3.4)-(3.5) implies that $(\partial x/\partial \alpha)(\alpha_0,\beta)$ is not identically zero. Hence we have Lemma 3.2. Exchanging α and β in the above, we obtain Lemma 3.3.   □

Proof   Integrating the first and the third equations of (3.4), we have

$$p + \Lambda(q) = \psi_1(\beta) \quad \text{and} \quad p - \Lambda(q) = \psi_2(\alpha) \qquad (3.6)$$

where $\Lambda'(q) = \lambda(q)$, $\psi_1(\beta) = z_1(\beta) + \Lambda(z_0'(\beta))$ and $\psi_2(\alpha) = z_1(\alpha) - \Lambda(z_0'(\alpha))$.

As $\Lambda'(q) > 0$, we see that an inverse function of $\Lambda(q)$ is smooth. Therefore p and q are obtained as smooth functions of (α, β) defined in the whole space $\mathbf{R}^2$. On the other hand, x and y are solutions of the following system:

$$\begin{cases} \lambda(q)\dfrac{\partial x}{\partial \alpha} + \dfrac{\partial y}{\partial \alpha} = 0, \\[3mm] -\lambda(q)\dfrac{\partial x}{\partial \beta} + \dfrac{\partial y}{\partial \beta} = 0. \end{cases} \tag{3.7}$$

As this is equivalent to a system of linear wave equations concerning x and y, we can get the solutions $x = x(\alpha, \beta)$ and $y = y(\alpha, \beta)$ as smooth functions of (α, β) defined in the whole space $\mathbf{R}^2 = \{(\alpha, \beta)\}$. The function $z = z(\alpha, \beta)$ is uniquely determined by the contact relation $dz = pdx + qdy$ and the initial conditions (3.2), that is to say,

$$\begin{cases} \dfrac{\partial z}{\partial \alpha} = p\dfrac{\partial x}{\partial \alpha} + q\dfrac{\partial y}{\partial \alpha}, \quad \dfrac{\partial z}{\partial \beta} = p\dfrac{\partial x}{\partial \beta} + q\dfrac{\partial y}{\partial \beta}, \\[3mm] z(\xi, \xi) = z_0(\xi), \quad \xi \in \mathbf{R}^1. \end{cases}$$

We can easily see that $z = z(\alpha, \beta)$ is a smooth function defined in the whole space $\mathbf{R}^2 = \{(\alpha, \beta)\}$. For the existence of the regular geometric solution, we must prove that it is regular, that is to say,

$$\text{rank} \begin{pmatrix} \partial x/\partial \alpha & \partial y/\partial \alpha & \partial z/\partial \alpha & \partial p/\partial \alpha & \partial q/\partial \alpha \\ \partial x/\partial \beta & \partial y/\partial \beta & \partial z/\partial \beta & \partial p/\partial \beta & \partial q/\partial \beta \end{pmatrix} = 2. \tag{3.8}$$

Taking the derivatives of (3.6) with respect to α and β, we get

$$2\dfrac{\partial p}{\partial \alpha} = \psi_2'(\alpha), \quad 2\dfrac{\partial p}{\partial \beta} = \psi_1'(\beta), \quad 2\lambda(q)\dfrac{\partial q}{\partial \alpha} = -\psi_2'(\alpha), \quad 2\lambda(q)\dfrac{\partial q}{\partial \beta} = \psi_1'(\beta).$$

Therefore, for the proof of (3.8), we must consider whether $\psi_1'(\beta)$ and $\psi_2'(\alpha)$ are 0 or not. Concerning this problem, we use the above two lemmas. Then we see that (3.8) holds at any point $(\alpha, \beta) = (\alpha_0, \beta_0)$.   □

In a domain where the Jacobian $D(x,y)/D(\alpha, \beta)$ does not vanish, we can uniquely solve $x = x(\alpha, \beta)$ and $y = y(\alpha, \beta)$ with respect to (α, β). We write $\alpha = \alpha(x, y)$ and $\beta = \beta(x, y)$. Then $z(x, y) = z(\alpha(x, y), \beta(x, y))$ is a classical solution of (3.1)-(3.2), because $(\partial z/\partial x)(x, y) = p(\alpha(x, y), \beta(x, y))$ and $(\partial z/\partial y)(x, y) = q(\alpha(x, y), \beta(x, y))$. Next we will prove the explosion of classical solutions at points where the Jacobian vanishes.

Theorem 3.4 *Assume* $\dfrac{D(x,y)}{D(\alpha, \beta)}(\alpha^0, \beta^0) = 0$. *If a point* (x, y) *goes to* $(x(\alpha^0, \beta^0), y(\alpha^0, \beta^0))$ *along a curve in the existence domain of the classical solution* $z = z(x, y)$, *then* (r, s, t) *tends to* ∞.

Proof In a domain where there exists a classical solution, we have

$$
\begin{cases}
\dfrac{\partial p}{\partial \alpha} = r\dfrac{\partial x}{\partial \alpha} + s\dfrac{\partial y}{\partial \alpha}\;, & \dfrac{\partial p}{\partial \beta} = r\dfrac{\partial x}{\partial \beta} + s\dfrac{\partial y}{\partial \beta}\;, \\[3mm]
\dfrac{\partial q}{\partial \alpha} = s\dfrac{\partial x}{\partial \alpha} + t\dfrac{\partial y}{\partial \alpha}\;, & \dfrac{\partial q}{\partial \beta} = s\dfrac{\partial x}{\partial \beta} + t\dfrac{\partial y}{\partial \beta}\;,
\end{cases}
\tag{3.9}
$$

where r, s and t are the second order derivatives of $z = z(x,y)$ introduced in (2.1). For (3.1)-(3.2), the Jacobian is written down by $D(x,y)/D(\alpha,\beta) = 2\lambda(q)(\partial x/\partial \alpha)(\partial x/\partial \beta)$. If the Jacobian does not vanish, then it follows from (3.9) that

$$
r = \frac{1}{4}\Big\{\frac{\psi_1'(\beta)}{\frac{\partial x}{\partial \beta}(\alpha,\beta)} + \frac{\psi_2'(\alpha)}{\frac{\partial x}{\partial \alpha}(\alpha,\beta)}\Big\}, \quad s = \frac{1}{4\lambda(q)}\Big\{\frac{\psi_1'(\beta)}{\frac{\partial x}{\partial \beta}(\alpha,\beta)} - \frac{\psi_2'(\alpha)}{\frac{\partial x}{\partial \alpha}(\alpha,\beta)}\Big\}, \quad t = \frac{1}{\lambda(q)^2}r.
\tag{3.10}
$$

Applying Lemma 3.2 and Lemma 3.3 to (3.10), we can get the above result. □

Remark We explain the meaning of Theorem 3.4. Sometimes, even if the Jacobian may vanish, there exists a classical solution. This phenomenon happens even for nonlinear first order partial differential equations (see M. Tsuji [21]). What Theorem 3.4 insists is that, for the Cauchy problem (3.1)-(3.2), a classical solution always blows up at a point where the Jacobian vanishes. Therefore we can exactly determine the life-span of the classical solution by the information on zeros of the Jacobian. Concerning the life-span of classical solutions, various kinds of results have been published, for example [28], [13], [24], etc, etc.

Example The Cauchy problem (3.1)-(3.2) has sometimes a global classical solution. Assume that the initial data satisfy $z_1(y) = \Lambda(z_0'(y))$. Moreover, if we choose a function $z_0(y)$ so that it holds

$$z_0''(y)\ \lambda'(z_0'(y)) \le 0,$$

then we see that the Jacobian $D(x,y)/D(\alpha,\beta)$ does not vanish for any (x,y) in $\{(x,y); x \ge 0, y \in \mathbf{R}^1\}$. Therefore the Cauchy problem (3.1)-(3.2) has uniquely a classical solution in the whole domain $\{(x,y); x \ge 0, y \in \mathbf{R}^1\}$. If the above condition is violated, then we can easily see by Theorem 3. 4 that the solution blows up in finite time.

As we have stated in the above, our principal interest is to extend a solution beyond the singularities. Therefore we introduce the notion of "solution with singularities" which is called "weak solution", though we do not yet arrive at the final decision on the definition of weak solutions. Let us introduce the most typical definition which is corresponding to P. D. Lax's one [14] introduced for systems of conservation laws.

Definition 3.5 *A function $z = z(x,y)$ is called a weak solution of (3.1)-(3.2) if the following conditions (i) and (ii) are satisfied:*
(i) The function $z = z(x,y)$ is continuous with $z(0,y) = z_0(y)$; and its derivatives in the

sense of distributions, $(\partial z/\partial x)(x,y)$ and $(\partial z/\partial y)(x,y)$, are bounded and measurable,
(ii) The function $z = z(x,y)$ satisfies the Cauchy problem (3.1)-(3.2) in the sense of distributions, that is to say, it holds for any $\varphi(x,y) \in C_0^\infty(\mathbf{R}^2)$ that

$$\int_{\mathbf{R}_+^2} \{ \frac{\partial z}{\partial x}\frac{\partial \varphi}{\partial x} - f(\frac{\partial z}{\partial y})\frac{\partial \varphi}{\partial y} \} dx dy + \int_{\mathbf{R}^1} z_1(y)\varphi(0,y)dy = 0. \tag{3.11}$$

Let $z = z(x,y)$ be a weak solution of (3.1)-(3.2) in the sense of Definition 3.5. Assume that $(\partial z/\partial x)(x,y) \equiv p(x,y)$ and $(\partial z/\partial y)(x,y) \equiv q(x,y)$ have jump discontinuities along a smooth curve $y = \gamma(x)$. As $z = z(x,y)$ is continuous, it holds that $z(x, \gamma(x) + 0) = z(x, \gamma(x) - 0)$. Differentiating this with respect to x, we get

$$[p] + [q]\dot{\gamma} = 0. \tag{3.12}$$

where $[\ ]$ means the quantity of difference, i.e., $[p] = p(x, \gamma(x) + 0) - p(x, \gamma(x) - 0)$. Since $z = z(x,y)$ satisfies (3.11), we obtain

$$[p]\dot{\gamma} + [f(q)] = 0. \tag{3.13}$$

This means that the curve $y = \gamma(x)$ must satisfy two kinds of differential equations, (3.12) and (3.13). This suggests us the following:

Theorem 3.6 *Assume that $f'(q) > 0$ and $f''(q) \neq 0$. Then we can not generally construct a weak solution of the Cauchy problem (3.1)-(3.2) in the sense of Definition 3.5 by projecting the above geometric solution to the base space and cutting off some part of the multi-valued projected solution so that it would become a single-valued solution of (3.1)-(3.2).*

Remark In this theorem, we have used the word "generally" to state that there exists the case in which we cannot construct a weak solution by the method explained in the theorem. Therefore we do not deny the possibility that there may exist a case where we can do so.

Proof Supposing that we could construct a weak solution of the Cauchy problem (3.1)-(3.2) by the geometric method stated in the theorem, we will show that we would be led to a contradiction. We consider the case where the initial data satisfy $\psi_1(\beta) = 0$ or $\psi_2(\alpha) = 0$. Here we assume $\psi_2(\alpha) = 0$. Then it holds that $p - \Lambda(q) = 0$ for all $(\alpha, \beta) \in \mathbf{R}^2$. Moreover, solving (3.6) and (3.7), we get

$$y = \beta - \lambda(q)x, \quad p = z_1(\beta), \quad q = z_0'(\beta), \quad \text{and} \quad x(\alpha, \beta) = \frac{1}{2\sqrt{\lambda(z_0'(\beta))}}\int_\alpha^\beta \frac{d\tau}{\sqrt{\lambda(z_0'(\tau))}}.$$

As z_0' is bounded, $x(\alpha, \beta)$ tends to $\pm\infty$ when $\beta - \alpha$ goes to $\pm\infty$ respectively. Therefore, when (α, β) moves in the whole space $\mathbf{R}^2$, so does $(x(\alpha,\beta), y(\alpha,\beta))$. As $z = z(\alpha, \beta)$ is defined by $dz = pdx + qdy$, we have $\partial z/\partial x = p$ and $\partial z/\partial y = q$ in the domain

$\{(x(\alpha,\beta),y(\alpha,\beta)); \ D(x,y)/D(\alpha,\beta) \neq 0\}$. As, by Sard's theorem, the measure of the set $\{(x(\alpha,\beta),y(\alpha,\beta)); \ D(x,y)/D(\alpha,\beta) = 0\}$ is zero, we define the values of $\partial z/\partial x$ and $\partial z/\partial y$ at a point where the Jacobian vanishes by the limits of $p = p(\alpha,\beta)$ and $q = q(\alpha,\beta)$ respectively. Then $(\partial z/\partial x)(x,y)$ and $(\partial z/\partial y)(x,y)$ become multi-valued functions defined in the whole space $\mathbf{R}^2$, and $z = z(x,y)$ turns out to satisfy the following Cauchy problem:

$$\frac{\partial z}{\partial x} - \Lambda(\frac{\partial z}{\partial y}) = 0 \quad \text{in} \quad \{x > 0, y \in \mathbf{R}^1\} \equiv \mathbf{R}_+^2, \tag{3.14}$$

$$z(0,y) = z_0(y) \quad \text{in} \quad \{x = 0, y \in \mathbf{R}^1\}. \tag{3.15}$$

The characteristic differential equations for (3.14)-(3.15) are written by

$$\begin{cases} \dfrac{dy}{dx} = -\lambda(q), \quad \dfrac{dz}{dx} = p - \lambda(q)q, \quad \dfrac{dp}{dx} = \dfrac{dq}{dx} = 0, \\[2mm] y(0) = \xi, \quad z(0) = z_0(\xi), \quad p(0) = \Lambda(z_0'(\xi)), \quad q(0) = z_0'(\xi). \end{cases} \tag{3.16}$$

We can immediately solve (3.16). Moreover, in the case where $z_1(\xi) - \Lambda(z_0'(\xi)) = 0$, we can easily see that the surface $\{(x,y,p,q); y = \xi - \lambda(z_0'(\xi))x, p = z_1(\xi), q = z_0'(\xi)\}$, i.e. the (x,y,p,q)-components of the solution of (3.16), is the same as the solution surface of (3.4)-(3.5). Here we recall the result of [20, 21]. As $(\partial y/\partial \xi)(x,\xi) = 1 - \lambda'(z_0'(\xi))z_0''(\xi)x$ does not vanish in a neighbourhood of $x = 0$, we can uniquely solve the equation $y = y(x,\xi)$ with respect to ξ and denote it by $\xi = \xi(x,y)$. Then we can get a classical solution of (3.1)-(3.2) by $z = z(x,\xi(x,y))$ where $z = z(x,\xi)$ is the z-component of the solution of (3.16). Here we put $h(\xi) = \lambda'(z_0'(\xi))z_0''(\xi)$ and assume that $h = h(\xi)$ takes its positive maximum at $\xi = \xi_0$ with $h''(\xi_0) > 0$. We write $M = h(\xi_0)$, $x_0 = 1/M$ and $y_0 = y(x_0,\xi_0)$. Then the function $\xi = \xi(x,y)$ takes three values for $x > x_0$ in a neighbourhood of (x_0,y_0), and so does the solution $z = z(x,y)$. In [20] we have proved that we can choose only one from these three values of $z = z(x,y)$ so that the solution becomes a single-valued continuous solution of (3.14)-(3.15), and that it automatically satisfies the conditions for generalized solutions introduced in the case of Hamilton-Jacobi equations. Therefore, if a continuous solution of (3.1)-(3.2) may be obtained by the geometric method, it must coincide with the continuous solution of (3.14)-(3.15) constructed as above. In the situation under consideration, its derivatives have jump discontinuity along some smooth curve $y = \gamma(x)$. Combining (3.12) and (3.13), we get

$$\frac{[f(q)]}{[p]} = \frac{[p]}{[q]}, \quad \text{i.e.,} \quad \frac{[f(q)]}{[q]} = \left(\frac{[\Lambda(q)]}{[q]}\right)^2 \tag{3.17}$$

The following Lemma 3.7 assures us that (3.17) does not hold. This is a contradiction. Hence we get Theorem 3.6. $\qquad\qquad\qquad\qquad\qquad\qquad\qquad\qquad\qquad\qquad\qquad\qquad\qquad\Box$

Lemma 3.7 *Assume that $f'(q) > 0$ and $f''(q) \neq 0$. Then, if $q_1 \neq q_2$ and $q_2 - q_1$ is sufficiently small, we have*

$$\frac{f(q_2) - f(q_1)}{q_2 - q_1} \neq \left(\frac{\Lambda(q_2) - \Lambda(q_1)}{q_2 - q_1}\right)^2$$

Proof   Keeping q_1 fixed, we expand the both sides in the Taylor series with respect to q_2 about $q_2 = q_1$. Then the coefficient of $(q_2 - q_1)^2$ of the left-hand side is $f^{(3)}(q_1)/6$, while that of the right-hand side is equal to $f^{(3)}(q_1)/6 - (f''(q_1))^2/48f'(q_1)$. This means the above conclusion.   □

If we may change the definition of weak solution, we shall be able to get various results of another type. For example, the following definition of "weak solution" is also possible.

Definition 3.8 *Let $z = z(x,y)$, $(\partial z/\partial x)(x,y)$ and $(\partial z/\partial y)(x,y)$ be bounded and measurable. Moreover $z = z(x,y)$ is continuous as a function of x with values in $L^1_{loc}(\mathbf{R}^1)$. The function $z = z(x,y)$ is a weak solution of (3.1)-(3.2) if it satisfies equation (3.1) in the distribution sense, i.e.,*

$$\int_{\mathbf{R}^2_+} \left\{ \frac{\partial z}{\partial x}\frac{\partial \varphi}{\partial x} - f(\frac{\partial z}{\partial y})\frac{\partial \varphi}{\partial y} \right\} dxdy + \int_{\mathbf{R}^1} z_1(y)\varphi(0,y)\, dy = 0$$

for all $\varphi(x,y) \in C^\infty_0(\mathbf{R}^2)$, and $z = z(x,y)$ tends to $z_0(y)$ in $L^1_{loc}(\mathbf{R}^1)$ when x goes to $+0$.

Then we get the following:

Theorem 3.9 *In the same situation as Theorem 3.7, we can get a weak solution of the Cauchy problem (3.1)-(3.2) in the sense of Definition 3.8 by projecting the above geometric solution to the base space and cutting off appropriate parts of that projected surface.*

Proof   Let us begin our discussion from (3.14)-(3.15). As it holds that $p - \Lambda(q) = 0$, $p(x,y) = \partial z/\partial x$ satisfies the following Cauchy problem:

$$\begin{cases} \dfrac{\partial p}{\partial x} - \dfrac{\partial}{\partial y}g(p) = 0, \\[2mm] p(0) = z_1(y). \end{cases}$$

where $g(p) = f(\Lambda^{-1}(p))$. We consider this Cauchy problem in the same situation as in the proof of Theorem 3.7. Then we can uniquely construct a shock curve $y = \gamma(x)$ for $p = p(x,y)$ which starts from the point (x_0,y_0) (refer to [23]). We see easily that the jump condition for $p = p(x,y)$ is the same as (3.13). Next we construct a weak solution $z = z(x,y)$ of (3.1)-(3.2) by using the solution of (3.16). Then we can show that $z = z(x,y)$ is not continuous across the curve $y = \gamma(x)$. In fact, if so, it follows that $z(x,\gamma(x)-0) = z(x,\gamma(x)+0)$. Hence we get (3.12). As we have already proved in Lemma 3.7 that the conditions (3.12) and (3.13) are not compatible, we see that the above solution $z = z(x,y)$ can not become continuous across the curve $y = \gamma(x)$. Therefore it is not a weak solution in the sense of Definition 3.5, but it is so in the sense of Definition 3.8.   □

Remark 1 In Definition 3.8, $\partial z/\partial x$ and $\partial z/\partial y$ are the derivatives of $z = z(x, y)$ in the classical sense, not in the distribution sense. Hence, even if $z = z(x, y)$ may have jump discontinuities, Dirac's measure does not appear in $\partial z/\partial x$ and $\partial z/\partial y$. But we insist that equation (3.1) is satisfied in the distribution sense. Therefore we guess that Definition 3.8 would not be accepted by many people. But we have shown in [24] that it works well in some case. As an example in [24], we have considered the well-known equation appeared in N. J. Zabusky [28].

Remark 2 Recently a notion of "viscosity solution" has been introduced even for second order nonlinear partial differential equations. For this purpose the equations must satisfy the maximum and minimum principle. But, as hyperbolic equations do not have this property, we can not apply the theory of viscosity solutions to our problem.

4 Systems of conservation laws

In this section we will consider a certain hyperbolic system of conservation laws which is related with equation (3.1). Let $z = z(x, y)$ be a solution of (3.1), and write $p = \partial z/\partial x$, $q = \partial z/\partial y$, $U(x, y) = (p, q)$, $F(U) = (f(q), p)$ and $U_0(y) = (z_1(y), z_0'(y)) \equiv (p_0(y), q_0(y))$. Then we get

$$\frac{\partial}{\partial x}U - \frac{\partial}{\partial y}F(U) = 0 \qquad \text{in } \{x > 0, y \in \mathbf{R}^1\}, \tag{4.1}$$

$$U(0, y) = U_0(y) \qquad \text{on } \{x = 0, y \in \mathbf{R}^1\}. \tag{4.2}$$

The system of the form (4.1) is called "p-system". As we have done for hyperbolic Monge-Ampère equations in §3, we will introduce a notion of "geometric solution" for (4.1)-(4.2). To do so, we will rewrite the equations by using differential forms. Then system (4.1) is represented by

$$\begin{cases} dp \wedge dy + df(q) \wedge dx = 0, \\ \\ dq \wedge dy + dp \wedge dx = 0. \end{cases} \tag{4.3}$$

Definition 4.1 *A regular geometric solution of* (4.1) *is a submanifold of dimension 2 defined in* $\mathbf{R}^4 = \{(x, y, p, q)\}$ *on which system* (4.3) *is satisfied.*

For getting a geometric solution in the above sense, we will decompose the equations of differential forms as a product of one forms just as in §3. Then we can easily see that system (4.3) is equivalent to the following system:

$$\begin{cases} (dp + \lambda(q)dq) \wedge (\lambda(q)dx + dy) = 0, \\ \\ (dp - \lambda(q)dq) \wedge (-\lambda(q)dx + dy) = 0, \end{cases} \tag{4.4}$$

where $\lambda(q) = \sqrt{f'(q)}$. We will repeat the same discussion as in §3 for solving system (4.4). Let us represent a geometric solution by

$$x = x(\alpha, \beta), \ y = y(\alpha, \beta), \ p = p(\alpha, \beta), \ q = q(\alpha, \beta).$$

A sufficient condition so that it is a geometric solution of (4.1)-(4.2) is given by

$$\begin{cases} \dfrac{\partial p}{\partial \alpha} + \lambda(q)\dfrac{\partial q}{\partial \alpha} = 0, \\[2mm] \lambda(q)\dfrac{\partial x}{\partial \alpha} + \dfrac{\partial y}{\partial \alpha} = 0, \\[2mm] \dfrac{\partial p}{\partial \beta} - \lambda(q)\dfrac{\partial q}{\partial \beta} = 0, \\[2mm] -\lambda(q)\dfrac{\partial x}{\partial \beta} + \dfrac{\partial y}{\partial \beta} = 0. \end{cases} \qquad (4.5)$$

The above system is just the same as (3.4). The initial condition corresponding to (4.2) is

$$x(\xi,\xi) = 0, \ \ y(\xi,\xi) = \xi, \ \ p(\xi,\xi) = p_0(\xi), \ \ q(\xi,\xi) = q_0(\xi), \ \ \xi \in \mathbf{R}^1. \qquad (4.6)$$

As discussed in §3, we can prove that the Cauchy problem (4.5)-(4.6) has a unique solution defined for all $(\alpha, \beta) \in \mathbf{R}^2$, and it satisfies

$$p + \Lambda(q) = \psi_1(\beta) \quad \text{and} \quad p - \Lambda(q) = \psi_2(\alpha)$$

where $\Lambda'(q) \equiv \lambda(q)$ and $\psi_1(\beta) = p_0(\beta) + \Lambda(q_0(\beta))$, and $\psi_2(\alpha) = p_0(\alpha) - \Lambda(q_0(\alpha))$. This solution determines a regular geometric solution of (4.1)-(4.2) in the large.

It is well known that, even if the initial data are sufficiently smooth, singularities generally appear in the solution of (4.1)-(4.2). Therefore we will construct a "solution with singularities" called "weak solution". Let us recall here the definition of weak solutions of (4.1)-(4.2) introduced by P. D. Lax [14].

Definition 4.2 *A bounded and measurable 2-vector function $U = U(x,y)$ is a weak solution of (4.1)-(4.2) if it satisfies (4.1)-(4.2) in the weak sense, i.e.,*

$$\int_{\mathbf{R}^2_+} \{U(x,y)\frac{\partial \Phi}{\partial x}(x,y) - F(U)\frac{\partial \Phi}{\partial y}(x,y)\}dxdy + \int_{\mathbf{R}^1} U_0(y)\Phi(0,y)dy = 0 \qquad (4.7)$$

for any 2-vector function $\Phi(x,y) \in C_0^\infty(\mathbf{R}^2)$.

Putting $p = \partial z/\partial x$ and $q = \partial z/\partial y$, we arrive at Definition 4.2 from Definition 3.5. Contrarily we have shown in [26] that, if the Cauchy problem (4.1)-(4.2) has a weak solution in the sense of Definition 4.2, then the Cauchy problem (3.1)-(3.2) admits a weak solution in the sense of Definition 3.5.

If $U = U(x, y)$ is a weak solution of (4.1) which has jump discontinuity along a smooth curve $y = \gamma(x)$, we get jump conditions of Rankine-Hugoniot as follows:

$$[p]\dot{\gamma} + [f(q)] = 0,$$

$$[q]\dot{\gamma} + [p] = 0.$$

Therefore the jump discontinuity must satisfy two kinds of differential equations. Using this property, we can show that Theorem 3.6 is rewritten in the following form, though we do not write the proof because it is almost similar to that of Theorem 3.6.

Theorem 4.3 *Assume that $f'(q) > 0$ and $f''(q) \neq 0$. Then we can not generally construct a weak solution of the Cauchy problem (4.1)-(4.2) in the sense of Definition 4.2 by cutting off some part of the above projected geometric solution so that it would become a single-valued solution of (4.1)-(4.2).*

Remark For Riemann's problem to (4.1), the same result has been already proved in [1]. But, in the case where the initial data are smooth, the above result has been announced as a conjecture (see p. 59 in [1]). The above result is also written a little in [24] which is the Proceeding of a meeting held in September 1996. At that time we did not know the paper [1]. We thank Y. Machida that he informed us the existence of [1].

Concerning single first order partial differential equations, we could construct weak solutions by the above method. For example, see M. Tsuji [20, 21], S. Nakane [17, 18], S. Izumiya [10], S. Izumiya and G. T. Kossioris [11, 12], etc.

In this note we insist that, for second order hyperbolic equations or hyperbolic systems of conservation laws, we can not construct weak solutions by the same method as for single first order partial differential equations. But, if we may change the definition of weak solutions, we can get a weak solution by cutting off some part of the above projected geometric solution. For example, if we may introduce a new definition of weak solutions of (4.1)-(4.2) corresponding to Definition 3.8, we can get an affirmative answer which is corresponding to Theorem 3.9. In a forthcoming paper, we will discuss more precisely what kind of solutions we can get by projecting the geometric solutions to the base space.

Acknowledgement This lecture is dedicated to Professor Jean Vaillant in the occasion of his retirement. The author thanks him for his kind and continuous support. Especially, when the author begun to work on this subject, he could spend one year from 1993 to 1994 with the group of Jean Vaillant and Bernard Gaveau at the University of Paris VI. This was the nice period for the author to study the subject of this talk. In this sense he has summerized here our recent results on the above subject.

References

[1] R. Bryant, P. Griffiths and L. Hsu, *Toward a geometry of differential equations*, in "Geometry, Topology, and Physics for Raoul Bott" edited by S.-T. Yau (International Press, 1994), 1-76.

[2] R. Courant and D. Hilbert, *Method of mathematical physics*, vol. 2, Interscience, New York, 1962.

[3] G. Darboux, *Leçon sur la théorie générale des surfaces*, tome 3, Gauthier-Villars, Paris, 1894.

[4] B. Gaveau, *Evolution of a shock for a single conservation law in 2+1 dimensions*, Bull. Sci. math. **113** (1989), 407-442.

[5] E. Goursat, *Leçons sur l'intégration des équations aux dérivées partielles du second ordre*, tome 1, Hermann, Paris, 1896.

[6] E. Goursat, *Cours d'analyse mathématique*, tome 3, Gauthier-Villars, Paris, 1927.

[7] J. Guckenheimer, *Solving a single conservation law*, Lecture Notes in Math. (Springer-Verlag) **468** (1975), 108-134.

[8] J. Hadamard, *Le problème de Cauchy et les équations aux dérivées partielles linéaires hyperboliques*, Hermann, Paris, 1932.

[9] Ha Tien Ngoan, D. Kong and M. Tsuji, *Integration of Monge-Ampère equations and surfaces with negative Gaussian curvature*, Ann. Scuola Normale Sup. Pisa **27** (1999), 309-330.

[10] S. Izumiya, *Geometric singularities for Hamilton-Jacobi equations*, Adv. Studies in Math. **22** (1993), 89-100.

[11] S. Izumiya and G. T. Kossioris, *Semi-local classification of geometric singularities for Hamilton-Jacobi equations*, J. Diff. Eq. **118** (1995), 166-193.

[12] S. Izumiya and G. T. Kossioris, *Formation of singularities for viscosity solutions of Hamilton-Jacobi equations*, Banach Center Publications **33** (1996), 127-148.

[13] P. D. Lax, *Development of singularities of solutions of nonlinear hyperbolic partial differential equations*, J. Math. Physics **5** (1964), 611-613.

[14] P. D. Lax, *Hyperbolic systems of conservation laws II*, Comm. Pure Appl. Math. **10** (1957), 537-566.

[15] H. Lewy, *Über das Anfangswertproblem einer hyperbolischen nichtlinearen partiellen Differentialgleichung zweiter Ordnung mit zwei unabhängigen Veränderlichen*, Math. Ann. **98** (1928), 179-191.

[16] V. V. Lychagin, *Contact geometry and non-linear second order differential equations*, Russian Math. Surveys **34** (1979), 149-180.

[17] S. Nakane, *Formation of shocks for a single conservation law*, SIAM J. Math. Anal. **19** (1988), 1391-1408.

[18] S. Nakane, *Formation of singularities for Hamilton-Jacobi equations in several space variables*, J. Math. Soc. Japan **43** (1991), 89-100.

[19] R. Thom, *Stabilité structuelle et morphogénèse*, W. A. Benjamin, Massachusetts, 1972.

[20] M. Tsuji, *Formation of singularities for Hamilton-Jacobi equation II*, J. Math. Kyoto Univ. **26** (1986), 299-308.

[21] M. Tsuji, *Prolongation of classical solutions and singularities of generalized solutions*, Ann. Inst. H. Poincarè - Analyse nonlinéaire **7** (1990), 505-523.

[22] M. Tsuji, *Formation of singularities for Monge-Ampère equations*, Bull. Sci. math. **119** (1995), 433-457.

[23] M. Tsuji, *Monge-Ampère equations and surfaces with negative Gaussian curvature*, Banach Center Publications **39** (1997), 161-170.

[24] M. Tsuji, *Geometric approach to blow-up phenomena in nonlinear problems*, "Real analytic and algebraic singularities", Pitman Research Notes in Math. **381** (Longman, 1998), 164-180.

[25] M. Tsuji and T. S. Nguyen Duy, *Geometric solutions of nonlinear second order hyperbolic equations*, to appear in "Acta Math. Vietnamica".

[26] T. D. Van, M. Tsuji and N. D. Thai Son, *The characteristic method and its generalizations for first-order non-linear partial differential equations*, Chapman & Hall/CRC (USA), 1999.

[27] H. Whitney, *On singularities of mappings of Euclidean spaces I*, Ann. Math. **62** (1955), 374-410.

[28] N. J. Zabusky, *Exact solution for vibrations of a nonlinear continuous model string*, J. Math. Physics **3** (1962), 1028-1039.

Causal evolution for Einsteinian gravitation

YVONNE CHOQUET-BRUHAT
Université Paris 6, 4 Place Jussieu, 75152 Paris

Abstract

The integration of the Einstein equations split into the solution of constraints on an initial space like 3 - manifold, an essentially elliptic system, and a system which will describe the dynamical evolution, modulo a choice of gauge. We prove in this paper that the simplest gauge choice leads to a system which is causal, but hyperbolic non strict in the sense of Leray - Ohya. We review some strictly hyperbolic systems obtained recently.

1 INTRODUCTION.

The Einstein equations equate the Ricci tensor of a pseudo riemannian 4-manifold (V, g), of lorentzian signature, the spacetime, with a phenomelogical tensor which describes the sources which we take here to be zero (vacuum case). The Einstein equations are a geometric system, invariant by diffeomorphisms of V and the associated isometries of g. From the analyst point of view they constitute a system of second order quasilinear partial differential equations which is over determined, Cauchy data must satisfy constraints, and underdetermined, the characteristic determinant is identical to zero. An important problem for the study of solutions, their physical interpretation and numerical computation is the choice of a gauge, i.e. a priori hypothesis for instance on coordinates choice, such that the evolution of initial data satisfying the constraints is well posed.

The geometric initial data are a 3-manifold M endowed with a riemaniann metric and a symmetric 2-tensor which will be the extrinsic curvature of M embedded in $V = M \times R$. A natural gauge choice seemed to be the data on V of the time lines by their projection N on R and β on M. Such a choice has been extensively used in numerical computation, though the evolution system $R_{ij} = 0$ obtained with such a choice was not known to be well posed. In this article we will show that the system is indeed hyperbolic in the sense of Leray-Ohya, in the Gevrey class $\gamma = 2$, and is causal, i.e. the domain of dependence of its solutions is determined by the light cone. When the considered evolution system is satisfied the constraints are preserved through a symmetric first order evolution system. Consideration of a system of order 4 obtained previously by combination of the equations $R_{ij} = 0$ with the constraints give the same result.

In section 8 we recall how the old harmonic gauge, interpreted now as conditions on N and β, gives a strictly hyperbolic evolution system and the larger functional spaces where local existence and global geometrical uniqueness of solutions are known.

In recent years, since the paper of C-B and York 1995, there has been a great interest in formulating the evolution part of Eintein equations as a first order symmetric hyperbolic system for geometrically defined unknowns. Several such systems have been devised. Particularly interesting are those constructed with the Weyl tensor (H. Friedrich, see review article 1996) or the Riemann tensor (Anderson, C-B and York 1997) because they lead to estimates of the geometrically defined Bel-Robinson energy used in some global existence proofs (Christodoulou and Klainerman 1989). We recall in the last section the symmetric hyperbolic Einstein-Bianchi system and the corresponding Bel-Robinson energy.

2 EINSTEIN EQUATIONS.

The spacetime of general relativity is a pseudo riemannian manifold (V, g),of lorentzian signature (- + + +). The Einstein equations link its Ricci tensor with a phenomelogical stress energy tensor which describes the sources. They read

$$Ricci(g) = \rho$$

that is, in local coordinates x^λ, $\lambda = 0, 1, 2, 3$, where $g = g_{\lambda\mu}dx^\lambda dx^\mu$,

$$R_{\alpha\beta} \equiv \frac{\partial}{\partial x^\lambda}\Gamma^\lambda_{\alpha\beta} - \frac{\partial}{\partial x^\alpha}\Gamma^\lambda_{\beta\lambda} + \Gamma^\lambda_{\alpha\beta}\Gamma^\mu_{\lambda\mu} - \Gamma^\lambda_{\alpha\mu}\Gamma^\mu_{\beta\lambda} = \rho_{\alpha\beta}$$

where the $\Gamma's$ are the Christoffel symbols:

$$\Gamma^\lambda_{\alpha\beta} = \frac{1}{2}g^{\lambda\mu}(\frac{\partial}{\partial x^\alpha}g_{\beta\mu} + \frac{\partial}{\partial x^\beta}g_{\alpha\mu} - \frac{\partial}{\partial x^\mu}g_{\alpha\beta})$$

The source ρ is a symmetric 2-tensor given in terms of the stress energy tensor T by

$$\rho_{\alpha\beta} \equiv T_{\alpha\beta} - \frac{1}{2}g_{\alpha\beta}trT, \quad \text{with} \quad trT \equiv g^{\lambda\mu}T_{\lambda\mu}$$

Due to the Bianchi identities the left hand side of the Einstein equations satisfies the identities, with ∇_α the covariant derivative in the metric g

$$\nabla_\alpha(R_{\alpha\beta} - \frac{1}{2}g_{\alpha\beta}\,R) \equiv 0, \quad R \equiv g^{\lambda\mu}R_{\lambda\mu}$$

The stress energy tensor of the sources satisfies the conservation laws which make the equations compatible

$$\nabla_\alpha T^{\alpha\beta} = 0$$

In vacuum the stress energy tensor is identically zero. We will consider here only this case. The presence of sources brings up new problems specific to various types of sources.

The Einstein equations (in vacuum) are a geometric system, invariant by diffeomorphisms of V and the associated isometries of g. From the analyst point of view they constitute a system of second order quasilinear partial differential equations which is both undetermined (the characteristic determinant is identical to zero) and overdetermined (one cannot give arbitrarily Cauchy data).

3   GEOMETRIC CAUCHY PROBLEM.

Due to the geometric nature of Einstein's equations it is appropriate to consider a Cauchy problem also in geometric form. The definition follows. .

An initial data set is a triple $(M, \bar{g}_0, K_0)$ where M is a 3 dimensional manifold, $\bar{g}_0$ a riemannian metric on M and K_0 a symmetric 2 tensor.

An extension of an initial data set is a spacetime (V, g) such that there exists an immersion $i : M \to M_0 \subset V$ with $i\overset{*}{\bar{g}}_0$ and i^*K_0 equal respectively to the metric induced by g on M_0 and the extrinsic curvature of M_0 as submanifold of (V, g).

We say that (V, g) is an einsteinian extension if g satisfies the Einstein (vacuum) equations on V.

A spacetime (V, g) is said to be globally hyperbolic if the set of timelike curves, between two arbitrary points is relatively compact in the Frechet topology of curves on V. This definition given by Leray 1952 has been shown by Geroch 1970 to be equivalent to the fact that (V, g) possesses a Cauchy surface, i.e. a spacelike submanifold M_0 such that each inextendible timelike or null curve cuts M_0 exactly once.

A development is a globally hyperbolic einsteinian extension.

4   SPACE AND TIME SPLITTING.

To link geometry with analysis one performs a space and time splitting of the Einstein equations. We consider a manifold V of the type $M \times R$ (the support of a development will always be of this type). We denote by $x^i, \in i = 1, 2, 3$ local coordinates in M, we set $x^0 = t \in R$. We choose a moving frame on V such that at a point (x, t) its axes e_i coincide with the axis of the natural frame, tangent to $M_t \equiv M \times \{t\}$, and its axis e_0 is orthogonal to M_t, with associated coframe such that $\theta^0 = dt$. A generic lorentzian metric on V with the M_t's spacelike reads in the associated coframe

$$g = -N^2 dt^2 + \bar{g}_{ij}\theta^i\theta^j, \quad \text{with} \quad \theta^i \equiv dx^i + \beta^i dt$$

The coefficients are time dependent geometric objects on M. The scalar N is called lapse, the space vector β is called shift, $\bar{g}$ is a riemannian metric. These elements are linked with the metric coefficients $g_{\alpha\beta}$ in the natural frame by the relations:

$$g_{ij} \equiv \bar{g}_{ij}, \quad N^2 = (-g^{00})^{-1}, \quad \beta^i \equiv N^2 g^{0i}$$

We denote by $\bar{\nabla}$ the covariant derivative in the metric $\bar{g}$. We have

$$\partial_i = \frac{\partial}{\partial x^i}, \quad \partial_0 = \partial_t - \beta^i\partial_i, \quad \text{with} \quad \partial_t = \frac{\partial}{\partial t}$$

and we denote

$$\hat{\partial}_0 = \frac{\partial}{\partial t} - L_\beta$$

with L_β the Lie derivative with respect to β, an operator wich maps a time dependent tensor field on M into another such tensor field.

We denote by K the extrinsic curvature of $M_t \equiv M \times \{t\}$ as submanifold of (V, g), i.e. we set:

$$K_{ij} = -\frac{1}{2N}\hat{\partial}_0 g_{ij}$$

A straightforward calculation (C-B 1956) gives the fundamental identities, written in the coframe $\theta^0 = dt, \theta^i = dx^i + \beta^i dt$, for the Ricci tensor of g :

$$R_{ij} \equiv \bar{R}_{ij} - \frac{\hat{\partial}_0 K_{ij}}{N} - 2K_{jh}K_i^h + K_{ij}K_h^h - \frac{\bar{\nabla}_j \partial_i N}{N}$$

$$R_{0i} \equiv N(-\bar{\nabla}_h k_i^h + \bar{\nabla}_i k_h^h)$$

$$R_{00} \equiv N(\partial_0 K_h^h - N K_{ij}K^{ij} + \bar{\nabla}^i \partial_i N)$$

5   CONSTRAINTS AND CONSERVATION.

The following part of the Einstein equations do not contain second derivatives of g neither first derivatives of K transversal to the spacelike manifolds M_t. They are the constraints. They read, with $S_{\alpha\beta} \equiv R_{\alpha\beta} - \frac{1}{2}g_{\alpha\beta}R$,

Momentum constraint

$$C_i \equiv \frac{1}{N}R_{0i} \equiv -\bar{\nabla}_h K_i^h + \bar{\nabla}_i K_h^h = 0$$

Hamiltonian constraint

$$C_0 \equiv \frac{2}{N^2}S_{00} \equiv \bar{R} - K_j^i K_i^j + (K_h^h)^2 = 0$$

These constraints are transformed into a system of elliptic equations on each submanifold M_t, in particular on M_0 for $\bar{g} = g_0, K = K_0$, by the conformal method .

The equations

$$R_{ij} \equiv \bar{R}_{ij} - \frac{\hat{\partial}_0 K_{ij}}{N} - 2K_{jh}K_i^h + K_{ij}K_h^h - \frac{\bar{\nabla}_j \partial_i N}{N} = 0$$

together with the definition

$$\hat{\partial}_0 g_{ij} = -2NK_{ij}$$

determine the derivatives transversal to M_t of $\bar{g}$ and K when these tensors are known on M_t as well as the lapse N and shift β. It is natural to look at these equations as evolution equations determining $\bar{g}$ and K, while N and β, projections of the tangent to the time line respectively on the normal and the tangent space to M_t, are considered as gauge variables. This point of view is conforted by the following theorem (Anderson and York 1997, previously given for sources in C-B and Noutcheguéme 1988)

THEOREM. When $R_{ij} = 0$ the constraints satisfy a linear homogeneous first order symmetric hyperbolic system, they are satisfied if satisfied initially.

Proof. When $R_{ij} = 0$ we have, in the privileged frame,

$$R = -N^2 R^{00}$$

hence

$$S^{00} = \frac{1}{2} R^{00} \quad \text{and} \quad R = -2N^2 S^{00} = 2S_0^0$$

and

$$S^{ij} = -\frac{1}{2} \bar{g}^{ij} R = -\bar{g}^{ij} S_0^0$$

the Bianchi identities give therefore a linear homogeneous system for S_0^i and S_0^0 with principal parts

$$N^{-2} \partial_0 S_0^i + \bar{g}^{ij} \partial_j S_0^0, \quad \text{and} \quad \partial_0 S_0^0 + \partial_i S_0^i$$

This system is symetrizable hyperbolic, it has a unique solution, zero if the initial values are zero. The characteristics, which determine the domain of dependence, are the light cone.

6 LERAY-OHYA HYPERBOLICITY OF R_{ij} $= 0$

An evolution part of Einstein equations should exhibit causal propagation, i.e. with domain of dependence determined by the light cone of the spacetime metric. It has been a long standing problem to decide it was the case for the equations $R_{ij} = 0$ when N and β are known. These equations are a second order differential system | for g_{ij}. The hyperbolicity of a quasilinear system is defined through the linear differential operator obtained by replacing in the coefficients the unknown by given values. In our case for given N, β and g_{ij} the principal part of this operator acting on a symmetric 2-tensor γ_{ij} is

$$\frac{1}{2}\{(N^{-2}\partial_{00}^2 - g^{hk}\partial_{hk}^2)\gamma_{ij} + \partial^k\partial_j\gamma_{ik} + \partial^k\partial_i\gamma_{jk} - g^{hk}\partial_i\partial_j\gamma_{hk}\}$$

The characteristic matrix at a point of spacetime is the linear operator obtained by replacing the derivation ∂ by a covariant vector ξ. The characteristic determinant is the determinant of this linear operator. We take as independent unknown $\gamma_{12}, \gamma_{23}, \gamma_{31}, \gamma_{11}, \gamma_{22}, \gamma_{33}$ and consider the 6 equations $R_{ij} = 0$,same indices.

To simplify the writing we compute this matrix in a coframe orthonormal for the given spacetime metric (N, β, g_{ij}). We denote by (t, x, y, z) the components of ξ in such a coframe. The characteristic matrix $\mathcal{M}$ reads then (up to multiplication by 2):

$$\mathcal{M} \equiv \begin{pmatrix} t^2 - z^2 & zx & zy & 0 & 0 & -xy \\ xz & t^2 - x^2 & xy & -yz & 0 & 0 \\ yz & xy & t^2 - y^2 & 0 & -xz & 0 \\ 2xy & 0 & 2zx & t^2 - y^2 - z^2 & -x^2 & -x^2 \\ 2xy & 2yz & 0 & -y^2 & t^2 - x^2 - z^2 & -y^2 \\ 0 & 2yz & 2xz & -z^2 & -z^2 & t^2 - x^2 - y^2 \end{pmatrix}$$

The characteristic polynomial is the determinant of this matrix. It is found to be

$$Det\mathcal{M} = b^6 a^3, \quad \text{with} \quad b = t, \quad a = t^2 - x^2 - y^2 - z^2$$

The characteristic cone is the dual of the cone defined in the cotangent plane by annulation of the characteristic polynomial.For our system the characteristic cone splits into the light cone of the given spacetime metric and the normal to its space slice. Since these charateristics appear as multiple and the system is non diagonal it is not hyperbolic in the usual sense. We will prove the following theorem

THEOREM. When $N > 0$ and β are given, arbitrary, the system $R_{ij} = 0$ is a system hyperbolic non strict in the sense of Leray Ohya for g_{ij}, in the Gevrey class $\gamma = 2$, as long as g_{ij} is properly riemannian. If the Cauchy data as well as N and β are in such a Gevrey class the Cauchy problem has a local in time solution, with domain of dependence determined by the light cone.

Proof.

The product $ab^2 \mathcal{M}^{-1}$, with $\mathcal{M}^{-1}$ the inverse of the characteristic matrix $\mathcal{M}$ is computed to be:

$$
\begin{matrix}
t^2 - x^2 - y^2 & -zx & -zy & 0 & 0 & xy \\
-zx & t^2 - y^2 - z^2 & -xy & zy & 0 & 0 \\
-zy & -xy & t^2 - x^2 - z^2 & 0 & zx & 0 \\
-2xy & 0 & -2zx & t^2 - x^2 & x^2 & x^2 \\
-2xy & -2zy & 0 & y^2 & t^2 - y^2 & y^2 \\
0 & -2zy & -2zx & z^2 & z^2 & t^2 - z^2
\end{matrix}
$$

We see that the elements of the matrix $ab^2 \mathcal{M}^{-1}$ are polynomials in x, y, z. The product of this matrix by $\mathcal{M}$ is a diagonal matrix with elements ab^2 in the diagonal. Consider now the differential operator R_{ij} acting on g_{ij}. Multiply it on the left by the differential operator defined by replacing in $ab^2 \mathcal{M}^{-1}$ the variables x, y, z by the derivatives $\partial_1, \partial_2, \partial_3$. The resulting operator is quasi diagonal with principal operator $\partial^\lambda \partial_\lambda \partial_0^2$. It is the product of two strictly hyperbolic operators, $\partial^\lambda \partial_\lambda \partial_0$ and ∂_0. The result follows from the Leray-Ohya general theory.

7   LERAY-OHYA HYPERBOLIC 4th ORDER SYSTEM.

LEMMA. The following combination of derivatives of components of the Ricci tensor of an arbitrary spacetime :

$$
\Lambda_{ij} \equiv \hat{\partial}_0 \hat{\partial}_0 R_{ij} - \hat{\partial}_0 \bar{\nabla}_{(i} R_{j)0} + \bar{\nabla}_j \partial_i R_{00}
$$

reads, when g is known, as a third order quasi diagonal hyperbolic system for the extrinsic curvature K_{ij}.

$$\Lambda_{ij} \equiv \hat{\partial}_0 \mathcal{D} K_{ij} + \hat{\partial}_0 \hat{\partial}_0 (H K_{ij} - 2 K_{im} K_j^m) - \hat{\partial}_0 \hat{\partial}_0 (N^{-1} \bar{\nabla}_j \partial_i N) +$$
$$\hat{\partial}_0 (-\bar{\nabla}_{(i}(K_{j)h} \partial^h N) - 2N \bar{R}_{::ijm}^h K_h^m - N \bar{R}_{m(i} K_{j)}^m + H \bar{\nabla}_j \partial_i N) +$$
$$\bar{\nabla}_j \partial_i (N \bar{\Delta} N - N^2 K.K) + \mathcal{C}_{ij}$$

with

$$\mathcal{D} K_{ij} \equiv -\hat{\partial}_0 (N^{-1} \hat{\partial}_0 K_{ij}) + \bar{\nabla}^h \bar{\nabla}_h (N K_{ij}), \quad \bar{\Delta} = \bar{\nabla}_h \bar{\nabla}^h, \quad H \equiv K_h^h$$

and

$$\mathcal{C}_{ij} \equiv \bar{\nabla}_j \partial_i (N \partial_0 H) - \hat{\partial}_0 (N \bar{\nabla}_j \partial_i H)$$

Proof. A straightforward computation shows that $\mathcal{C}_{ij}$ contains terms of at most second order in K (and also in N) and first order in $\bar{g}$ (replace $\hat{\partial}_0 g_{ij}$ by $-2N K_{ij}$). The other terms of Λ_{ij}, except for $\hat{\partial}_0 \mathcal{D} K_{ij}$ are second order in K. All terms of Λ_{ij} are at most second order in $\bar{g}$ except for third order terms appearing through $\bar{\nabla}_j \partial_i (N \bar{\Delta} N)$. Because of these terms the system for $\bar{g}$ and K given by

$$\Lambda_{ij} = 0 \tag{1}$$

and

$$\hat{\partial}_0 g_{ij} = -2N K_{ij} \tag{2}$$

is not quasi diagonal. It is not hyperbolic in the usual sense of Leray, we will prove the following theorem.

THEOREM. The system (1), (2) with unknown $\bar{g}$, K is for any choice of lapse N and shift β equivalent to a system hyperbolic non strict in the sense of Leray-Ohya with local existence of solutions in Gevrey classes $\gamma = 2$ and domain of dependence determined by the light cone.

Proof. Replace in the equations $\Lambda_{ij} = 0$ the tensor K by $-(2N)^{-1} \hat{\partial}_0 \bar{g}$: this gives a quasi diagonal system for $\bar{g}$, but with principal operator $(\partial_0)^2 \partial^\lambda \partial_\lambda$. The result follows immediately from the Leray-Ohya theory.

The system for $\bar{g}$, K can be turned into a *hyperbolic system* by a *gauge choice* as follows.

THEOREM. Suppose that N satisfies the wave equation.

$$N^{-2}\partial_0\partial_0 N - \bar{\Delta}N = f \qquad (3)$$

with f an arbitrarily given smooth function on $M \times R$. The system (1),(2),(3), called S, is equivalent to a hyperbolic Leray system for $\bar{g}, K, N$, for arbitary shift.

Proof. We use the wave equation (3) to reduce the terms $-\hat{\partial}^2_{00}(N^{-1}\bar{\nabla}_j\partial_i N)+ \bar{\nabla}_j\partial_i(N\bar{\Delta}N)$ in Λ_{ij} to terms of third order in N, second order in g and K.

We replace the equation (3) by the also hyperbolic equation

$$\hat{\partial}_0(N^{-2}\partial_0\partial_0 N - \bar{\Delta}N) = \partial_0 f \qquad (4)$$

and replace $\hat{\partial}_0\bar{g}$ by $-2NK$ wherever it appears. Then the equation (4) is third order in N, while first order in $\bar{g}$ and K. We call S' the system thus modified.

We give to the equations and unknowns the Leray-Volevic indices:

$$m(1) = 0, \quad m(2) = 2, \quad m(3) = 1 \qquad (5)$$

$$n(g) = n(K) = 3, \quad n(N) = 4 \qquad (6)$$

The principal matrix of the system S' is then diagonal with elements the hyperbolic operators $\partial_0\partial^\alpha\partial_\alpha$ or ∂_0.

If the equation (3) is satisfied on the initial submanifold as well as S', the equation (3) and the system S are satisfied.

REMARK. The system $\Lambda_{ij} = 0$ has the additionnal property to satisfy a polarized null condition, that is the quadratic form defined by the second derivative of A at some given metric g, $\Lambda"_{ij}(g)(\gamma,\gamma)$, vanishes when $\gamma = \ell \otimes \ell$ with ℓ a null vector for the spacetime metric g such that γ is in the kernel of the first derivative of the Ricci tensor of spacetime at g (C-B 2000).

8   A STRICTLY HYPERBOLIC SYSTEM.

A variety of hyperbolic evolution systems for Einstein equations have been obtained, with a speed greatly increasing in recent years, by replacing the trivial gauge choice (which is the data of N and β on V) by more elaborate ones, together with combining the evolution equations with the constraints.

The hope in changing the gauge is to find systems either better suited to the study of global existence problems, or more stable under numerical codes. We give some references in the bibliography. We will return below to the original gauge choice (C-B 1952) in the perspective of conditions on the lapse and the shift.

The following identity was already known by De Donder, Lanczos and Darmois. It splits the Ricci tensor with components $R_{\alpha\beta}$ in the natural frame into a quasi linear quasi diagonal wave operator and 'gauge' terms, as follows:

$$R_{\alpha\beta} \equiv R_{\alpha\beta}^{(h)} + \frac{1}{2}(g_{\beta\lambda}\frac{\partial}{\partial x^{\alpha}}F^{\lambda} + g_{\alpha\lambda}\frac{\partial}{\partial x^{\beta}}F^{\lambda})$$

with

$$R_{\alpha\beta}^{(h)} \equiv -\frac{1}{2}g^{\lambda\mu}\frac{\partial^2}{\partial x^{\lambda}\partial x^{\mu}}g_{\alpha\beta} + H_{\alpha\beta} = 0$$

where $H_{\alpha\beta}$ is a quadratic form in first derivatives of g with coefficients polynomials in g and its contravariant associate.

The F^{λ} are given by

$$F^{\lambda} \equiv g^{\alpha\beta}\Gamma_{\alpha\beta}^{\lambda} \equiv \nabla^{\alpha}\nabla_{\alpha}x^{(\lambda)}$$

They are non tensorial quantities, result of the action of the wave operator of g on the coordinate functions. For this reason their vanishing is called 'harmonicity condition'.

The contravariant components of the Ricci tensor admit an analogous splitting, namely:

$$R^{\alpha\beta} \equiv R_{(h)}^{\alpha\beta} + \frac{1}{2}(g^{\alpha\lambda}\frac{\partial}{\partial x^{\lambda}}F^{\beta} + g^{\beta\lambda}\frac{\partial}{\partial x^{\lambda}}F^{\alpha})$$

with

$$R^{\alpha\beta} = \frac{1}{2}g^{\lambda\mu}\frac{\partial^2}{\partial x^{\lambda}\partial x^{\mu}}g^{\alpha\beta} + K^{\alpha\beta}$$

Let us denote by $R_{\alpha\beta}^{(\theta)}$ the components of the Ricci tensor in the previous frame $\theta^0 = dt$, $\theta^i = dx^i + \beta^i dt$, to distinguish them from the components in the natural frame now denoted $R_{\alpha\beta}$. These components are linked by the relations:

$$R_{ij}^{(\theta)} = R_{\alpha\beta}\frac{\partial(dx^\alpha)}{\partial\theta^i}\frac{\partial(dx^\beta)}{\partial\theta^j} = R_{ij},$$

$$R^{00} = R_{(\theta)}^{\alpha\beta}\frac{\partial(dt)}{\partial\theta^\alpha}\frac{\partial(dt)}{\partial\theta^\beta} = R_{(\theta)}^{00}$$

$$R_{(\theta)}^{0i} = R^{\alpha\beta}\frac{\partial(\theta^0)}{\partial(dx^\alpha)}\frac{\partial(\theta^i)}{\partial(dx^\beta)} = R^{00} + R^{0j}$$

The equations $R_{ij}^{(h)} = 0$ are a quasidiagonal second order system for g_{ij} when N and β are known. The equations $R_{(h)}^{00} = 0$ and $R_{(h)}^{0i}$ are quasilinear wave equations for N and β when the $g_{ij}'s$ are known: we interpret these equations as gauge conditions. The set of all these equations constitute a quasidiagonal second order system for g_{ij}, N and β, hyperbolic and causal as long as $N > 0$ and $\bar{g}$ is properly riemannian.

The Bianchi identities show that for a solution of these equations the quantities F^λ satisfy a linear homogeneous quasidiagonal hyperbolic and causal system. Its initial data can be made zero by choice of initial coordinates if and only if the geometric initial data $\bar{g}, K$ satisfy the constraints. A solution of our hyperbolic system satisfies then the full Einstein equations.

The following local existence and uniqueness theorem improves the differentiability obtained in the original theorem of C-B 1952 who used C^k spaces and a constructive (parametrix) method. The improvement to Sobolev spaces (one can endow M with a given smooth riemannian metric to define those spaces) with $s \geq 4$ for existence and $s \geq 5$ for geometric uniqueness is given in C-B 1968 using Leray's results. The improvement given in the theorem was sugested by Hawking and Ellis 1973, proved by semigroup methods by Hughes, Kato and Marsden 1978 and by energy methods by C-B, Christodoulou and Francaviglia 1979. The other hyperbolic systems constructed in the past twenty years did not lead, up to now, to further improvement on the regularity required of the Cauchy data. Such an improvement would be an important step.

THEOREM. Given an initial data set, $\bar{g}_0, K_0 \in H_s^{local}, H_{s-1}^{local}$ satisfying the constraints, there exists an einsteinian extension if $s \geq 3$.

The question of uniqueness is a geometrical problem. It is in general easy to prove that the solution is unique in the chosen gauge, for instance in the harmonic gauge recalled above. But two isometric spacetimes must be considered as identical. The following theorem (C-B and Geroch) gives this geometric uniqueness (maximal means inextendible).

THEOREM. The development of an initial data set is unique up to isometries in the class of maximal developments if $s \geq 4$. The domain of dependence is determined by the light cone of the spacetime metric.

The proof of C-B and Geroch considers smooth data and developments, the refined result is due to Chrusciel 1996. The geometric uniqueness in the case $s = 3$, even the local one, is still an open problem.

9   FIRST ORDER SYMMETRIC HYPER-BOLIC SYSTEM.

The Riemann tensor satisfies the identities

$$\nabla_\alpha R_{\beta\gamma,\lambda\mu} + \nabla_\beta R_{\gamma\alpha,\lambda\mu} + \nabla_\gamma R_{\alpha\beta,\lambda\mu} \equiv 0 \tag{7}$$

it holds therefore that, modulo the symmetries of the Riemann tensor

$$\nabla_\alpha R^\alpha_{..\mu,\beta\gamma} \equiv \nabla_\beta R_{\gamma\mu} - \nabla_\gamma R_{\beta\mu} \tag{8}$$

hence if the Ricci tensor $R_{\alpha\beta}$ satisfies the vacuum Einstein equations

$$R_{\alpha\beta} = 0 \tag{9}$$

it holds that

$$\nabla_\alpha R^\alpha_{..\mu,\beta\gamma} = 0. \tag{10}$$

The system (8), (11) splits as the Eintein equations into constraints, containing no time derivatives of curvature, namely in the frame used in the 3+1 splitting:

$$\nabla_i R_{jk,\lambda\mu} + \nabla_k R_{ij,\lambda\mu} + \nabla_j R_{ki,\lambda\mu} \equiv 0 \tag{11}$$

$$\nabla_\alpha R^\alpha_{..0,\beta\gamma} = 0. \tag{12}$$

and an evolution system

$$\nabla_0 R_{hk,\lambda\mu} + \nabla_k R_{0h,\lambda\mu} - \nabla_h R_{0k,\lambda\mu} = 0 \tag{13}$$

$$\nabla_0 R^0_{::::i,\lambda\mu} + \nabla_h R^h_{::::i,\lambda\mu} = 0 \tag{14}$$

This system has a principal matrix consisting of 6 identical 6 by 6 blocks around the diagonal, obtained by fixing a pair $\lambda, \mu, \lambda < \mu$. Each block is symmetrizable through the metric $\bar{g}$, and hyperbolic if $\bar{g}$ is properly Riemannian and $N > 0$ because the principal matrix M^0 for the derivatives ∂_0 was, up to product by N^{-1}, the unit matrix and the derivatives ∂_h do not contain $\partial/\partial t$.

The Bel-Robinson energy is the energy associated to this symmetric hyperbolic system.

REMARK. Following Bel one can introduce two pairs of gravitational "electric" and "magnetic" space tensors associated with the 3+1 splitting of the spacetime and the double two-form Riemann(g) :

$$N^2 E_{ij} \equiv R_{0i,0j}, \quad D_{ij} \equiv \frac{1}{4}\eta_{ihk}\eta_{jlm}R^{hk,lm}$$

$$N H_{ij} \equiv \frac{1}{2}\eta_{ihk}R^{hk}_{::::,0j}, \quad N B_{ji} \equiv \frac{1}{2}\eta_{ihk}A^{::::hk}_{0j,}$$

where η_{ijk} is the volume form of $\bar{g}$. The principal part of the evolution system ressemble then to the Maxwell equations, but contains an additional non principal part. Its explicit expression is given in Anderson,C-B and York 1997.

The Bianchi equations do not tell the whole story since they contain the spacetime metric g, which itself depends on the Riemann tensor.

A possibility to obtain a symmetric evolution system (Friedrich 1996 with the Weyl tensor) for both g and $Riemann(g)$ (Anderson, C-B and York 1997) is to introduce again the auxiliary unknown K and use $3 + 1$ identities, involving now not only the Ricci tensor but also the Riemann tensor. One can then obtain a symmetric first order hyperbolic system for K and $\bar{\Gamma}$, the space metric connection, modulo a choice of gauge, namely the integrated form of the harmonic time-slicing condition used before (C-B and Ruggeri 1983). The energy associated to this system has unfortunately no clear geometrical meaning. Determination of the metric from the Riemann tensor through elliptic equations seems more promising for the solution of global problems

(see Christodoulou and Klainerman 1989, Andersson and Moncrief, in preparation)

References.

Abrahams A., Anderson A., Choquet-Bruhat, Y., York, J.W. (1996) A non strictly hyperbolic system for Einstein equations with arbitrary lapse and shift. C.R. Acad. Sci. Paris Série 2b, 835-841.

Anderson A., Choquet-Bruhat, Y., York, J.W.(1997) Einstein Bianchi hyperbolic system for general relativity. Topol. Meth. Nonlinear Anal. 10, 353-373.

Anderson A., Choquet-Bruhat, Y., York, J.W. (2000) Einstein's equations and equivalent dynamical systems. In Cotsakis, S. (ed.) Mathematical and quantum aspects of relativity and cosmology. Cotsakis S., Gibbons G., (ed) Lecture Notes in Physics 537, Springer.

Andersson, L. , Moncrief, V. , private communication.

Choquet (Fourès)-Bruhat, Y. 1952 Théorème d'existence pour certains systèmes d'équations aux dérivées patielles non linéaires. Acta Matematica 88, 141-225.

Choquet (Fourès)-Bruhat, Y. 1956 Sur l'intégration des équations de la relativité générale.J. Rat. Mech. and Anal. 55, 951-966.

Choquet-Bruhat Y. 1968 Espaces temps einsteiniens généraux, chocs gravitationnels Ann. Inst. Poincaré 8, n°4 327-338.

Choquet-Bruhat Y. 2000 The null condition and asymptotic expansions for the Einstein equations. Ann. der Physik 9, 258-267.

Choquet-Bruhat Y., Christodoulou D., Francaviglia M., 1978 Cauchy data on a manifold. Ann. Inst. Poincaré A 23 241-250

Choquet-Bruhat Y., Geroch R. 1969 Global aspects of the Cauchy problem in general relativity. Comm. Math. Phys. 14 329-335.

Choquet-Bruhat, Y., Noutchegueme, N. 1986 Système hyperbolique pour les équations d'Einstein avec sources. C. R. Acad. Sci. Paris série 303, 259-263.

Choquet-Bruhat, Y., Ruggeri T. 1983 Hyperbolicity of the 3+1 system of Einstein equations. Comm. Math. Phys. 89, 269-275.

Choquet-Bruhat, Y., York, J.W. 1995 Geometrical well posed systems for the Einstein equations. C.R. Acad. Sci. Paris série 1 321, 1089-1095.

Christodoulou D., Klainerman S. 1989 The non linear stability of Minkowski space. Princeton University Press.

Chrusciel P. T. 1991 On the uniqueness in the large of solutions of Eintein's equations. Proc. Cent. Math. Anal. (ANU, Canberra) 20

Friedrich, H. 1996 Hyperbolic reductions for Einstein'equations. Class. Quant. Grav. 13, 1451-1459.

Hawking S., Ellis G. 1973 The global structure of spacetime Cambridge Univ. Press.

Leray J. 1953 Hyperbolic differential equations Lecture Notes, Princeton.

Leray J., Ohya Y. (1967) Equations et systèmes non linéaires hyperboliques non stricts. Math. Ann. 170 167-205.

Reula O. 1998 Hyperbolic methods for Einstein's equations. www.livingreviews.org/Articles/volume1/1998-3reula

On the Cauchy–Kowalevskaya theorem of Nagumo type for systems

WAICHIRO MATSUMOTO, MINORU MURAI AND TAKAAKI NAGASE

DEPARTMENT OF APPLIED MATHEMATICS AND INFORMATICS
FACULTY OF SCIENCE AND TECHNOLOGY
RYUKOKU UNIVERSITY
SETA
520-2194 OTSU
JAPAN
E-mail address: waichiro@rins.ryukoku.ac.jp

Dedicated to Professor Jean Vaillant

1. INTRODUCTION

W. Matsumoto and H. Yamahara [4] and [5] gave the necessary and sufficient condition for the Cauchy-Kowalevskaya theorem on systems of linear partial differential equations applying the theory of the normal form of systems obtained by W. Matsumoto[1, 3, 2]. On the other hand, M. Nagumo [6] showed that, for example in case of a higher order scalar equation, we can obtain the unique classical local solution for the analytic data in x if the coefficients are analytic in x and continuous in t and if the operator is Kowalevskian.
(We call the theorem of this type "the Cauchy-Kowalevskaya theorem of Nagumo type" = "the C-K theorem of N-type".) W. Matsumoto [2] gave an example which shows that we need some higher differentiability in t of the coefficients proportional to the order on $\frac{\partial}{\partial x}$ minus one and the size of system in order to obtain the C-K theorem of N-type for systems of first order on $\frac{\partial}{\partial t}$. Further, he gave a conjecture that W. Matsumoto and H. Yamahara's condition in [4, 5, 2] rests necessary and sufficient for the C-K theorem of N-type on systems when the coefficients are analytic in x and of C^∞ class in t. In this paper, we show that this conjecture is correct when the order on $\frac{\partial}{\partial t}$ is one, the order on $\frac{\partial}{\partial x}$ is two and the rank of the highest order part is at most one.

2. CONJECTURE

Let us consider the following Cauchy problem:

(2.1)
$$\begin{cases} Pu \equiv D_t u - A_d(t, x, D_x)u \equiv D_t u - \sum_{|\alpha| \le m} A_\alpha(t, x)D_x{}^\alpha u = f(t, x) \quad , \\ u|_{t=t_\circ} = u_\circ(x) \quad , \end{cases}$$

where A_α is a $N \times N$ matrix of the functions in $C^\infty(]T_1, T_2[; A(\Omega))$ ($|\alpha| \le m$, Ω is a domain in $\mathbf{R}^\ell$, $A(\Omega)$ is the set of the real analytic functions in Ω), u, $u_\circ$ and $f(t, x)$ are N-dimensional vectors, $D_t = \frac{1}{\sqrt{-1}}\frac{\partial}{\partial t}$ and $D_x = \frac{1}{\sqrt{-1}}\frac{\partial}{\partial x}$. If we can solve (2.1) with $u_\circ$ in $A(\Omega)$ and $f = 0$, we can also solve the equations with non-null right-hand sides analytic in x and of C^∞ class in t by Duhamel's principle. Then, from now on, we consider only

(2.2)
$$\begin{cases} Pu \equiv D_t u - A_d(t, x, D_x)u \equiv D_t u - \sum_{|\alpha| \le m} A_\alpha(t, x)D_x{}^\alpha u = 0 \quad , \\ u|_{t=t_\circ} = u_\circ(x) \quad , \end{cases}$$

We use the calculus of the formal symbols, the normal form and p-determinant of systems given in [1, 3, 2].

By the relation $D_t u = A_d(t, x, D_x)u$, $D_t^i u$ is represented by a linear combination of the derivatives on x of u. Let us set

(2.3) $$D_t^i u = A[i](t, x, D_x)u \quad, \quad (i \geq 0).$$

$\{A[i]\}_{i=0}^{\infty}$ satisfies the following recurrence formula:

(2.4)
$$\begin{cases} A[0] = I_N \quad, \\ \\ A[i] = A[i-1] \circ A_d + (A[i-1])_t \quad, \quad (i \geq 1) \quad, \end{cases}$$

where $A \circ B$ is the operator product of A and B and $(A)_t$ is the matrix obtained from A operating D_t on its coefficients.

Conjecture.   *The following conditions are equivalent.*

1) The Cauchy-Kowalevskaya theorem of Nagumo type for $P(t, x, D_t, D_x)$ holds in Ω.

2) The lower order terms in the normal form of P in Corollary 2 in [2] satisfy

(2.5) $$\operatorname{ord} b_k(h) \leq 1 - m(n_k - h), \quad (1 \leq h \leq n_k, 1 \leq k \leq d).$$

3) There exists an open conic dense set O in $T^*(]T_1, T_2[\times\Omega)$ and $P(t, x, D_t, D_x)$ is reduced to a first order system through a similar transformation in the formal symbol class on O.

4) 1-det P is of degree N : the size of P

5) P is Kowalevskian in our sense, that is, p-evolutive for $0 \leq p \leq 1$.

6) There exists a natural number $i_\circ$ such that

(2.6) $$\operatorname{ord} A[i](t, x, D_x) \leq i + i_\circ, \quad (i \in \mathbf{Z}_+).$$

The equivalence between 2), 3), 4), 5) and 6) and " 1) $\Rightarrow$ 2)" are shown by the same way as W. Matsumoto [2] and W. Matsumoto and H. Yamahara [5]. In order that Condition 6) holds, the highest order part of $A_d(t, x, D_x)$ must be nilpotent if $m > 1$. Let us show "6) $\Rightarrow$ 1)," under the following assumption.

Assumption 1. $m = 2$ and $\operatorname{rank} \sum_{|\alpha|=2} A_\alpha(t,x)\xi^\alpha \leq 1$.

3. Expression of Condition 6) by the symbol of P_d

We denote the symbol of the homogeneous part of second order of $A_d(t,x,D_x)$ by $A(t,x,\xi)$, that of first order by $B(t,x,\xi)$ and that of order zero by $C(t,x)$. Assumption 1 implies the following lemma.

Lemma 3.1. *Under Assumption 1, there exist an open dense subset O of $]T_1, T_2[\times \Omega \times \mathbf{R}^\ell$, vector valued functions $\mathbf{a}(t,x,\xi)$ and $\mathbf{r}(t,x,\xi)$ real analytic in x and ξ and of C^∞ class in t in O and homogeneous on ξ such that*

(3.1) $$A(t,x,\xi) = \mathbf{a}(t,x,\xi)^t \mathbf{r}(t,x,\xi) \qquad in \ O .$$

Proof. Let us set $O_1 = \{(t,x,\xi) \in]T_1, T_2[\times \Omega \times \mathbf{R}^\ell \,|\, the \ first \ row \ vector \ {}^t(a_{11}(t,x,\xi),$ $\cdots, a_{N1}(t,x,\xi))$ of $A(t,x,\xi) = (a_{ij}(t,x,\xi))_{1\leq i,j\leq N}$ does not vanish $\}$. We denote the first row vector of A in O_1 by $\mathbf{a}(t,x,\xi)$ and set $\mathbf{r}(t,x,\xi) = {}^t(1, a_{i2}(t,x,\xi)/a_{i1}(t,x,\xi), \cdots,$ $a_{iN}(t,x,\xi)/a_{i1}(t,x,\xi))$ in O_1 using some $a_{i1}(t,x,\xi) \neq 0$. The definition of $\mathbf{r}(t,x,\xi)$ does not depend on the choice of i and holds $A(t,x,\xi) = \mathbf{a}(t,x,\xi)^t \mathbf{r}(t,x,\xi)$ in O_1.

Next, we set $O_2 = \{(t,x,\xi) \in]T_1, T_2[\times \Omega \times \mathbf{R}^\ell \backslash \bar{O}_1 \,|\, the \ second \ row \ vector \ of \ A(t,x,\xi)$ does not vanish $\}$, denote the second row vector of $A(t,x,\xi)$ in O_2 by $\mathbf{a}(t,x,\xi)$ and set $\mathbf{r}(t,x,\xi) = {}^t(0, 1, a_{i3}(t,x,\xi)/a_{i2}(t,x,\xi), \cdots, a_{iN}(t,x,\xi)/a_{i2}(t,x,\xi))$ in O_2 using some $a_{i2}(t,x,\xi) \neq 0$. The relation $A(t,x,\xi) = \mathbf{a}(t,x,\xi)^t \mathbf{r}(t,x,\xi)$ in O_2 holds.

Repeating this procedure, we can obtain O_k ($1 \leq k \leq N$), $\mathbf{a}$ and $\mathbf{r}$ in $\cup_{1\leq k\leq N} O_k$.

Further, we set $O_0 =]T_1, T_2[\times \Omega \times \mathbf{R}^\ell \backslash \cup_{1\leq k\leq N} \bar{O}_k$, where $A = 0$. We take $\mathbf{a} = \mathbf{r} = 0$ there. Finally, we set $O = \cup_{0\leq k\leq N} O_k$.

$\qquad\qquad\qquad\qquad\qquad\qquad\qquad\qquad\qquad\qquad\qquad\qquad\qquad\qquad\qquad\qquad\qquad$ □

Remark 1. In our consideration in this section, every property is trivial on O_0, where A vanishes. Further, as each property in Proposition 3.4 below is continuous on (t,x,ξ), we need discuss all in $\cup_{1\leq k\leq N} O_k$, which is dense in $\{(t,x,\xi)\,|\,A(t,x,\xi) \neq 0\}$

We set

$$\hat{B} = B - \sum_{i=1}^\ell \mathbf{a}^{(i)\,t} \mathbf{r}_{(i)}$$

Further, we define the operators R and L by

$$F(t,x,\xi)R = \sum_{i=1}^\ell F^{(i)} A_{(i)} + FB , \qquad LF(t,x,\xi) = \sum_{i=1}^\ell A^{(i)} F_{(i)} + BF .$$

where $F^{(i)} = \frac{\partial}{\partial \xi_i}F(t,x,\xi)$ and $F_{(i)} = D_{x_i}F(t,x,\xi)$.

Lemma 3.2. *If* $(AR^h)A = 0$ *for* $0 \le h \le i-1$,

(3.2) $$(AR^i)A = A(L^i A).$$

Further,

(3.3) $$(AR^h)A = 0 \text{ for } 0 \le h \le i \iff A(L^h A) = 0 \text{ for } 0 \le h \le i.$$

Proof.   We show Lemma 3.2 by the induction on i.
The case of $i = 0$.   The both sides are same.
The general case.   We assume that we have seen the equivalence of $(AR^h)A = 0$ *for* $0 \le h \le i-1$ and $A(L^h A) = 0$ *for* $0 \le h \le i-1$. We have the following calculation:

$$A(L^i A) = A\left\{ \sum_{p=1}^{\ell} A^{(p)} (L^{(i-1)} A)_{(p)} + B(L^{(i-1)} A) \right\}$$

$$= A^{(p)} \left\{ \sum_{p=1}^{\ell} A_{(p)} (L^{(i-1)} A) + B(L^{(i-1)} A) \right\}$$

$$= (AR)(L^{i-1} A)$$

$$\vdots$$

$$= (AR^i)A.$$

Thus, we see that $(AR^i)A = 0$ is equivalent to $A(L^i A) = 0$.   □

Lemma 3.3.
(1) If $(AR^h)A = 0$ *holds for* $0 \le h \le i-1$, *there exist matrices* $F_h^{<j>}(t,x,\xi)$ *and* $\hat{F}_h^{<j>}(t,x,\xi)$ *(* $0 \le j \le h$, $0 \le h \le i$ *) real analytic in* x *and* ξ *and of* C^∞ *in* t *and homogeneous of degree* j *in* ξ *in* O *such that*

(3.4)
$$\begin{cases} AR^h = \sum_{j=0}^{h} F_h^{<h-j>} A\hat{B}^j, \\[2mm] F_h^{<0>} = I \quad \text{in } O \qquad\qquad (0 \le h \le i), \end{cases}$$

(3.5)
$$\begin{cases} A\hat{B}^h = \sum_{j=0}^{h} \hat{F}_h^{<h-j>} AR^j, \\[2mm] \hat{F}_h^{<0>} = I \quad \text{in } O \qquad\qquad (0 \le h \le i). \end{cases}$$

Further, $(AR^h)A = 0$ *(* $0 \le h \le i$ *) is equivalent to* ${}^t r \hat{B}^h \mathbf{a} = 0$ *in* O *(* $0 \le h \le i$ *).*

(2) If $A(L^h A) = 0$ *holds for* $0 \le h \le i-1$, *there exist matrices* $G_h^{<j>}(t,x,\xi)$ *and* $\hat{G}_h^{<j>}(t,x,\xi)$ *(* $0 \le j \le h$, $0 \le h \le i$ *) real analytic in* x *and* ξ *and of* C^∞ *in* t *and homogeneous of degree* j *in* ξ *in* O *such that*

$$(3.6) \quad \begin{cases} L^h A = \sum_{j=0}^{h} \hat{B}^j A G_h^{<h-j>}, \\[2mm] G_h^{<0>} = I \quad in \ O \qquad (0 \le h \le i), \end{cases}$$

$$(3.7) \quad \begin{cases} \hat{B}^h A = \sum_{j=0}^{h} L^j A \hat{G}_h^{<h-j>}, \\[2mm] \hat{G}_h^{<0>} = I \quad in \ O \qquad (0 \le h \le i). \end{cases}$$

Further, $A(L^h A) = 0$ ($0 \le h \le i$) is equivalent to ${}^t r \hat{B}^h \mathbf{a} = 0$ in O ($0 \le h \le i$).

Proof. As (2) is provable by the same way as the proof of (1), we only show (1). We show (1) by the induction on i.

The case of $i = 0$. Trivial,

The general case. We assume that Lemma 3.3 (1) holds for $i-1$. Then, as $(AR^h)A = 0$ ($0 \le h \le i-1$) holds and then ${}^t r \hat{B}^h \mathbf{a} = 0$ in O ($0 \le h \le i-1$) holds, we see

$$AR^i = \sum_{p=1}^{\ell} (\sum_{j=0}^{i-1} F_{i-1}^{<i-1-j>} A\hat{B}^j)^{(p)} A_{(p)} + \sum_{j=0}^{i-1} F_{i-1}^{<i-1-j>} A\hat{B}^j B$$

$$= \sum_{j=0}^{i-1} \sum_{p=1}^{\ell} \{ (F_{i-1}^{<i-1-j>} \mathbf{a})^{(p)\,t} r \hat{B}^j \mathbf{a}_{(p)}{}^t r + F_{i-1}^{<i-1-j>} \mathbf{a} ({}^t r \hat{B}^j)^{(p)} \mathbf{a}_{(p)}{}^t r$$

$$+ F_{i-1}^{<i-1-j>} \mathbf{a} ({}^t r \hat{B}^j)^{(p)} \mathbf{a}^t r_{(p)} \} + \sum_{j=0}^{i-1} F_{i-1}^{<i-1-j>} A\hat{B}^j B$$

$$= \sum_{j=0}^{i-1} \sum_{p=1}^{\ell} \{ ({}^t r \hat{B}^j)^{(p)} \mathbf{a}_{(p)} F_{i-1}^{<i-1-j>} - (F_{i-1}^{<i-1-j>} \mathbf{a})^{(p)} ({}^t r \hat{B}^j)_{(p)} \} A$$

$$+ \sum_{j=0}^{i-1} F_{i-1}^{<i-1-j>} A\hat{B}^{j+1}$$

Setting $F_i^{<j>} = F_{i-1}^{<j>}$ ($0 \le j \le i-1$) and $F_i^{<i>} = \sum_{j=0}^{i-1} \sum_{p=1}^{\ell} \{ ({}^t r \hat{B}^j)^{(p)} \mathbf{a}_{(p)} F_{i-1}^{<i-1-j>} - (F_{i-1}^{<i-1-j>} \mathbf{a})^{(p)} ({}^t r \hat{B}^j)_{(p)} \}$, (3.4) holds for i.

As $F_h^{<0>} = I$, we obtain (3.5) immediately from (3.4).

As $(AR^h)A = 0$ ($0 \le h \le i-1$) and ${}^t r \hat{B}^h \mathbf{a} = 0$ ($0 \le h \le i-1$) are equivalent, $(AR^h)A = 0$ ($0 \le h \le i$) also equivalent to ${}^t r \hat{B}^h \mathbf{a} = 0$ in O ($0 \le h \le i$) by virtue of the representation (3.4).

$\square$

By lemmas 3.2 and 3.3, we can represent Condition 6) by the relations between the coefficients of A_d

Proposition 3.4.   *The following equivalences hold.*

$$6) \iff 7) \ (AR^h)A = 0 \ (h \geq 0) \iff 7') \ (AR^h)A = 0 \ (0 \leq h \leq N-1)$$

$$\iff 8) \ A(L^hA) = 0 \ (h \geq 0) \iff 8') \ A(L^hA) = 0 \ (0 \leq h \leq N-1)$$

$$\iff 9) \ A\hat{B}^hA = 0 \ (h \geq 0) \iff 9') \ A\hat{B}^hA = 0 \ (0 \leq h \leq N-1) \ .$$

Proof   Lemmas 3.2 and 3.3 give the equivalences between 7), 8) and 9) and between 7'), 8') and 9'). Thus, we need show the equivalences between 6) and 7) and between 9) and 9'). It is trivial that 9) implies 9').

Proof of 9') $\Rightarrow$ 9).   By Cayley-Hamilton theorem, $\hat{B}^N$ is a linear combination of $\{\hat{B}^h\}_{h=0}^{N-1}$ and then $\hat{B}^i$ $(i \geq N)$ is also a linear combination of $\{\hat{B}^h\}_{h=0}^{N-1}$. Therefore, 9') implies 9).

Proof of 6) $\Rightarrow$ 7).   We show this by the induction on h.
The case of $h = 0$.   As we remarked at the last of Section 2, $A(t,x,\xi)$ must be nilpotent. Further, as rank $A \leq 1$, $A(t,x,\xi)^2$ vanishes.
The general case.   We assume that $(AR^h)A = 0$ hold for $0 \leq h \leq i-1$ and consider the case of $h = i$.

Lemma 3.5.   *If $(AR^h)A = 0$ holds for $0 \leq h \leq i-1$, there exist matrices $\{A_{i+1}^{<k>}\}_{k=0}^i$ real analytic in x and ξ and of C^∞ class in t and homogeneous of degree k on ξ such that*

$$(3.8) \quad \begin{cases} A[i+1] \equiv \sum_{k=0}^i A_{i+1}^{<i-k>}(AR^k) & \text{mod ord} \quad i+1, \\ \\ A_{i+1}^{<0>} = I. \end{cases}$$

Proof.   We show Lemma 3.5 by the induction on i.
The case of $i = 0$.   Under no condition, $A[1] = A + B + C$ and it has the form (3.8) with $i = 0$.
The general case.   We assume the following relation:

$$A[i] \equiv \sum_{k=0}^{i-1} A_i^{<i-1-k>}(AR^k) + B_i \quad \text{mod ord} \quad i-1, \quad A_i^{<0>} = I,$$

where B_i is homogeneous of degree i on ξ. By the formula (2.4), we have

$$A[i+1] = A[i] \circ (A + B + C) + (A[i])_t$$

$$\equiv \sum_{k=0}^{i-1} \{ \sum_{q=1}^{\ell} (A_i^{<i-1-k>}(AR^k))^{(q)} A_{(q)} + A_i^{<i-1-k>}(AR^k)B \} + B_i A$$

$$\equiv \sum_{k=0}^{i-1} A_i^{<i-1-k>}(AR^{k+1}) - \sum_{k=0}^{i-1}\sum_{q=1}^{\ell} A_i^{<i-1-j>(q)}(AR^k)_{(q)} A + B_i A$$

$$\text{mod ord } i+1 .$$

Setting $A_{i+1}^{<k>} = A_i^{<k>}$ for $0 \le k \le i-1$ and $A_{i+1}^{<i>} = B_i - \sum_{j=0}^{i-1}\sum_{q=1}^{\ell} A_i^{<i-1-k>(q)}(AR^k)_{(q)}$, we arrive at (3.8).

$\square$

We return to the proof of Proposition 3.4. As $(AR^h)A = 0$ ($0 \le h \le i-1$), by virtue of Lemmas 3.3 and 3.5, we have

$$A[p(i+1)] = D_t^{p(i+1)}$$

$$\equiv D_t^{(p-1)(i+1)} \circ (\sum_{k=0}^{i} A_{i+1}^{<i-k>}(AR^k))$$

$$\equiv D_t^{(p-1)(i+1)} \circ (\sum_{k=0}^{i} A_{i+1}^{<i-k>}(\sum_{j=0}^{k} F_k^{<k-j>} A\hat{B}^j))$$

$$\vdots$$

$$\equiv \{\sum_{j=0}^{i}(\sum_{k=j}^{i} A_{i+1}^{<i-k>} F_k^{<k-j>})A\hat{B}^j\}^p \qquad \text{mod ord } p(i+2) - 1.$$

By the relations $(AR^h)A = 0$ ($0 \le h \le i-1$) and Lemma 3.3 and 3.5, it follows that

$$A[p(i+1)+1] \equiv \{\sum_{j=0}^{i}(\sum_{k=j}^{i} A_{i+1}^{<i-k>} F_k^{<k-j>})A\hat{B}^j\}^p A \qquad \text{mod ord } p(i+2) + 1$$

$$= \{\sum_{j=0}^{i}(\sum_{k=j}^{i} A_{i+1}^{<i-k>} F_k^{<k-j>})A\hat{B}^j\}^{p-1} \mathbf{a}(^t\mathbf{r}\hat{B}^i\mathbf{a})^t\mathbf{r}$$

$$= (^t\mathbf{r}\hat{B}^i\mathbf{a})\{\sum_{j=0}^{i}(\sum_{k=j}^{i} A_{i+1}^{<i-k>} F_k^{<k-j>})A\hat{B}^j\}^{p-1} A$$

$$\vdots$$

$$= ((^t\mathbf{r}\hat{B}^i\mathbf{a}))^p A.$$

If $^t\mathbf{r}\hat{B}^i\mathbf{a} \neq 0$, this implies that the order of $A[p(i+1)+1]$ is $p(i+2)+2$ and it contradicts to 6). Then. 6) implies 7) for i.

Proof of 7) $\Rightarrow$ 6).   As (3.8) holds for all i under 7), the order of $A[i]$ is at most $i+1$ and 6) holds.

Q.E.D.

4. MAIN THEOREMS

When $A(t, x, \xi)$ is polynomial in ξ with real analytic coefficients in $]T_1, T_2[\times\Omega$, we can find an scalar-valued function $\alpha(t, x, \xi)$ and vector-valued functions $\mathbf{a}(t, x, \xi)$ and $\mathbf{r}(t, x, \xi)$ polynomial in ξ with real analytic coefficients in $]T_1, T_2[\times\Omega$ such that $A = \alpha \mathbf{a}^t\mathbf{r}$ and the elements of $\mathbf{a}$ and $\mathbf{r}$ are relatively prime, respectively. (α, $\mathbf{a}$ and $\mathbf{r}$ are essentially uniquely determined.) On the other hand, if A is not real analytic, the possibility of such decomposition in full $T^*(]T_1, T_2[\times\Omega)$ is not clear in spite of Lemma 3.1. Then, we introduce the following additional condition.

Assumption 2.   There exist two vectors $\mathbf{a}$ and $\mathbf{r}$, homogeneous polynomial in ξ with coefficients real analytic in x and of C^∞ class in t in $]T_1, T_2[\times\Omega$ such that

$$A(t, x, \xi) = \mathbf{a}(t, x, \xi)^t\mathbf{r}(t, x, \xi).$$

By what we have remarked at the last of Section 2 and Proposition 3.4, we only need prove that 7′) or 8′) implies 1). We have the following proposition.

Proposition 4.1.   *We assume Assumptions 1, 2 and Condition 6).*

(1)   There exists a differential operator with coefficients real analytic in x and of C^∞ class in t in $]T_1, T_2[\times\Omega$:

(4.1) $\displaystyle Q = D_t^N + \sum_{i=1}^{N}(\sum_{k=i}^{i+1} C_{ik}(t, x, D_x))D_t^{N-i}$,   ord $C_{ik} = k$ *and homogeneous* ,

such that

(4.2) $\displaystyle P \circ Q = D_t^{N+1} + \sum_{i=1}^{N+1} B_i(t, x, D_x)D_t^{N+1-i}$ ord $B_i = i$.

(2)   There exists a differential operator with coefficients real analytic in x and of C^∞ class in t in $]T_1, T_2[\times\Omega$:

(4.3) $\displaystyle \hat{Q} = D_t^N + \sum_{i=1}^{N}(\sum_{k=i}^{i+1} \hat{C}_{ik}(t, x, D_x))D_t^{N-i}$,   ord $\hat{C}_{ik} = k$ *and homogeneous* , *such that*

$$(4.4) \qquad \hat{Q} \circ P = D_t^{N+1} + \sum_{i=1}^{N+1} \hat{B}_i(t, x, D_x) D_t^{N+1-i} \qquad \text{ord } \hat{B}_i = i.$$

Proof As (2) is provable by the same way as the proof of (1), we only show (1). By the simple calculation, we have

$$P \circ Q \equiv D_t^{N+1} - \sum_{i=1}^{N} AC_{i\,i+1} D_t^{N-i}$$

$$- AD_t^N + \sum_{i=1}^{N} C_{i\,i+1} D_t^{N-i+1} - \sum_{i=1}^{N} \sum_{p=1}^{\ell} A^{(p)} C_{i\,i+1\,(p)} D_t^{N-i}$$

$$- \sum_{i=1}^{N} BC_{i\,i+1} D_t^{N-i} - \sum_{i=1}^{N} AC_{i\,i} D_t^{N-i} \qquad \text{mod ord } N+1.$$

In order that $P \circ Q$ becomes of order $N+1$, the following equalities must be hold.

$$(4.5) \qquad AC_{i\,i+1} = 0 \qquad (1 \leq i \leq N),$$
$$(4.6) \qquad C_{12} = A,$$
$$(4.7) \qquad C_{i+1\,i+2} = LC_{i\,i+1} + AC_{ii} \qquad (1 \leq i \leq N-1),$$
$$(4.8) \qquad LC_{N\,N+1} + AC_{NN} = 0.$$

Lemma 4.2. *If (4.6) and (4.7) hold, (4.5) is automatically satisfied.*

Proof. Because of Proposition 3.4 and the following calculation, (4.5) follows:

$$AC_{i+1\,i+2} = A(LC_{i\,i+1})$$
$$= A\{ L(LC_{i-1\,i} + AC_{i-1\,i-1}) \}$$
$$= A(L^2 C_{i-1\,i}) + A(LA)C_{i-1\,i-1} - \sum_{p=1}^{\ell} A^2 A^{(p)} C_{i-1\,i-1\,(p)}$$
$$= A(L^2 C_{i-1\,i})$$
$$\vdots$$
$$= A(L^i C_{12}) = A(L^i A) = 0 \qquad (0 \leq i \leq N-1).$$

$\square$

Let us represent $C_{i\,i+1}(t, x, \xi)$ ($2 \leq i \leq N$) by $\{C_{jj}(t, x, \xi)\}_{j=1}^{N}$.

Lemma 4.3. *We assume that $A(L^h A) = 0$ holds for ($0 \le h \le N - 1$). There exist matrices $\{H_i^{<k>}(t, x, \xi; D_x)\}_{0 \le k \le i, \, 1 \le i \le N}$ of differential operators of D_x with coefficients homogeneous polynomial in ξ, real analytic in x and of C^∞ class in t in $]T_1, T_2[\times \Omega \times \mathbb{C}^\ell$ such that*

(4.9)

$$
\begin{cases}
C_{i\,i+1} = \sum_{h=0}^{i-1} \sum_{j=0}^{i-h-1} (L^h A) H_{i-h-1}^{<i-h-j-1>} C_{jj} \\[2mm]
\text{ord}_{D_x}\, H_{i-1}^{<k>} = k\,, \qquad \deg_\xi\, H_{i-1}^{<k>} = k \\[2mm]
C_{00} = H_{i-1}^{<0>} = I \qquad (0 \le k \le i,\ 1 \le i \le N+1) \\[2mm]
H_i^{<i>} = 0 \qquad\qquad (1 \le i \le N)
\end{cases}
$$

where $C_{N+1\,N+2}$ is defined by $LC_{N\,N+1} + AC_{N\,N}$.

Proof.   We show this lemma by the induction on i.
The case of $i = 1$.   As $C_{12} = A$, taking $C_{00} = H_0^{<0>} = I$, (4.9) holds for $i = 1$.
The general case.   We assume that there exist $\{H_h^{<k>}(t, x, \xi; D_x)\}_{0 \le k \le h, \, 1 \le h \le i-1}$ and $C_{i\,i+1} = \sum_{h=0}^{i-1} \sum_{j=0}^{i-h-1} (L^h A) H_{i-h-1}^{<i-h-j-1>} C_{jj}$ holds. Let us consider $C_{i+1\,i+2}$ ($1 \le i \le N$). By (4.7), we have

$$
\begin{aligned}
C_{i+1\,i+2} &= LC_{i\,i+1} + AC_{ii} \\
&= \sum_{h=0}^{i-1} \sum_{j=0}^{i-h-1} \Big[\sum_{p=1}^{\ell} A^{(p)} \{(L^h A) H_{i-h-1}^{<i-h-j-1>} C_{jj}\}_{(p)} + B(L^h A) H_{i-h-1}^{<i-h-j-1>} C_{jj} \Big] \\
&\quad + AC_{ii} \\
&= \sum_{h=0}^{i-1} \sum_{j=0}^{i-h-1} (L^{h+1} A) H_{i-h-1}^{<i-h-j-1>} C_{jj} \\
&\quad - A \sum_{j=0}^{i-1} \sum_{h=0}^{i-j-1} \sum_{p=1}^{\ell} (L^h A)^{(p)} \{H_{i-h-1}^{<i-h-j-1>} D_{x_p} + H_{i-h-1}^{<i-h-j-1>}{}_{(p)}\} C_{jj} + AC_{ii}.
\end{aligned}
$$

We set $H_i^{<k>} = -\sum_{h=0}^{k-1} \sum_{p=1}^{\ell} (L^h A)^{(p)} \{H_{i-h-1}^{<k-h-1>} D_{x_p} + H_{i-h-1}^{<k-h-1>}{}_{(p)}\}$ ($1 \le k \le i$) and $H_i^{<0>} = I$. As $H_i^{<i>} = -\sum_{p=1}^{\ell} (L^{i-1} A)^{(p)} D_{x_p}$ and $C_{00} = I$, $H_i^{<i>} C_{00}$ vanishes and we can reset $H_i^{<i>} = 0$. Thus, we have obtained (4.9) replaced i by $i + 1$. $\square$

We return to the proof of Proposition 4.1. By virtue of Lemma 4.3, (4.8) means that $C_{N+1\,N+2} = 0$, that is,

(4.10)
$$\sum_{h=0}^{N}\sum_{j=0}^{N-h}(L^h A)H_{N-h}^{<N-h-j>}C_{jj}=0$$

Here, by Lemma 3.3 (3.6),

$$L^N A = \sum_{k=0}^{N}\hat{B}^k A G_N^{<N-k>}$$

and by Cayley-Hamilton theorem, there exist scalar functions $\alpha_k(t,x,\xi)$ $(0\le k\le N-1)$ homogeneous polynomial of degree k in ξ with coefficients real analytic in x and of C^∞ class in t in $]T_1, T_2[\times\Omega$ such that

$$\hat{B}^N = \sum_{k=0}^{N-1}\alpha_{N-k}\hat{B}^k .$$

Further, by Lemma 3.3 (3.7), we obtain

$$L^N A = \sum_{h=0}^{N-1}(L^h A)\Big\{\sum_{k=h}^{N-1}\hat{G}_k^{<k-h>}(G_N^{<N-k>}+\alpha_{N-k}I)\Big\} .$$

We set $K_{N-h}=\sum_{k=h}^{N-1}\hat{G}_k^{<k-h>}(G_N^{<N-k>}+\alpha_{N-k}I)$ $(\,0\le h\le N-1\,)$. Finally, we arrive at

$$\sum_{h=0}^{N-1}(L^h A)\Big\{\sum_{j=1}^{N-h}H_{N-h}^{<N-h-j>}C_{jj}+K_{N-h}\Big\}=0 .$$

We take

$$\sum_{j=1}^{N-h}H_{N-h}^{<N-h-j>}C_{jj}+K_{N-h}=0 \qquad (0\le h\le N-1).$$

that is, applying $H_k^{<0>}=I$,

$$C_{kk}=-\sum_{j=1}^{k-1}H_k^{<k-j>}C_{jj}-K_k \qquad (1\le k\le N) .$$

Taking k from 1 to N, we can determine $\{C_{jj}\}_{j=1}^N$.

Q.E.D.

Theorem 1. *Under Assumptions 1 and 2, Conditions 1) and the others in Conjecture are equivalent.*

Proof

Solvability   By virtue of the original C-K theorem of N-type (See M. Nagumo[6]) for

$$\begin{cases} (P \circ Q)v = 0, \\ D_t^N v|_{t=t_o} = u_o(x), \quad D_t^j v|_{t=t_o} = 0 \ (0 \le j \le N-1), \end{cases}$$

we have a local solution $v(t, x)$. $u = Qv$ gives the solution of (2.2).

Uniqueness   Also by virtue of the original C-K theorem of N-type for

$$\begin{cases} (\hat{Q} \circ P)v = 0, \\ D_t^j v|_{t=t_o} = 0 \quad (0 \le j \le N), \end{cases}$$

the solution $v(t, x)$ is unique. Thus, the solution $u(t, x)$ of (2.2) also must be unique.

Q.E.D.

In case of $\ell = 1$, we can obtain the above theorem without the decomposition $A = \mathbf{a}^t\mathbf{r}$ and need not Assumption 2.

Corollary 2.   *We assume $\ell = 1$ and Assumption 1. Conditions 1) and the others in Conjecture are equivalent. Further, they are equivalent to*

$$9'') \qquad AB^h A = 0 \qquad (0 \le h \le N-1).$$

REFERENCES

[1] W. Matsumoto ; *Normal form of systems of partial differential and pseudo-differential operators in formal symbol classes* , Jour. Math. Kyoto Univ. **34** (1994), 15-40.

[2] ———; *The Cauchy problem for systems – through the normal form of systems and the theory of the weighted determinant –*, Séminaire de E. D. P. Exposé no XVIII, Ecole Polytechnique, (1998-99).

[3] ———; *Direct proof of the perfect block diagonalization of systems of pseudo-differential operators in the ultradifferentiable classes*, Jour. Math. Kyoto Univ. **40**, no. 3 (2000), 541-566.

[4] W. Matsumoto and H. Yamahara; *On the Cauchy-Kowalevskaya theorem for systems*, Proc. Japan Acad., **67**, Ser. A (1991), 181-185.

[5] ———; *The Cauchy-Kowalevskaya theorem for systems*, (to appear).

[6] M. Nagumo; *Über des Anfangswertproblem partieller Differentialgleichungen* Japan Jour. Math. **18** (1941-43), 41-47.

Differential analysis on stratified spaces

B.-W. Schulze   N. Tarkhanov

(B.-W. Schulze) UNIVERSITÄT POTSDAM, INSTITUT FÜR MATHEMATIK, POSTFACH 60 15 53, 14415 POTS-
DAM, GERMANY
E-mail address: schulze@math.uni-potsdam.de

(N. Tarkhanov) UNIVERSITÄT POTSDAM, INSTITUT FÜR MATHEMATIK, POSTFACH 60 15 53, 14415 POTS-
DAM, GERMANY
E-mail address: tarkhanov@math.uni-potsdam.de

Abstract

We describe a calculus of pseudodifferential operators on a stratified
space.

1   Introduction

It is well-known that parametrices of elliptic partial differential equations on
C^∞ manifolds can be expressed by pseudodifferential operators. This implies
the elliptic regularity in terms of the standard Sobolev spaces that are natural
domains of pseudodifferential operators.

The same problem is extremely interesting on manifolds or, more generally, on the Thom-Mather stratified spaces, with piecewise smooth geometry,
e.g., with conical points, edges, corners of higher order or non-compact 'exits'.
Analytically this corresponds to operators with 'degenerate' symbols.

In the literature there are several approaches to the analysis on manifolds
with corners.

Boundary value problems in domains with conical points on the boundary
were studied quite thoroughly by Kondrat'ev [Kon67].

Maz'ya and Plamenevskii [MP77] treat elliptic boundary value problems
for differential equations on manifolds with singularities of a sufficiently general nature. As singularities they admit edges of different dimensions and their
various intersections at non-zero angles. The same sets are regarded as carriers
of discontinuities of the coefficients. However, the treatment falls short of pro-
viding a pseudodifferential algebra where the parametrices to elliptic elements
are available.

Melrose [Mel87] studies so-called b-pseudodifferential operators on manifolds with corners. While originating from geometry his theory does not apply,
however, to many interesting elliptic operators, e.g., the Laplace operator in
the corner $(\bar{\mathbb{R}}_+)^n$, $n \geq 3$.

As far as we know the problem of representing parametrices of differential
operators near corners in terms of symbols of operators in an algebra and of

obtaining corner asymptotics by means of the parametrices was first treated by
Schulze [Sch92]. The experience with simpler problems with singularities, e.g.,
mixed elliptic ones such as the Zaremba problem, shows that the analogous
questions lead at once to corresponding algebras of rather generality. For
proving an analogue of the Atiyah-Singer index theorem such an approach
would be necessary, anyway.

Plamenevskii and Senichkin [PS95] discuss the C^*-algebras generated by
pseudodifferential operators of order zero with discontinuous symbols. The
symbols may have discontinuities along some submanifolds of the unit sphere
intersecting at non-zero angles. The purpose is to describe the spectrum of such
algebras, i.e., the set of all equivalence classes of irreducible representations
endowed with a natural topology (the so-called Jacobson topology). Thus,
they confine themselves to homogeneous symbols and L^2-spaces.

In recent years this area of problems found growing interest in the literature
while the applications in differential geometry, topology and natural sciences
are often classical. The vast variety of special investigations suggests a general
approach for sufficiently wide classes of such problems.

The goal of this paper is a calculus of pseudodifferential operators on manifolds with higher order corners including ellipticity and Fredholm property.
The style of exposition is dictated by the desire to formulate the theory on the
whole and to confirm the expectation that in spite of complexity of tools there
does actually exist an operator algebra containing the ideas of Agranovich's
[AV64, Agr65, AD62], Vishik and Eskin's [VE65, VE67] and Boutet de Monvel's [BdM71] theories as well as higher order operator structures for conical
and edge singularities.

2   Spaces with point singularities

Recall that by a "manifold" with conical singularities we understand a Hausdorff topological space B with a discrete subset B_0 of 'conical points', such that
$B \setminus B_0$ is a C^∞ manifold, and every point $v \in B_0$ has a neighbourhood O in
B homeomorphic to the topological cone over a C^∞ compact closed manifold
X. Thus,

$$O \xrightarrow{\cong} \frac{[0,1) \times X}{\{0\} \times X}, \qquad (2.1)$$

the manifold X being referred to as the *base* of the cone close to v. Moreover,
we require these local homeomorphisms to restrict to diffeomorphisms of open
sets

$$O \setminus \{v\} \xrightarrow{\cong} (0,1) \times X.$$

Any two homeomorphisms h_1 and h_2 are said to be *equivalent* if the composition
$h_2 \circ h_1^{-1}$ extends to a diffeomorphism of $[0,1) \times X$. This gives B a C^∞ structure

Fig. 1: A corner.

with singular points.

According to (2.1), local coordinates in a punctured neighbourhood of any conical point v in B split as (r, x), $r \in \mathbb{R}_+$ being the cone axis variable and $x \in X$.

3   Spaces with corners

The cone is a special case of a corner, the base of the latter being itself a manifold with conical points. Some elements of the cone theory are to be applied for the general corners again, cf. Schulze [Sch98].

A "manifold" with corners is a Hausdorff topological space C along with closed subspaces C_0 and C_1, such that C_0 is a discrete subset of C_1 consisting of the 'corners', $C_1 \setminus C_0$ is a C^∞ manifold of dimension 1 consisting of the edges which emanate from the corners, and $C \setminus C_1$ is a C^∞ manifold of dimension $n + 2$. Every $v \in C_0$ has an open neighbourhood O in C homeomorphic to the topological cone over a C^∞ compact closed manifold with conical singularities B,

$$O \overset{\cong}{\hookrightarrow} \frac{[0, 1) \times B}{\{0\} \times B}, \tag{3.1}$$

B being the *base* of the corner close to v (see Fig. 1).

We require these homeomorphisms to restrict to diffeomorphisms of open sets

$$O \cap (C_1 \setminus C_0) \overset{\cong}{\hookrightarrow} (0, 1) \times B_0,$$
$$O \setminus C_1 \overset{\cong}{\hookrightarrow} (0, 1) \times (B \setminus B_0).$$

Once again we specify classes of *equivalent* homeomorphisms by requiring suitable compositions to preserve the differentiability up to $t = 0$. This gives C a C^∞ structure with corners near v.

By (3.1), local coordinates in a deleted neighbourhood of any corner $v \in C$ split as (t, p), $t \in \mathbb{R}_+$ being the corner axis variable and $p \in B$. In the theory of Thom-Mather stratified spaces B is known as the *link* of the stratum C through v, cf. [GM88].

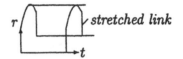

Fig. 2: A stretched corner.

4   Spaces with edges

A further assumption on C is that every point $p \in C_1 \setminus C_0$ has a neighbourhood O in C which is homeomorphic to a wedge

$$O \overset{\cong}{\to} \Omega \times \frac{[0,1) \times X}{\{0\} \times X}, \tag{4.1}$$

Ω being an open interval on the real axis and X a C^∞ compact closed manifold. We confine ourselves to those homeomorphisms (4.1) which restrict to diffeomorphisms

$$O \setminus C_1 \overset{\cong}{\to} \Omega \times (0,1) \times X,$$

and the various mappings are assumed to be compatible in an obvious way over the intersections of neighbourhoods. This leads to a C^∞ structure with edges near $C_1 \setminus C_0$ on C.

It is easy to introduce the categories of spaces with conical points, edges, corners and evident notions of morphisms, in particular, isomorphisms.

The space in Fig. 1 has one corner v, the base B of the corner has one conical point.

The analysis near v will take place in the coordinates (t, r, x) of the open stretched corner $\mathbb{R}_+ \times \mathbb{R}_+ \times X$ (cf. Fig. 2).

5   Typical differential operators

For a C^∞ manifold M, we denote by $\mathrm{Diff}^m(M)$ the space of all differential operators of order m with C^∞ coefficients on M. It is equipped with a natural Fréchet topology. In particular, we look at operators $A \in \mathrm{Diff}^m(C \setminus C_1)$ of the form

$$A = \frac{1}{t^m} \sum_{j=0}^{m} A_j(t) \, (tD_t)^j$$

in the coordinates from (3.1) on $O \setminus C_1$, where $A_j(t)$ is a C^∞ function on $[0,1)$ with values in $\mathrm{Diff}^{m-j}(B \setminus B_0)$. We require every $A_j(t)$ to be of Fuchs type on

B, i.e., A takes the form

$$A = \frac{1}{(tr)^m} \sum_{j+k \leq m} A_{jk}(t,r)\,(trD_t)^j\,(rD_r)^k \tag{5.1}$$

in the coordinates from (2.1) close to any point $v \in B_0$, with $A_{jk}(t,r)$ being a C^∞ function on $[0,1) \times [0,1)$ with values in $\mathrm{Diff}^{m-(j+k)}(X)$. The factors t^{j-m} are no longer important away from the corners where t is non-zero. Hence, near any point on the smooth part of an edge in $C_1 \setminus C_0$ we can rewrite (5.1) in the form

$$A = \frac{1}{r^m} \sum_{j+k \leq m} \tilde{A}_{jk}(t,r)\,(rD_t)^j\,(rD_r)^k$$

in the coordinates from (4.1), $\tilde{A}_{jk}(t,r)$ being a C^∞ function on $\Omega \times [0,1)$ with values in $\mathrm{Diff}^{m-(j+k)}(X)$.

Operators of the form (5.1) will be referred to as the *typical* differential operators on manifolds with corners. By the above, they bear both cone and edge degeneracy as well as more general corner degeneracy.

As but one example of corner-degenerate operators we mention the Laplace-Beltrami operator with respect to the Riemannian metric on C that is of the form

$$
\begin{aligned}
dt^2 + t^2\,dp^2 &= dt^2 + t^2\left(dr^2 + r^2\,dx^2\right) \\
&= (tr)^2\left(\left(\frac{dt}{tr}\right)^2 + \left(\frac{dr}{r}\right)^2 + dx^2\right)
\end{aligned}
$$

in the local coordinates from (3.1), the first equality being close to a conical point of the link B (cf. Fig. 2).

Now the program of the calculus is to introduce symbol structures, pseudodifferential operators and adequate Sobolev spaces, such that the elliptic operators are Fredholm and possess parametrices within the calculus. Here, the ellipticity of an operator A means the invertibility of certain symbols related to A, while the Fredholm property refers to an adequate scale of Sobolev spaces. Moreover, it is desirable to study subspaces with corner asymptotics, such that the elliptic regularity still remains valid therein.

The solution requires an analogous theory on manifolds with conical points and edges. The calculus for edges is known to recover the case of boundary value problems where the edge is the boundary and the model cone is the inner normal $\bar{\mathbb{R}}_+$, cf. Ch. 4 in [Sch98]. Here, for getting the Fredholm property additional edge conditions are posed, satisfying an analogue of the Lopatinskii condition. In general, they are of trace and potential type just as in Boutet de Monvel's theory, cf. [BdM71].

Note that C can in turn be regarded as the base of a "third order" corner, namely

$$\frac{[0,1) \times C}{\{0\} \times C},$$

etc. The theory for this singular configuration encompasses the problems of quarter plane type as well as boundary value problems in domains whose boundary bears "second order" corners. The axiomatic ideas contain formal procedures to obtain from a given operator algebra on some singular variety a new one in the cone over that base by means of a machinery called "conification". A suitable conified algebra on a corresponding infinite model cone then serves as a starting object for another procedure called "edgification". These concepts are elaborated in Schulze [Sch98].

6   Cuspidal singularities

We now illustrate a few aspects of the analysis on manifolds with edges which intersect each other at zero angles, cf. [RST00]. The underlying space looks locally like a wedge $\mathbb{R}^q \times B^\wedge$, where $B^\wedge = \mathbb{R}_+ \times B$ is the semicylinder over a manifold with singularities B. The Riemannian metric on $\mathbb{R}^q \times B^\wedge$ is of the form

$$dy^2 + dt^2 + \left(\frac{1}{\delta'(t)}\right)^2 dp^2 = \left(\frac{1}{\delta'(t)}\right)^2 \left(\left(\frac{d}{dt}e^{\delta(t)}\right)^2 (e^{-\delta(t)}dy)^2 + (d\delta(t))^2 + dp^2\right),$$

where dy^2 and dp^2 are Riemannian metrics on $\mathbb{R}^q$ and B, respectively, and $T = \delta(t)$ is a diffeomorphism of a neighbourhood of $t = 0$ to a neighbourhood of $T = -\infty$.

In the case of transversal intersections we have $\delta(t) = \log(t)$. In contrast to this, $\delta(t)$ behaves like $-1/t^p$, $p > 0$, in the case of power-like cuspidal singularities. The "conification" on $B^\wedge$ and the "edgification" on $\mathbb{R}^q \times B^\wedge$ have to complete an operator algebra over B by functions of the typical vector fields

$$\frac{\partial}{\partial \delta(t)} := \frac{1}{\delta'(t)}\frac{\partial}{\partial t} \quad \text{and} \quad e^{\delta(t)}\frac{\partial}{\partial y_j}, \quad j = 1, \ldots, q,$$

respectively.

Under the coordinate $s = e^{\delta(t)}$ on the cone axis, these vector fields are written as $s\,\partial/\partial s$ and $s\,\partial/\partial y_j$, these latter occur in the case of transversal intersections. This is not surprising because any cuspidal singularity can topologically be transformed to a conical one. However, the change of variables $s = e^{\delta(t)}$ fails to preserve the C^∞ structure close to the edge $t = 0$. Hence it follows that "smooth coefficients" near $t = 0$ are pushed forward to "singular ones" near $s = 0$. The analysis on manifolds with cuspidal singularities thus reduces to that for the case of conical singularities, but with singular "coefficients".

The required nature of symbols near the edge $t = 0$ depends on the function spaces to be domains of operators in the algebra. In fact, the only natural requirement stems from the general observation that the domain of an operator should be a module over the space of its "coefficients". Since pseudodifferential operators on manifolds with singularities are intended to act in spaces with asymptotics, the coefficients themselves should bear appropriate asymptotic expansions.

As usual, the asymptotic expansions correspond to Euler solutions to equations with coefficients constant in t, which are obtained from given equations by freezing the "coefficients" at $t = 0$. This behaviour of symbols near $t = 0$ is then inherited under any change of variables $s = e^{\delta(t)}$. Thus, our results apply as well to problems on manifolds with cuspidal corners, unless we leave the setting of Euler asymptotics. For more details we refer the reader to [ST99].

7 Twisted calculus

Recall the definition of abstract edge Sobolev spaces. Let V be a Banach space and $(\kappa_\lambda)_{\lambda>0}$ be a fixed representation of $\mathbb{R}_+$ in $\mathcal{L}(V)$, i.e., $\kappa_\lambda \in C(\mathbb{R}_+, \mathcal{L}_\sigma(V))$ satisfies $\kappa_1 = \mathrm{Id}$ and $\kappa_\lambda \kappa_\mu = \kappa_{\lambda\mu}$, for all $\lambda, \mu \in \mathbb{R}_+$.

Choose a strictly positive C^∞ function $\tau \mapsto \langle \tau \rangle$ on $\mathbb{R}$ satisfying $\langle \tau \rangle = |\tau|$ for all $|\tau| \geq c$, the constant $c > 0$ being fixed, e.g., $c = 1$. We write $\mathcal{F}_{t\mapsto\tau}$ for the Fourier transform in $\mathbb{R}$.

Definition 7.1 *For $s \in \mathbb{R}$, the space $H^s(\mathbb{R}, \pi^*V)$ is defined to be the completion of $\mathcal{S}(\mathbb{R}, V)$ with respect to the norm*

$$\|u\|_{H^s(\mathbb{R},\pi^*V)} = \left(\int \langle \tau \rangle^{2s} \|\kappa_{\langle\tau\rangle}^{-1} \mathcal{F}_{t\mapsto\tau} u\|_V^2 \, d\tau \right)^{1/2}$$

If $\kappa_\lambda \equiv \mathrm{Id}_V$ for all $\lambda \in \mathbb{R}_+$, we recover the usual Sobolev spaces $H^s(\mathbb{R}, V)$ of V-valued functions on $\mathbb{R}$.

Yet another crucial example corresponds to the fibre space $V = H^{s,\gamma}(X^\wedge)$, a weighted Sobolev space on the stretched cone $X^\wedge$ with $s, \gamma \in \mathbb{R}$. The group action on V given by $\kappa_\lambda u(r, x) = \lambda^{(1+n)/2} u(\lambda r, x)$, for $\lambda > 0$, n being the dimension of X. Then, we obtain the *wedge Sobolev spaces*

$$H^{s,\gamma}(\mathbb{R} \times X^\wedge) = H^s(\mathbb{R}, \pi^* H^{s,\gamma}(X^\wedge))$$

of smoothness s and weight γ. When localised to compact subsets of the wedge $\mathbb{R} \times X^\wedge$, the space $H^{s,\gamma}(\mathbb{R} \times X^\wedge)$ is known to coincide with the usual Sobolev space $H^s_{\mathrm{loc}}(\mathbb{R} \times X^\wedge)$. Moreover, we have

$$H^{0,-\frac{n+1}{2}}(\mathbb{R} \times X^\wedge) \cong L^2(\mathbb{R} \times X^\wedge, r^n dt dr dx),$$

the measure $r^n dr dx$ corresponding to the cone Riemannian metric in the fibre $X^\wedge$ of the wedge.

The operator $I = \mathcal{F}^{-1}_{\tau \to t} \kappa^{-1}_{\langle \tau \rangle} \mathcal{F}_{t \to \tau}$ induces an isomorphism of $H^s(\mathbb{R}, \pi^* V)$ onto $H^s(\mathbb{R}, V)$. This allows us to extend the definition of $H^s(\mathbb{R}, \pi^* \Sigma)$ to the vector subspaces $\Sigma \subset V$ that are not necessarily preserved under κ_λ. Namely, we set

$$H^s(\mathbb{R}, \pi^* \Sigma) \cong I^{-1} H^s(\mathbb{R}, \Sigma), \tag{7.1}$$

for any Banach space Σ continuously embedded to V. Then (7.1) is again a Banach space in the topology induced by the bijection I.

The invariance of our definitions under diffeomorphisms $\Omega \overset{\cong}{\to} \tilde{\Omega}$ and the property $H^{s,\gamma}(\mathbb{R} \times X^\wedge) \hookrightarrow H^s_{\mathrm{loc}}(\mathbb{R} \times X^\wedge)$ give rise to the global weighted Sobolev spaces $H^{s,\gamma}_{\mathrm{loc}}(C \setminus C_0)$ and $H^{s,\gamma}_{\mathrm{comp}}(C \setminus C_0)$ as well as those with asymptotics $H^{s,\gamma}_{\mathrm{as,loc}}(C \setminus C_0)$ and $H^{s,\gamma}_{\mathrm{as,comp}}(C \setminus C_0)$.

Let V and $\tilde{V}$ be Hilbert spaces and $(\kappa_\lambda)_{\lambda \in \mathbb{R}_+}$ and $(\tilde{\kappa}_\lambda)_{\lambda \in \mathbb{R}_+}$ fixed group actions on V and $\tilde{V}$, respectively.

Definition 7.2. *Given any open set $\Omega \subset \mathbb{R}^Q$ and $m \in \mathbb{R}$, we denote by $\mathcal{S}^m(\Omega \times \mathbb{R}, \mathcal{L}(V, \tilde{V}))$ the set of all C^∞ functions $a(t,\tau)$ on $\Omega \times \mathbb{R}$ with values in $\mathcal{L}(V, \tilde{V})$, such that*

$$\left\| \tilde{\kappa}^{-1}_{\langle \tau \rangle} \left(D^\alpha_t D^\beta_\tau a(t,\tau) \right) \kappa_{\langle \tau \rangle} \right\|_{\mathcal{L}(V, \tilde{V})} \leq c_{K,\alpha,\beta} \langle \tau \rangle^{m-|\beta|}$$

for all $(t,\tau) \in K \times \mathbb{R}$ and $\alpha \in \mathbb{Z}^Q_+$, $\beta \in \mathbb{Z}_+$, where K is any compact subset of Ω.

Analogous notation is used for subspaces of classical symbols, indicated by "cl".

More precisely, $\mathcal{S}^m_{\mathrm{cl}}(\Omega \times \mathbb{R}, \mathcal{L}(V, \tilde{V}))$ is the subspace of $\mathcal{S}^m(\Omega \times \mathbb{R}, \mathcal{L}(V, \tilde{V}))$ formed by the symbols a that possess asymptotic expansions

$$a \sim \chi(\tau) \sum_{j=0}^{\infty} a_j$$

with $\chi(\tau)$ an excision function and $a_j \in C^\infty_{\mathrm{loc}}(\Omega \times (\mathbb{R} \setminus \{0\}), \mathcal{L}(V, \tilde{V}))$ homogeneous of degree $m - j$ with respect to the group actions in V and $\tilde{V}$ in the sense that

$$a_j(t, \lambda\tau) = \lambda^{m-j} \tilde{\kappa}_\lambda a_j(t, \tau) \kappa^{-1}_\lambda, \quad \lambda > 0;$$

for all $t \in \Omega$ and $\tau \in \mathbb{R} \setminus \{0\}$.

Any symbol $a \in S^m((\Omega \times \Omega) \times \mathbb{R}, \mathcal{L}(V, \tilde{V}))$ induces a canonical operator on Ω by

$$\mathrm{op}(a)u(t) = \frac{1}{2\pi} \iint e^{i(t-t')\tau} a(t, t', \tau)\, u(t')\, dt' d\tau,$$

first regarded as a continuous mapping $C^\infty_{\mathrm{comp}}(\Omega, V) \to C^\infty_{\mathrm{loc}}(\Omega, \tilde{V})$.

Denote by $\Psi^m(\Omega; V, \tilde{V})$ the set of all operators $\mathrm{op}(a)$ with arbitrary double symbols $a \in S^m((\Omega \times \Omega) \times \mathbb{R}, \mathcal{L}(V, \tilde{V}))$, and by $\Psi^m_{\mathrm{cl}}(\Omega; V, \tilde{V})$ the subspace of classical operators. Then by the Schwartz Kernel Theorem $\Psi^{-\infty}(\Omega; V, \tilde{V})$ can be identified with $C^\infty_{\mathrm{loc}}(\Omega \times \Omega, \mathcal{L}(V, \tilde{V}))$. For $a \in S^m((\Omega \times \Omega) \times \mathbb{R}, \mathcal{L}(V, \tilde{V}))$, we set

$$\sigma^m_{\mathrm{edge}}(a)(t, \tau) = a_0(t, t, \tau), \quad (t, \tau) \in T^*\Omega \setminus \{0\}, \tag{7.2}$$

and define the *principal homogeneous edge symbol* of an operator $A = \mathrm{op}(a)$ by $\sigma^m_{\mathrm{edge}}(A) = \sigma^m_{\mathrm{edge}}(a)$.

The kernels of operators $A = \mathrm{op}(a)$ live in the space of distributions on $\Omega \times \Omega$ with values in $\mathcal{L}(V, \tilde{V})$. Then, we may talk about the operators properly supported with respect to the (t, t')-variables. Every $A \in \Psi^m(\Omega; V, \tilde{V})$ extends to continuous operators $H^s_{\mathrm{comp}}(\Omega, \pi^*V) \to H^{s-m}_{\mathrm{loc}}(\Omega, \pi^*\tilde{V})$ for all $s \in \mathbb{R}$. We may write "comp" "loc" in both the domain and target space if A is properly supported.

In the sequel we use the letters W and $\tilde{W}$ to designate arbitrary C^∞ vector bundles over the edge $C_1 \setminus C_0$. When restricted to Ω, they are trivial, i.e.,

$$\begin{aligned} W &= \Omega \times \mathbb{C}^N, \\ \tilde{W} &= \Omega \times \mathbb{C}^{\tilde{N}}. \end{aligned}$$

Definition 7.3 *Let* $w = (\gamma, \gamma - m, \mathcal{I})$, *where* $m, \gamma \in \mathbb{R}$ *and* $\mathcal{I} = (-l, 0]$, $0 < l \leq \infty$. *Then,* $\Psi^{-\infty}_G(\Omega \times X^\wedge; W, \tilde{W}; w)$ *stands for the space of all operators*

$$\mathcal{G} \in \bigcap_{s \in \mathbb{R}} \mathcal{L}(H^{s,\gamma}_{\mathrm{comp}}(\Omega \times X^\wedge) \oplus H^s_{\mathrm{comp}}(\Omega, W), H^{\infty,\gamma-m}_{\mathrm{loc}}(\Omega \times X^\wedge) \oplus H^\infty_{\mathrm{loc}}(\Omega, \tilde{W}))$$

that map as

$$\begin{aligned} \mathcal{G} &: \ H^{s,\gamma}_{\mathrm{comp}}(\Omega \times X^\wedge) \oplus H^s_{\mathrm{comp}}(\Omega, W) \to H^{\infty,\gamma-m}_{\mathrm{as,loc}}(\Omega \times X^\wedge) \oplus H^\infty_{\mathrm{loc}}(\Omega, \tilde{W}), \\ \mathcal{G}^* &: \ H^{s,m-\gamma}_{\mathrm{comp}}(\Omega \times X^\wedge) \oplus H^s_{\mathrm{comp}}(\Omega, \tilde{W}) \to H^{\infty,-\gamma}_{\tilde{\mathrm{as}},\mathrm{loc}}(\Omega \times X^\wedge) \oplus H^\infty_{\mathrm{loc}}(\Omega, W) \end{aligned}$$

for all $s \in \mathbb{R}$ *and some asymptotic types* "as" *and* "ãs" *related to weight data* $(\gamma - m, \mathcal{I})$ *and* $(-\gamma, \mathcal{I})$, *respectively.*

Note that $\mathcal{G}^*$ is the formal adjoint of $\mathcal{G}$ with respect to a fixed scalar product in $H^{0,0}(\Omega \times X^\wedge) \oplus H^0(\Omega)$.

The operators of $\Psi^{-\infty}_G(\Omega \times X^\wedge; W, \tilde{W}; w)$ are called smoothing *Green operators* on the wedge $\Omega \times X^\wedge$. In contrast to the cone case, Green operators on the wedge are specified by their action along the edge, here along Ω. While being smoothing away from the edge, they may bear a finite order along Ω. In order to describe this new feature on the symbol level, one introduces Green edge symbols cf. [Sch98].

8   Calculi for higher order singularities

We are now in a position to single out three main features of our approach to constructing algebras of pseudodifferential operators on singular varieties.

The first feature is using special coordinates near singularities to identify coordinate patches with cone bundles over Euclidean spaces $\mathbb{R}^q$, $q \geq 0$. Under these coordinates the singularity itself is moved to a variety of points at infinity. This enables one to write the operators in a unified way, namely as Fourier pseudodifferential operators along $\mathbb{R}^q$ with operator-valued symbols, in spite of the diversity of possible operator representations (Mellin, etc.).

The analysis in weighted Sobolev spaces near singularities reduces to that in weighted Sobolev spaces over $\mathbb{R}^q$, with natural weight functions $e^{\gamma\langle y\rangle}$ and $\langle y\rangle^\mu$ to control the behaviour of functions at points at infinity.

The choice of two weights actually fits in the analysis on varieties with cuspidal singularities, cf. [RST97], because $\delta'(r) \sim \langle\delta(r)\rangle^{p/(p-1)}$ as $r \to 0$, $t = \delta(r)$ being a local coordinate close to the cusp.

The second feature is the idea of edge Sobolev spaces $H^s(\mathbb{R}^q, \pi^*V)$ whose definition relies on group actions in the fibres over the edge $\mathbb{R}^q$, cf. Definition 7.1. The property

$$H^s(\mathbb{R}^{q_1}, \pi^* H^s(\mathbb{R}^{q_2}, \pi^*V)) = H^s(\mathbb{R}^{q_1+q_2}, \pi^*V), \qquad (8.1)$$

$H^s(\mathbb{R}^{q_2}, \pi^*V)$ being endowed with the group action $u \mapsto \kappa_\lambda(\lambda^{q_2/2}u(\lambda y))$, cf. [Sch98, p. 115], allows one to ensure compatibility of definitions over different strata.

Finally, the third feature is invoking operator-valued Fourier symbols over $\mathbb{R}^q$ which admit asymptotic expansions in "twisted" homogeneous symbols, the "twisted" homogeneity referring to the group actions. This makes the theory of pseudodifferential operators on stratified varieties quite analogous to Boutet de Monvel's theory of pseudodifferential boundary value problems, cf. [BdM71].

9   Thom-Mather stratified spaces

More precisely, let V be a *smoothly stratified space with local cone bundle neighbourhoods*, cf. [GM88]. We shall introduce this notion by defining inductively for each $N \geq 1$ a category LCB(N) of smoothly stratified spaces

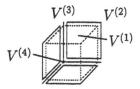

Fig. 3: A smooth stratification.

with local cone bundle neighbourhoods with at most N strata. A smoothly stratified space with local cone bundle neighbourhoods is simply an object in $\mathrm{LCB} = \cup_N \mathrm{LCB}(N)$. For each of these categories there will be natural notions of product with a smooth manifold and of the boundary of an object. Also, underlying each object in $\mathrm{LCB}(N)$ will be a stratified space with at most N strata. The category $\mathrm{LCB}(1)$ is the category of smooth manifolds and smooth mappings. Assume that $\mathrm{LCB}(j)$ has been defined for all $j < N$. For any closed finite-dimensional object B in $\mathrm{LCB}(N-1)$ we denote by $C_t(B)$ the cone on the topological space underlying B. Its top point is v, and the open cone is identified with $(0,1] \times B$. If Ω is an open set in a Euclidean space, we give $\Omega \times C_t(B)$ a stratification in which $\Omega \times \{v\}$ is the bottom stratum and the others are of the form Ω times the open cone over strata of B. An $\mathrm{LCB}(N)$ *coordinate chart* for a stratified space V is a strata-preserving homeomorphism h from an open subset O of V to $\Omega \times C_t(B)$, where Ω is an open subset of a Euclidean space and B is a closed finite-dimensional object in $\mathrm{LCB}(N-1)$. Two coordinate charts $h_\nu : O_\nu \to \Omega_\nu \times C_t(B_\nu)$, $\nu = 1,2$, are said to be *compatible* if the composition $h_2 \circ h_1^{-1}$ induces local diffeomorphisms both from Ω_1 to Ω_2 and from $\Omega_1 \times ((0,1] \times B_1)$ to $\Omega_2 \times ((0,1] \times B_2)$, the former being a "limit" of the latter. An object in $\mathrm{LCB}(N)$ is a stratified space with a maximal atlas of compatible $\mathrm{LCB}(N)$ coordinate charts covering the space. A *morphism* in this category is a strata-preserving mapping f which is C^∞ on each stratum and behaves properly with respect to local cone bundle structures. It is well known that any *Whitney stratified subspace* of a smooth manifold admits a natural structure of a smoothly stratified space with local cone bundle neighbourhoods, cf. *ibid.*

Write $V^{(1)}, \ldots, V^{(N)}$ for the strata of V, $V^{(j)}$ being of codimension n_j (see Fig. 3). A stratum need not be connected, hence we may assume without loss of generality that $0 = n_1 < \ldots < n_N \leq \dim V$. By definition, each $V^{(j)}$ is a smooth manifold, and the closure of $V^{(j)}$ in V lies in $V^{(j)} \cup \ldots \cup V^{(N)}$ for all $j = 1, \ldots, N$.

We may formally write

$$V^{(j)} = V^{(J)} \times \frac{V^{(j)}}{V^{(J)}}, \tag{9.1}$$

for $1 \leq j \leq J \leq N$, and our next goal is to give a precise meaning to this equality. We shall interpret it locally in a neighbourhood O of any point $p \in V^{(J)}$. If O is small enough, then $V^{(j)}$ has the structure of a cone bundle on $V^{(J)}$ within O, the fibre being $C_t(B)$. If non-empty, B is a smoothly stratified space with local cone bundle neighbourhoods with at most $(J-j)$ strata. We call B the link of $V^{(j)}$ over $V^{(J)}$, it is actually the same over each connected component of $V^{(J)}$. The fibre $C_t(B)$ is invariant under the group action $(t, x) \mapsto (\lambda t, x)$, $\lambda > 0$; this latter specifies a singular fibre structure by itself. To not exclude artificial stratifications, we allow the fibre $C_t(B)$ to bear a smooth structure, too. In this case $C_t(B)$ is locally identified with an Euclidean space (namely, $\mathbb{R}^{n_J - n_j}$) and endowed with the group action $z \mapsto \lambda z$, $\lambda > 0$. Now, by $V^{(j)}/V^{(J)}$ in (9.1) is just meant the local fibre of $V^{(j)}$ over $V^{(J)}$.

10   Weighted Sobolev spaces

In case the strata meet each other at non-zero angles the typical differential operators on V are those of Fuchs-type. This suggests a canonical transformation of the coordinate along the cone axis of $C_t(B)$, which maps the vertex $t = 0$ to the point at infinity. More precisely, the new coordinates $(T = \log t, x)$ for small $t > 0$ allow one to identify $C_t(B)$ with the cylinder $\mathbb{R} \times B$ over B, the conical point corresponding to $T = -\infty$. Note that B itself has a finite covering by local cone bundle neighbourhoods and that a smooth manifold is locally a cone over the unit sphere. We are thus led to the problem of defining weighted Sobolev spaces in the wedge $\mathbb{R}^q \times C_t(B)$, which properly control the behaviour of functions up to points at infinity. We start with the scale $H^s(B)$, $s \in \mathbb{R}$, of usual Sobolev spaces on a C^∞ manifold B, which possesses a finite coordinate covering. Such a manifold has one stratum, the cone over B has two ones. We proceed by induction and assume that the relevant Sobolev spaces have been introduced for all B with at most $(J-j)$ strata. Any stratum contributes by additional weight exponents, so we have a scale of Sobolev spaces $H^{s,w'}(B)$ on B, parametrised by smoothness $s \in \mathbb{R}$ and a tuple of weights $w' = (w_1, \ldots, w_{J-j-1})$, every component being a weight function on $\mathbb{R}_+$ usually identified with a quadruple of real numbers. By the above, B has a finite covering (O_ν) by coordinate charts, each O_ν being diffeomorphic to a wedge by $h_\nu : O_\nu \to \Omega_\nu \times C_t(B_\nu)$. Pick a C^∞ partition of unity on B subordinate to this covering, (φ_ν). The space $H^{s,w'}(B)$ is glued together from the local spaces $H^s(\mathbb{R}^{q_\nu}, \pi^* H^{s,w'}(C_t(B_\nu)))$ in the sense that $u \in H^{s,w'}(B)$ if and only if $(h_\nu)_* (\varphi_\nu u) \in H^s(\mathbb{R}^{q_\nu}, \pi^* H^{s,w'}(C_t(B_\nu)))$ for all ν. We have $C_t(B) = \cup_\nu C_t(O_\nu)$, and the diffeomorphisms $1 \times h_\nu$ take $C_t(O_\nu)$ to the cones over $\mathbb{R}^{q_\nu} \times C_t(B_\nu)$. Note that we think of $C_t(B)$ as being infinite and we regard the infinity as a conical point. Choose the covering of $\bar{\mathbb{R}}_+$ by the intervals $[0, 1)$ and $(1/2, \infty)$, and a partition of unity on $\bar{\mathbb{R}}_+$ subordinate to this covering, $(\omega, 1 - \omega)$. We make use of the cut-off function ω to introduce a weight function along the cone axis,

$$w_{J-j}(t) = t^{-\gamma} \left(\log 1/t\right)^{\mu} \omega(t) + t^{\delta} \left(\log t\right)^{\nu} \left(1 - \omega(t)\right),$$

with $\gamma, \mu, \delta, \nu \in \mathbb{R}$, which is typical for the analysis on a cone. For the tuple $w = (w', w_{J-j})$, the space $H^{s,w}(C_t(B))$ is defined to consist of all functions $u(t, x)$, such that

$$\begin{aligned}
(\log t \times h_{\nu})_* \left(\omega \varphi_{\nu} w_{J-j} u\right) &\in H^s(\mathbb{R}^{1+q_{\nu}}, \pi^* H^{s,w'}(C_t(B_{\nu}))), \\
(1 \times h_{\nu})_* \left((1 - \omega) \varphi_{\nu} w_{J-j} u\right) &\in H^s(\mathbb{R}^{1+q_{\nu}}, \pi^* H^{s,w'}(C_t(B_{\nu})))
\end{aligned}$$

for all ν.

We might certainly express this in a unified way by introducing a diffeomorphism $T = \delta(t)$ of $\mathbb{R}_+$ onto $\mathbb{R}$ with the property that $\delta(t) = \log t$, for small t, and $\delta(t) = t$, for large t.

Lemma 10.1 *As defined above, the space $H^{s,w}(C_t(B))$ is invariant under the group action $(K_{\lambda} u)(t, x) = \lambda^N u(\lambda t, x)$, $\lambda > 0$.*

Proof. When topologising $H^{s,w}(C_t(B))$ under the natural norm, we get, by (8.1),

$$\|K_{\lambda} u\|^2_{H^{s,w}(C_t(B))} = \sum_{\nu} \|(\delta \times h_{\nu})_* \varphi_{\nu} w_{J-j} K_{\lambda} u\|^2_{H^s(\mathbb{R}, \pi^* V_{\nu})}$$

for all $\lambda > 0$, where $V_{\nu} = H^s(\mathbb{R}^{q_{\nu}}, \pi^* H^{s,w'}(C_t(B_{\nu})))$. We will restrict our attention to those functions u which are supported in a fixed chart O_{ν}. Then we may omit, by abuse of notation, the index ν and both φ_{ν} and h_{ν}. Moreover, it will cause no confusion if we write $u(t)$ for u and $w(t)$ for w_{J-j}. It follows that

$$\|K_{\lambda} u\|^2_{H^{s,w}(C_t(B))} = \lambda^{2N} \int_{\mathbb{R}} \langle \tau \rangle^{2s} \|\kappa_{\langle \tau \rangle}^{-1} \mathcal{F}_{T \to \tau} \delta_* \left(w(t) u(\lambda t)\right)\|^2_V \, d\tau,$$

(κ_{λ}) being the group action in the fibre V. Changing the variable by

$$T = \delta \left(\frac{1}{\lambda} \delta^{-1}(S)\right),$$

we obtain

$$\mathcal{F}_{T \to \tau} \delta_* \left(w(t) u(\lambda t)\right) = \int_{\mathbb{R}} e^{-i\tau \delta \left(\frac{1}{\lambda} \delta^{-1}(S)\right)} w \left(\frac{1}{\lambda} \delta^{-1}(S)\right) u \left(\delta^{-1}(S)\right) d\delta \left(\frac{1}{\lambda} \delta^{-1}(S)\right)$$

whence

$$\mathcal{F}_{T \mapsto \tau} \delta_* \left(w(t) u(\lambda t) \right) = \frac{1}{\lambda} \int_{\mathbb{R}} K\left(\lambda; \tau, \sigma\right) \mathcal{F}_{S \mapsto \sigma} \delta_* \left(wu \right) d\sigma,$$

$$K\left(\lambda; \tau, \sigma\right) = \frac{1}{2\pi} \int_{\mathbb{R}} e^{-i\tau \delta\left(\frac{1}{\lambda}\delta^{-1}(S)\right) + i\sigma S} \frac{\left(\delta' w\right) \left(\frac{1}{\lambda}\delta^{-1}(S)\right)}{\left(\delta' w\right) \left(\delta^{-1}(S)\right)} dS,$$

the latter integral being understood in the sense of distributions. Applying Schwarz's inequality yields

$$\|\kappa_{\langle\tau\rangle}^{-1} \mathcal{F}_{T \mapsto \tau} \delta_* (w(t) u(\lambda t))\|_V^2 \le \left(\frac{1}{\lambda^2} \int_{\mathbb{R}} \langle \sigma \rangle^{-2s} |K|^2 \left\| \kappa_{\frac{\langle\tau\rangle}{\langle\sigma\rangle}}^{-1} \right\|_{\mathcal{L}(V)}^2 d\sigma \right) \|u\|_{H^{s,w}(C_t(B))}^2,$$

and so

$$\|K_\lambda u\|_{H^{s,w}(C_t(B))}^2 \le \left(\lambda^{2N-2} \iint_{\mathbb{R}^2} \left(\frac{\langle\tau\rangle}{\langle\sigma\rangle} \right)^{2s} \left\| \kappa_{\frac{\langle\tau\rangle}{\langle\sigma\rangle}}^{-1} \right\|_{\mathcal{L}(V)}^2 |K|^2 d\tau d\sigma \right) \|u\|_{H^{s,w}(C_t(B))}^2.$$

To complete the proof it suffices to observe that the expression in the round brackets is dominated by

$$\lambda^{2N-2} \iint_{\mathbb{R}^2} \langle \tau - \sigma \rangle^{2Q} |K\left(\lambda; \tau, \sigma\right)|^2 d\tau d\sigma$$

which is due to a well-known property of abstract group actions and Peetre's inequality.

$\square$

Having defined the space $H^{s,w}(C_t(B))$ along with the group action, we can invoke Definition 7.1 to introduce the space $H^s(\mathbb{R}^q, \pi^* H^{s,w}(C_t(B)))$, for any q, thus completing the step of induction. What is still lacking is an explicit value of N in dependence of q and the norm in $H^{s,w}(C_t(B))$. The criterion for the choice of N is that the space $H^s(\mathbb{R}^q, \pi^* H^{s,w}(C_t(B)))$ be locally equivalent to the usual Sobolev space H^s_{loc} away from the singularities of $\mathbb{R}^q \times C_t(B)$, "locally" meaning "on compact sets". As usual, taking $\nu = 0$ and $N = \delta + 1/2$ fills the bill.

The underlying idea of our "resolution of singularities" consists of moving the singular points to infinity and using the Fourier analysis in a Euclidean space under strong control by weight functions. This idea goes back at least as far as [Rab69].

11   Pseudodifferential edge problems

We now turn to an arbitrary smoothly stratified space with local cone bundle neighbourhoods V. In order to introduce pseudodifferential operators on V, we proceed by induction. Suppose we have already defined a pseudodifferential calculus on all spaces with at most $N-1$ strata. Given any $B \in \text{LCB}(N-1)$, by a pseudodifferential operator on $C_t(B)$ we mean $\delta^\sharp \mathcal{A}$, where $\mathcal{A}$ is a pseudodifferential operator on $\mathbb{R}$ whose symbol takes its values in the algebra of pseudodifferential operators on B, $T = \delta(t)$ is a diffeomorphism of $\mathbb{R}_+$ onto $\mathbb{R}$, as above, and $\delta^\sharp \mathcal{A} = \delta^* \mathcal{A} \delta_*$ is the operator pull-back of $\mathcal{A}$ under δ. Obviously, the operators obtained this way are of the Mellin type close to the conical point of $C_t(B)$. Let now $V \in \text{LCB}(N)$. By definition, V has a finite covering (O_ν) by coordinate charts diffeomorphic to model wedges by $h_\nu : O_\nu \to \Omega_\nu \times C_t(B_\nu)$, where Ω_ν is an open subset of $\mathbb{R}^{q_\nu}$. We pick a C^∞ partition of unity on V subordinate to this covering, (φ_ν), and a system of C^∞ functions (ψ_ν) which covers (φ_ν). The pseudodifferential operators on the stratified space V are of the form

$$\mathcal{A} = \sum_\nu \varphi_\nu \left(h_\nu^\sharp \mathcal{A}_\nu \right) \psi_\nu,$$

where $\mathcal{A}_\nu$ is a pseudodifferential operator along $\mathbb{R}^{q_\nu}$ with an operator-valued symbol taking its values in the algebra on $C_t(B_\nu)$, and $h_\nu^\sharp \mathcal{A}_\nu = h_\nu^* \mathcal{A}_\nu (h_\nu)_*$ is the pull-back of $\mathcal{A}_\nu$ under h_ν.

Another way of stating the pseudodifferential calculus on V is to cover V by local cylinder bundle neighbourhoods (O_ν). This requires diffeomorphisms $\Delta_\nu : O_\nu \to \Omega_\nu \times ((-\infty, T_\nu) \times B_\nu)$, for every ν, together with a precise control of the behaviour at $T = -\infty$. In the case of transversal intersections, we carry over the point at infinity to a finite point by the diffeomorphism $t = e^T$, thus transplanting the C^∞ structure from $t = 0$ to $T = -\infty$. For cuspidal intersections, other diffeomorphisms are used, whose explicit nature is prescribed by the geometry. Note that a cylinder $\mathbb{R} \times B$ bears the group action $(T, x) \mapsto (T + \log \lambda, x)$, for $\lambda \in \mathbb{R}$, which has actually been used in the definition of twisted Sobolev spaces. Under this approach, pseudodifferential operators in O_ν are of the form $\Delta_\nu^\sharp \mathcal{A}_\nu$, with $\mathcal{A}_\nu$ a pseudodifferential operator on $\mathbb{R}^{q_\nu+1}$ whose symbol takes the values in the algebra on B_ν. They can be treated in the framework of a calculus of pseudodifferential operators in $\mathbb{R}^Q$ with slowly varying operator-valued symbols, cf. [RST97]. However, it is more convenient to have specified the exit to infinity by choosing a relevant variable, here T.

The pseudodifferential operators to be introduced are intended to act in weighted Sobolev spaces on V as

$$\mathcal{A} : \bigoplus_{j=1}^N H^{s,(w_{j+1},\ldots,w_N)}(V^{(j)}, F_j) \to \bigoplus_{j=1}^N H^{s-m,(w_{j+1},\ldots,w_N)-m}(V^{(j)}, \tilde{F}_j) \quad (11.1)$$

for all $s \in \mathbb{R}$, where $w = (w_2, \ldots, w_N)$ is a tuple of weights, w_j corresponding to the stratum $V^{(j)}$, and F_j, $\tilde{F}_j$ are C^∞ vector bundles over $V^{(j)}$. For $j = N$, the tuple $(w_{j+1}, \ldots, w_N)$ is empty, which causes no confusion because $V^{(N)}$ is a C^∞

compact closed manifold. By (11.1), $\mathcal{A}$ can be specified as an $(N \times N)$-matrix of operators

$$\mathcal{A} = (A_{ij})_{\substack{i=1,\dots,N \\ j=1,\dots,N}},$$

with

$$A_{ij} \in \bigcap_{s \in \mathbb{R}} \mathcal{L}(H^{s,(w_{j+1},\dots,w_N)}(V^{(j)}, F_j), H^{s-m,(w_{i+1},\dots,w_N)-m}(V^{(i)}, \tilde{F}_i)).$$

The entries A_{ij} with $i > j$ have the meaning of *trace operators*, the "trace" standing for restriction from $V^{(j)}$ to $V^{(i)}$. On the other hand, the entries A_{ij} with $i < j$ have the dual meaning of *potential operators*. To handle such operators within an algebra, we should also add compositions of trace and potential operators. They contribute to the diagonal entries A_{jj} and are known as *Green operators* on the corresponding strata. In fact, all the entries are pseudodifferential operators and can be specified through their operator-valued symbols.

To this end, it is sufficient to describe the action of $\mathcal{A}$ close to any point $p \in V$. For definiteness, consider $p \in V^{(J)}$ where $1 \leq J \leq N$. Then we have $\mathcal{A} = \mathcal{F}_{\eta \to y}^{-1} a^{(J)}(y, \eta) \mathcal{F}_{y \to \eta}$ in local coordinates $y \in \mathbb{R}^{q_J}$ of $V^{(J)}$ near the point p, where

$$a^{(J)} = \begin{pmatrix} a_{11}^{(J)} & 0 & \dots & 0 & a_{1J} \\ 0 & a_{22}^{(J)} & \dots & 0 & a_{2J} \\ \dots & \dots & \dots & \dots & \dots \\ 0 & 0 & \dots & a_{J-1,J-1}^{(J)} & a_{J-1,J} \\ a_{J1} & a_{J2} & \dots & a_{J,J-1} & a_{JJ}^{(J)} \end{pmatrix} \quad (11.2)$$

acts as

$$a^{(J)} : \pi^* \bigoplus_{j=1}^{J} H^{s,(w_{j+1},\dots,w_J)}\left(\frac{V^{(j)}}{V^{(J)}}, F_j\right) \to \pi^* \bigoplus_{j=1}^{J} H^{s-m,(w_{j+1},\dots,w_J)-m}\left(\frac{V^{(j)}}{V^{(J)}}, \tilde{F}_j\right)$$

for all $s \in \mathbb{R}$, π standing for the canonical projection $T^*V^{(J)} \to V^{(J)}$. To encompass all strata, we identify these matrices within those with $(N \times N)$ entries in an obvious way.

Write $\kappa_{j,\lambda}^{(J)}$ for the group action in $H^{s,(w_{j+1},\dots,w_J)}(V^{(j)}/V^{(J)})$, and gather them to a matrix

$$\kappa_{\lambda}^{(J)} = \bigoplus_{j=1}^{J} \kappa_{j,\lambda}^{(J)}$$

to act in $\bigoplus_{j=1}^{J} H^{s,(w_{j+1},\dots,w_J)}(V^{(j)}/V^{(J)}, F_j)$. By the above, the usual choice for $\kappa_{j,\lambda}^{(J)}$ is

$$u(t) \mapsto \lambda^{\frac{n_J-n_j}{2}} u(\lambda t),$$

which meets (8.1). Analogously, we denote by $\tilde{\kappa}_{\lambda}^{(J)}$, $\lambda > 0$, the group action in $\bigoplus_{j=1}^{J} H^{s-m,(w_{j+1},\dots,w_J)-m}(V^{(j)}/V^{(J)}, \tilde{F}_j)$.

Thus, every stratum $V^{(J)}$ gives rise to group actions $\kappa_\lambda^{(J)}$ and $\tilde{\kappa}_\lambda^{(J)}$ in weighted Sobolev spaces in the fibres of V over $V^{(J)}$. Associated to these group actions are spaces of operator-valued symbols satisfying "twisted" symbol estimates, cf. Definition 7.2, and a concept of homogeneity. This allows one to repeat all the steps in the construction of the algebra of pseudodifferential operators on a manifold with smooth edges, as described in [Sch98]. It begins with typical differential operators on V whose form can be read off from that of the Laplace operator with respect to a Riemannian metric on the "smooth part" of V. Near $V^{(J)}$ the symbols of typical operators contain the covariables along $V^{(J)}$ through the aggregates $(r_1 \ldots r_{J-1})\eta$ where $r_1, \ldots, r_{J-1}$ are defining functions of the faces whose intersection gives $V^{(J)}$. Hence any differentiation in the covariables results not only in decreasing the order by 1 but also in an additional factor $(r_1 \ldots r_{J-1})$ which vanishes on $V^{(J)}$ and thus leads to a gain in the weight. Since the gains in both order and weight provide compactness in the relevant weighted Sobolev spaces, the symbols of typical operators belong to the class of symbols of compact fibre variation introduced in [Luk72]. To include the broadest interesting operator classes we reduce our standing assumptions on the symbols under consideration to the only requirement of compact fibre variation.

When summing up over all strata, we arrive at the matrixes of operators which look like

$$
\mathcal{A} = \begin{pmatrix}
\sum_{J=1}^{N} A_{11}^{(J)} & A_{12} & \cdots & A_{1,N-1} & A_{1N} \\
A_{21} & \sum_{J=2}^{N} A_{22}^{(J)} & \cdots & A_{2,N-1} & A_{2N} \\
\cdots & \cdots & \cdots & \cdots & \cdots \\
A_{N-1,1} & A_{N-1,2} & \cdots & \sum_{J=N-1}^{N} A_{N-1,N-1}^{(J)} & A_{N-1,N} \\
A_{N1} & A_{N2} & \cdots & A_{N,N-1} & A_{NN}^{(N)}
\end{pmatrix},
$$

the stratum $V^{(J)}$ contributing by $A_{jj}^{(J)}$ to any entry (j,j) with $j \leq J$. Only the summand $A_{jj}^{(j)}$ of the entry (j,j) is a usual pseudodifferential operator with scalar-valued symbol on $V^{(j)}$. The summands $A_{jj}^{(J)}$ with $j < J$ bear operator-valued symbols living on the cotangent bundle of $V^{(J)}$. They are known as *Green operators* on $V^{(j)}$ associated to the stratum $V^{(J)}$, cf. Definition 7.3 and elsewhere.

12 Ellipticity

Let $\Psi^m(V; F, \tilde{F}; w)$ denote the space of all operators (11.1) on V, as described above. Any operator $\mathcal{A} \in \Psi^m(V; F, \tilde{F}; w)$ has N principal homogeneous symbols of order m,

$$
\sigma(\mathcal{A}) = (\sigma_J^m(\mathcal{A}))_{J=1,\ldots,N},
$$

$\sigma_J^m(\mathcal{A})$ corresponding to the stratum $V^{(J)}$. For $(y, \eta) \in T^*V^{(J)}$, we actually have

$$\sigma_J^m(\mathcal{A})(y, \eta) = \lim_{\lambda \to \infty} \lambda^{-m} \, \tilde{\kappa}_{\lambda^{-1}}^{(J)} \, \mathfrak{a}^{(J)}(y, \lambda\eta) \, \kappa_\lambda^{(J)} \tag{12.1}$$

where $\mathfrak{a}^{(J)}$ is a local symbol of $\mathcal{A}$ along $V^{(J)}$, cf. (11.2). The passage to the limit automatically includes freezing the coefficients at the vertex of each cone $V^{(j)}/V^{(J)}$, $1 \le j \le J$, in a special manner.

It is worth pointing out that formula (12.1) gives a symbol $\sigma_N^m(\mathcal{A})$ even if $\dim V^{(N)} = 0$. In this case $V^{(N)}$ consists of isolated points, hence its cotangent bundle is also discrete. Taking $\sigma_N^m(\mathcal{A})$ over a point $v \in V^{(N)}$ just amounts to freezing the "coefficients" of $\mathcal{A}$ at v.

If all the symbols $\sigma_1^m(\mathcal{A}), \ldots, \sigma_N^m(\mathcal{A})$ vanish identically, then the operator (11.1) is compact. Hence we may invoke the tuple $\sigma(\mathcal{A})$ to introduce a concept of elliptic operators on V.

Definition 12.1 *An operator $\mathcal{A} \in \Psi^m(V; F, \tilde{F}; w)$ is said to be elliptic if, for every $J = 1, \ldots, N$, the symbol $\sigma_J^m(\mathcal{A})$ is invertible away from the zero section of $T^*V^{(J)}$.*

The condition on the invertibility of $\sigma_N^m(\mathcal{A})$ away from the zero section of $T^*V^{(N)}$ is vague in case $V^{(N)}$ is zero-dimensional. To define it more exactly, we observe that what we really need is the Fredholm property of the complete operator-valued symbol over all of $T^*V^{(N)}$ along with its invertibility outside a compact subset of $T^*V^{(N)}$. Hence a proper substitute for the invertibility will be the Fredholm property of $\sigma_N^m(\mathcal{A})$ on all of $T^*V^{(N)}$, provided that $V^{(N)}$ is of dimension 0. If $\mathcal{A}$ is pseudodifferential close to $V^{(N)}$, the Fredholm property of $\sigma_N^m(\mathcal{A})$ is in turn equivalent to the invertibility of its Fourier symbol over a suitable horizontal line, which acts in weighted Sobolev spaces on the link of $V^{(1)}$ over $V^{(N)}$.

Let us have look from this viewpoint at the analysis on manifolds with conical points. Obviously, these correspond to smoothly stratified spaces V with local cone bundle neighbourhoods with at most 2 strata. In fact, $V^{(1)}$ is the "smooth" part of the manifold, and $V^{(2)}$ is the discrete set of conical points. Assume for simplicity that $V^{(2)}$ consists of only one point v, and write X for the link of $V^{(1)}$ over $V^{(2)}$. When localised to $V^{(1)}$, the space $\Psi^m(V; F, \tilde{F}; w)$ is nothing but $\Psi_{cl}^m(V^{(1)}; F_1, \tilde{F}_1)$, i.e., the space of classical pseudodifferential operators of type $F_1 \to \tilde{F}_1$ and order m, acting in the usual Sobolev spaces on $V^{(1)}$ as $H_{comp}^s(V^{(1)}, F_1) \to H_{loc}^{s-m}(V^{(1)}, \tilde{F}_1)$. Further, v possesses a neighbourhood O on V with local coordinates $h: O \to C_t(X)$. Since $V^{(2)}$ is zero-dimensional, the restriction of $\mathcal{A} \in \Psi^m(V; F, \tilde{F}; w)$ to O is identified with its operator-valued symbol $\mathfrak{a}^{(2)}(y, \eta)$ living on $T^*V^{(2)} \cong \{0\}$. This symbol is a (2×2)-matrix acting as

$$H^{s,w}(C_t(X),F_1) \qquad H^{s-m,w-m}(C_t(X),\tilde{F}_1)$$
$$\oplus \qquad \rightarrow \qquad \oplus$$
$$F_2 \qquad\qquad \tilde{F}_2$$

for all $s \in \mathbb{R}$, where $w(t) = t^{-\gamma}(\log 1/t)^{\mu}\omega(t) + t^{\delta}(\log t)^{\nu}(1 - \omega(t))$ is a weight function on the cone axis. The symbol $\sigma_2^m(\mathcal{A})$ is obtained from $\mathfrak{a}^{(2)}$ by freezing the "coefficients" at $t = 0$. Since both F_2 and $\tilde{F}_2$ are finite-dimensional, the Fredholm property of $\sigma_2^m(\mathcal{A})$ just amounts to that of its entry $(1,1)$, for which we write $\sigma_2^m(a_{11}^{(2)})$. As described above, the change of variables $T = \log t$ yields local coordinates $\Delta : O \rightarrow (-\infty, T) \times X$ near v, v itself corresponding to $\{-\infty\} \times X$. In these coordinates, we have

$$a_{11}^{(2)} = \Delta^{\sharp}\mathcal{F}_{\tau \to T}^{-1}a(T, \tau - i\gamma)\mathcal{F}_{T \to \tau}$$

where $a(T, \tau - i\gamma)$ takes its values in $\cap_{s \in \mathbb{R}}\mathcal{L}(H^s(X,F_1), H^{s-m}(X,\tilde{F}_1))$. It follows that

$$\sigma_2^m(a_{11}^{(2)}) = \Delta^{\sharp}\mathcal{F}_{\tau \to T}^{-1}a(-\infty, \tau - i\gamma)\mathcal{F}_{T \to \tau},$$

and so $\sigma_2^m(a_{11}^{(2)})$ is Fredholm if and only if the family $a(-\infty, \tau - i\gamma)$ is invertible. Clearly, $a(-\infty, \tau - i\gamma)$ is the conormal symbol of $\mathcal{A}$ at v. We conclude that for spaces with point singularities Definition 12.1 reduces to the concept of ellipticity in the cone algebra of [Sch98].

13 Fredholm property

We finish the paper by characterising Fredholm operators in the calculus on a smoothly stratified space with local cone bundle neighbourhoods.

Theorem 13.1 *Suppose* $\mathcal{A} \in \Psi^m(V; F, \tilde{F}; w)$ *is an elliptic edge problem. Then the operator* (11.1) *is Fredholm, for each* $s \in \mathbb{R}$, *and it has a parametrix in* $\Psi^{-m}(V; \tilde{F}, F; w^{-1})$.

Proof. By assumption, the symbol $\sigma_1^m(\mathcal{A})$ is invertible outside the zero section of $T^*V^{(J)}$. Using the standard Leibniz product argument, we find an elliptic operator $\mathcal{R}_1 \in \Psi^{-m}(V; \tilde{F}, F; w^{-1})$ such that

$$\sigma_1^{-m}(\mathcal{R}_1) = (\sigma_1^m(\mathcal{A}))^{-1}.$$

Since the symbol mappings behave naturally under composition of operators, it follows readily that $\mathcal{R}_1\mathcal{A} \in \Psi^0(V; F; w^{-1} \circ w)$ is an elliptic operator and

$$\sigma_1^0(\mathcal{R}_1\mathcal{A}) = \sigma_1^{-m}(\mathcal{R}_1)\sigma_1^m(\mathcal{A}) = 1.$$

We now find an elliptic operator $\mathcal{R}_2 \in \Psi^0(V; F; w^{-1} \circ w)$ with the property that

$$\begin{aligned} \sigma_1^0(\mathcal{R}_2) &= 1, \\ \sigma_2^0(\mathcal{R}_2) &= \left(\sigma_2^0\left(\mathcal{R}_1\mathcal{A}\right)\right)^{-1}. \end{aligned}$$

Arguing as above, we conclude that $\mathcal{R}_2\mathcal{R}_1\mathcal{A} \in \Psi^0(V; F; w^{-1} \circ w)$ is an elliptic operator and

$$\begin{aligned} \sigma_1^0\left(\mathcal{R}_2\mathcal{R}_1\mathcal{A}\right) &= 1, \\ \sigma_2^0\left(\mathcal{R}_2\mathcal{R}_1\mathcal{A}\right) &= 1. \end{aligned}$$

We continue by induction. Suppose $\mathcal{R}_1, \ldots, \mathcal{R}_J$, $1 \le J < N$, have already been constructed. Find an elliptic operator $\mathcal{R}_{J+1} \in \Psi^0(V; F; w^{-1} \circ w)$ such that

$$\sigma_1^0\left(\mathcal{R}_{J+1}\right) = 1,$$
$$\cdots \quad \cdots \quad \cdots$$
$$\sigma_J^0\left(\mathcal{R}_{J+1}\right) = 1,$$
$$\sigma_{J+1}^0\left(\mathcal{R}_{J+1}\right) = \left(\sigma_{J+1}^0\left(\mathcal{R}_J \ldots \mathcal{R}_1\mathcal{A}\right)\right)^{-1}.$$

Then

$$\sigma_1^0\left(\mathcal{R}_{J+1}\mathcal{R}_J \ldots \mathcal{R}_1\mathcal{A}\right) = 1,$$
$$\cdots \quad \cdots \quad \cdots$$
$$\sigma_J^0\left(\mathcal{R}_{J+1}\mathcal{R}_J \ldots \mathcal{R}_1\mathcal{A}\right) = 1,$$
$$\sigma_{J+1}^0\left(\mathcal{R}_{J+1}\mathcal{R}_J \ldots \mathcal{R}_1\mathcal{A}\right) = 1,$$

which completes the step of induction.

Set

$$\mathcal{R} = \mathcal{R}_N \ldots \mathcal{R}_1,$$

then $\mathcal{R} \in \Psi^{-m}(V; \tilde{F}, F; w^{-1})$ and

$$\sigma_J^0\left(\mathcal{R}\mathcal{A} - 1\right) = 0$$

for all $J = 1, \ldots, N$. Hence it follows that $\mathcal{R}\mathcal{A} - 1$ is a compact operator, i.e., $\mathcal{R}$ is a left parametrix of $\mathcal{A}$.

In the same way we prove the existence of a right parametrix, thus showing that actually $\mathcal{R}$ is a parametrix.

$\square$

References

[Agr65]   M. S. Agranovich, *Elliptic singular integro-differential operators*, Uspekhi Mat. Nauk **20** (1965), no. 5, 3–120.

[AD62]   M. S. Agranovich and A. S. Dynin, *General boundary value problems for elliptic systems in a multidimensional domain*, Dokl. Akad. Nauk SSSR **146** (1962), 511–514.

[AV64]   M. S. Agranovich and M. I. Vishik, *Elliptic problems with parameter and parabolic problems of general type*, Uspekhi Mat. Nauk **19** (1964), no. 3, 53–160.

[BdM71]   L. BOUTET DE MONVEL, *Boundary problems for pseudo-differential operators*, Acta Math. **126** (1971), no. 1–2, 11–51.

[Gel60]   I. M. GELFAND, *On elliptic equations*, Russian Math. Surveys **15** (1960), no. 3, 113–127.

[GM88]   M. GORESKY and R. MACPHERSON, *Stratified Morse Theory*, Springer-Verlag, Berlin et al., 1988.

[Kon67]   V. A. KONDRAT'EV, *Boundary value problems for elliptic equations in domains with conical points*, Trudy Mosk. Mat. Obshch. **16** (1967), 209–292.

[Luk72]   G. LUKE, *Pseudodifferential operators on Hilbert bundles*, J. Diff. Equ. **12** (1972), 566–589.

[Mel87]   R. B. MELROSE, *Pseudodifferential Operators on Manifolds with Corners*, Manuscript MIT, Boston, 1987.

[Mel96b]   R. B. MELROSE, *Fibrations, compactifications and algebras of pseudodifferential operators*, In: Partial Differential Equations and Mathematical Physics. The Danish-Swedish Analysis Seminar 1995 (L. Hörmander and A. Mellin, eds.), Birkhäuser, Basel et al., 1996, pp. 246–261.

[MP77]   V. G. MAZ'YA and B. A. PLAMENEVSKII, *Elliptic boundary value problems on manifolds with singularities*, Problems of Mathematical Analysis, Vol. 6, Univ. of Leningrad, 1977, pp. 85–142.

[Rab69]   V. S. RABINOVICH, *Pseudodifferential operators in non-bounded domains with conical structure at infinity*, Mat. Sb. **80** (1969), 77–97.

[RST97] V. RABINOVICH, B.-W. SCHULZE, and N. TARKHANOV, *A calculus of boundary value problems in domains with non-Lipschitz singular points*, Math. Nachr. **215** (2000), 45 pp.

[RST00] V. RABINOVICH, B.-W. SCHULZE, and N. TARKHANOV, *Elliptic boundary value problems in domains with intersecting cuspidal edges*, Russian Ac. Sci. Math. Dokl. **375** (2000), no. 6, 337–341.

[PS95]   B. A. PLAMENEVSKII and V. N. SENICHKIN, *Solvable operator algebras*, St. Petersburg Math. J. **6** (1995), no. 5, 895–968.

[Sch92] B.-W. SCHULZE, *The Mellin pseudodifferential calculus on manifolds with corners*, Symposium "Analysis on Manifolds with Singularities", Breitenbrunn, 1990. Teubner-Texte zur Mathematik 131, Teubner-Verlag, Leipzig, 1992, pp. 208–289.

[Sch98] B.-W. SCHULZE, *Boundary Value Problems and Singular Pseudodifferential Operators*, J. Wiley, Chichester, 1998.

[Sch01] B.-W. SCHULZE, *Operator Algebras with Symbol Hierarchies on Manifolds with Singularities*, In: J. Gil, D. Grieser, M. Lesch (eds), Approaches to Singular Analysis, Advances in Partial Differential Equations, Birkhäser Verlag, Basel, 2001, pp. 167–207.

[ST99]   B.-W. SCHULZE and N. TARKHANOV, *Ellipticity and parametrices on manifolds with cuspidal edges*, Contemporary Math. **242** (1999), 217–256.

[VE65]   M. I. VISHIK and G. I. ESKIN, *Convolution equations in a bounded region*, Uspekhi Mat. Nauk **20** (1965), no. 3, 89–152.

[VE67]   M. I. VISHIK and G. I. ESKIN, *Elliptic equations in convolution in a bounded domain and their applications*, Uspekhi Mat. Nauk **22** (1967), 15–76.

Edge Sobolev spaces, weakly hyperbolic equations, and branching of singularities

MICHAEL DREHER and INGO WITT

M. Dreher: Institute of Mathematics, University of Tsukuba, Tsukuba-shi, Ibaraki 305–8571, Japan, email: dreher@math.tsukuba.ac.jp; I. Witt: Institute of Mathematics, University of Potsdam, PF 60 15 53, D-14415 Potsdam, Germany, email: ingo@math.uni-potsdam.de.

Summary

Edge Sobolev spaces are proposed as a main new tool for the investigation of weakly hyperbolic equations. The well–posedness of the linear and the semilinear Cauchy problem in the class of such edge Sobolev spaces is proved. Applications to the propagation of singularities for solutions to semilinear problems are considered.

1 Introduction

We consider the two semilinear Cauchy problems

$$Lu = f(u), \qquad (\partial_t^j u)(0, x) = u_j(x), \quad j = 0, 1, \qquad (1.1)$$

$$Lu = f(u, \partial_t u, t^{l_*} \nabla_x u), \qquad (\partial_t^j u)(0, x) = u_j(x), \quad j = 0, 1, \qquad (1.2)$$

where L is the weakly hyperbolic operator

$$L = \partial_t^2 + 2 \sum_{j=1}^n \lambda(t) c_j(t) \partial_t \partial_{x_j} - \sum_{i,j=1}^n \lambda(t)^2 a_{ij}(t) \partial_{x_i} \partial_{x_j}$$

$$+ \sum_{j=1}^n \lambda'(t) b_j(t) \partial_{x_j} + c_0(t) \partial_t \qquad (1.3)$$

with coefficients a_{ij}, b_j, c_j belonging to $C^\infty([-T_0, T_0], \mathbb{R})$ and $\lambda(t) = t^{l_*}$ with some $l_* \in \mathbb{N}_+ = \{1, 2, 3, \dots\}$.

The variables t and x satisfy $(t,x) \in [0,T_0] \times \mathbb{R}^n$; in the end of this paper we will also consider the case $(t,x) \in [-T_0,T_0] \times \mathbb{R}^n$. The operator L is supposed to be weakly hyperbolic with degeneracy for $t = 0$ only, i.e.,

$$\left(\sum_{j=1}^{n} c_j(t)\xi_j\right)^2 + \sum_{i,j=1}^{n} a_{ij}(t)\xi_i\xi_j \geq \alpha_0|\xi|^2, \quad \alpha_0 > 0, \quad \forall(t,\xi).$$

The choice of the exponents of t in (1.3) reflects so–called Levi conditions which are necessary and sufficient conditions for the C^∞ well–posedness of the *linear* Cauchy problem, see [11], [13]. If, for instance, the t–exponent of the coefficient of ∂_{x_j} were less than $l_* - 1$, the linear Cauchy problem for that L would be well–posed only in certain Gevrey spaces, see [18].

We list some known results. The Cauchy problems (1.1), (1.2) are locally well–posed in $C^k([0,T], H^s(\mathbb{R}^n))$ for s large enough ([7], [12], [13], [15]) and $C^k([0,T], C^\infty(\mathbb{R}^n))$ ([3], [4], [5]).

Furthermore, singularities of the initial data may propagate in an astonishing way: in [14], it has been shown that the solution $v = v(t,x)$ of

$$Lv = v_{tt} - t^2 v_{xx} - (4m+1)v_x = 0, \quad m \in \mathbb{N}, \tag{1.4}$$

with initial data $v(0,x) = u_0(x)$, $v_t(0,x) = 0$ is given by

$$v(t,x) = \sum_{j=0}^{m} C_{jm}t^{2j}(\partial_x^j u_0)(x + t^2/2), \quad C_{jm} \neq 0. \tag{1.5}$$

This representation shows that singularities of the initial datum u_0 propagate only to the left.

Taniguchi and Tozaki discovered branching phenomena for similar operators in [19]. They have studied the Cauchy problem

$$v_{tt} - t^{2l_*}v_{xx} - bl_* t^{l_*-1}v_x = 0, \quad (\partial_t^j v)(-1,x) = u_j(x), \quad j = 0,1,$$

and assumed that the initial data have a singularity at some point x_0. Since the equation is strictly hyperbolic for $t < 0$, this singularity propagates, in general, along each of the two characteristic curves starting at $(-1,x_0)$. When these characteristic curves cross the line $t = 0$, they split, and the singularities then propagate along four characteristics for $t > 0$. However, in certain cases, determined by a discrete set of values for b, one or two of these four characteristic curves do not carry any singularities. A relation of this phenomenon to Stokes phenomena was described in [1].

The function spaces $C^k([0,T], H^s(\mathbf{R}^n))$ and $C^k([0,T], C^\infty(\mathbf{R}^n))$, for which local well–posedness could be proved, have the disadvantage that their elements have different smoothness with respect to t and x. We do not know any previous result concerning the weakly hyperbolic Cauchy problem stating that solutions belong to a function space that embeds into the Sobolev spaces $H^s_{\mathrm{loc}}((0,T) \times \mathbf{R}^n)$, for some $s \in \mathbf{R}$, under the assumption that the initial data and the right–hand side belong to appropriate function spaces of the same kind.

In this paper, solutions to (1.1) and (1.2) are sought in edge Sobolev spaces, a concept which has been initially invented in the analysis of elliptic pseudo-differential equations near edges, see [10], [17].

The operator L can be written as

$$L = t^{-\mu} P\left(t, t\partial_t, \Lambda(t)\partial_x\right),$$

where $\Lambda(t) = \int_0^t \lambda(t')\, dt'$ and $P(t,\tau,\xi)$ is a polynomial in τ, ξ of degree $\mu = 2$ with coefficients depending on t smoothly up to $t = 0$. Operators with such a structure arise in the investigation of edge pseudodifferential problems on manifolds with cuspidal edges, where cusps are described by means of the function $\lambda(t)$. The singularity of the manifold requires the use of adapted classes of Sobolev spaces, so–called *edge Sobolev spaces*.

We shall define edge Sobolev spaces $H^{s,\delta;\lambda}((0,T) \times \mathbf{R}^n)$, where $s \geq 0$ denotes the Sobolev smoothness with respect to (t,x) for $t > 0$ and $\delta \in \mathbf{R}$ is an additional parameter. More precisely,

$$H^s_{\mathrm{comp}}(\mathbf{R}_+ \times \mathbf{R}^n)\big|_{(0,T)\times\mathbf{R}^n} \subset H^{s,\delta;\lambda}((0,T) \times \mathbf{R}^n)$$
$$\subset H^s_{\mathrm{loc}}(\mathbf{R}_+ \times \mathbf{R}^n)\big|_{(0,T)\times\mathbf{R}^n}$$

with continuous embeddings.

The elements of the spaces $H^{s,\delta;\lambda}((0,T) \times \mathbf{R}^n)$ have different Sobolev smoothness at $t = 0$ in the following sense: There are traces τ_j, $\tau_j u(x) = (\partial_t^j u)(0,x)$, with continuous mappings

$$\tau_j : H^{s,\delta;\lambda}((0,T) \times \mathbf{R}^n) \to H^{s-\beta j + \beta\delta l_* - \beta/2}(\mathbf{R}^n), \quad \beta = \frac{1}{l_* + 1}$$

for all $j \in \mathbf{N}$, $j < s - 1/2$. This reflects the loss of Sobolev regularity observed when passing from the Cauchy data at $t = 0$ to the solution. Namely, (1.5) shows that $u_0 \in H^{s+m}(\mathbf{R})$ implies $v(t,.) \in H^s(\mathbf{R})$ only, since $C_{mm} \neq 0$.

This phenomenon has consequences for the investigation of the nonlinear problems (1.1), (1.2). The usual iteration procedure giving the existence of solutions for small times cannot be applied in the case of the standard function space $C([0,T], H^s(\mathbb{R}^n))$, since we have no longer a mapping which maps this Banach space into itself.

However, it turns out, that the iteration approach is applicable if we employ the specially chosen edge Sobolev spaces $H^{s,\delta;\lambda}((0,T) \times \mathbb{R}^n)$. Roughly speaking, the iteration algorithm does not feel the loss of regularity, because it has been absorbed in the function spaces. The idea to choose a special function space adapted to the weakly hyperbolic operator has also been used in [6], [8], and [16].

The paper is organized as follows. We construct the edge Sobolev spaces $H^{s,\delta;\lambda}((0,T) \times \mathbb{R}^n)$ and list their properties in Section 2. Then we show in Section 3 how the $H^{s,\delta;\lambda}((0,T) \times \mathbb{R}^n)$ well–posedness of (1.1), (1.2) can be proved. The proofs of these results can be found in [9]. In Section 4, we consider the hyperbolic equation from (1.1), but with data prescribed at $t = -T_0$, and show that the strongest singularities of the solution u propagate in the same way as the singularities of the solution v solving $Lv = 0$ and having the same initial data as u for $t = -T$. The propagation of the singularities of v was discussed in [19].

2   Edge Sobolev Spaces

In this section, we shall define Sobolev spaces on the cone $\mathbb{R}_+$ first. In a second step, edge Sobolev spaces on the manifold $\mathbb{R}_+ \times \mathbb{R}^n$ will be constructed. Then, the restriction of these Sobolev spaces to $(0,T) \times \mathbb{R}^n$ will give us the desired spaces $H^{s,\delta;\lambda}((0,T) \times \mathbb{R}^n)$. Details on the abstract approach to edge Sobolev spaces can be found, e.g., in [10], [17]. Proofs of the results listed here are given in [9].

2.1   Weighted Sobolev Spaces on $\mathbb{R}_+$

We say that $u = u(t)$ belongs to the Mellin Sobolev space $\mathcal{H}^{s,\delta}(\mathbb{R}_+)$, $s \in \mathbb{N}$, $\delta \in \mathbb{R}$, if

$$\|u\|_{\mathcal{H}^{s,\delta}(\mathbb{R}_+)}^2 = \sum_{k=0}^{s} \int_0^\infty \left| t^{-\delta}(t\partial_t)^k u(t) \right|^2 dt < \infty.$$

For arbitrary $s, \delta \in \mathbb{R}$, the Mellin Sobolev space $\mathcal{H}^{s,\delta}(\mathbb{R}_+)$ can be defined by means of interpolation and duality, or by the requirement that

$$\|u\|_{\mathcal{H}^{s,\delta}(\mathbb{R}_+)}^2 = \frac{1}{2\pi i} \int_{\mathrm{Re}\, z = 1/2 - \delta} \langle z \rangle^{2s} |Mu(z)|^2 \, dz < \infty,$$

where $Mu(z) = \int_0^\infty t^{z-1} u(t)\, dt$ denotes the Mellin transform. (Both norms coincide if $s \in \mathbb{N}$.) Recall that $M \colon L^2(\mathbb{R}_+) \to L^2(\{z \in \mathbb{C} \colon \operatorname{Re} z = 1/2\}; (2\pi i)^{-1} dz)$ is an isometry and

$$M\{(-t\partial_t)u\}(z) = z M u(z), \quad M\{t^{-\delta}u\}(z) = Mu(z - \delta).$$

Furthermore, the space $C^\infty_{\mathrm{comp}}(\mathbb{R}_+)$ is dense in $\mathcal{H}^{s,\delta}(\mathbb{R}_+)$.
We introduce the notations

$$H^s(\mathbb{R}_+ \times \mathbb{R}^n) = \{v|_{\mathbb{R}_+ \times \mathbb{R}^n} \colon v \in H^s(\mathbb{R}^{1+n})\}, \quad n \geq 0,$$

$$H^s_0(\overline{\mathbb{R}}_+ \times \mathbb{R}^n) = \{v \in H^s(\mathbb{R}^{1+n}) \colon \operatorname{supp} v \subseteq \overline{\mathbb{R}}_+ \times \mathbb{R}^n\}, \quad n \geq 0,$$

$$S(\overline{\mathbb{R}}_+ \times \mathbb{R}^n) = \{v|_{\mathbb{R}_+ \times \mathbb{R}^n} \colon v \in S(\mathbb{R}^{1+n})\}, \quad n \geq 0.$$

Example 2.1. For $s \geq 0$, $H^s_0(\overline{\mathbb{R}}_+) = \mathcal{H}^{0,0}(\mathbb{R}_+) \cap \mathcal{H}^{s,s}(\mathbb{R}_+)$.

Definition 2.2. Let $s \geq 0$, $\delta \in \mathbb{R}$ and $\omega \in C^\infty(\overline{\mathbb{R}}_+)$ be a cut–off function close to $t = 0$, i.e., $\operatorname{supp} \omega$ is bounded and $\omega(t) = 1$ for t close to 0. Then the *cone Sobolev spaces* $H^{s,\delta;\lambda}(\mathbb{R}_+)$, $H^{s,\delta;\lambda}_0(\overline{\mathbb{R}}_+)$ are defined by

$$H^{s,\delta;\lambda}(\mathbb{R}_+) = \{\omega u_0 + (1-\omega)u_1 \colon u_0 \in H^s(\mathbb{R}_+),\, u_1 \in \mathcal{H}^{s,\delta;\lambda}_\sharp(\mathbb{R}_+)\},$$

$$H^{s,\delta;\lambda}_0(\overline{\mathbb{R}}_+) = \{\omega u_0 + (1-\omega)u_1 \colon u_0 \in H^s_0(\overline{\mathbb{R}}_+),\, u_1 \in \mathcal{H}^{s,\delta;\lambda}_\sharp(\mathbb{R}_+)\},$$

where $\mathcal{H}^{s,\delta;\lambda}_\sharp(\mathbb{R}_+) = \mathcal{H}^{0,\delta l_*}(\mathbb{R}_+) \cap \mathcal{H}^{s,s(l_*+1)+\delta l_*}(\mathbb{R}_+)$. The space $H^{s,\delta;\lambda}(\mathbb{R}_+)$ is equipped with the norm

$$\|u\|^2_{H^{s,\delta;\lambda}(\mathbb{R}_+)} = \|\omega u_0\|^2_{H^s(\mathbb{R}_+)} + \|(1-\omega)u_1\|^2_{\mathcal{H}^{0,\delta l_*}(\mathbb{R}_+) \cap \mathcal{H}^{s,s(l_*+1)+\delta l_*}(\mathbb{R}_+)}.$$

Let us list some properties of these spaces.

Proposition 2.3. (a) *The spaces* $H^{s,\delta;\lambda}(\mathbb{R}_+)$, $H^{s,\delta;\lambda}_0(\overline{\mathbb{R}}_+)$ *do not depend on the choice of the cut–off function* ω, *up to the equivalence of norms.*

(b) $S(\overline{\mathbb{R}}_+)$ *is dense in* $H^{s,\delta;\lambda}(\mathbb{R}_+)$. *If* $s \notin 1/2 + \mathbb{N}$, *then* $H^{s,\delta;\lambda}_0(\overline{\mathbb{R}}_+)$ *is the closure of* $C^\infty_{\mathrm{comp}}(\mathbb{R}_+)$ *in* $H^{s,\delta;\lambda}(\mathbb{R}_+)$.

(c) *For fixed* $\delta \in \mathbb{R}$, $\{H^{s,\delta;\lambda}(\mathbb{R}_+) \colon s \geq 0\}$ *forms an interpolation scale with respect to the complex interpolation method.*

(d) *If* $l_* = 0$, *then* $H^{s,\delta;\lambda}(\mathbb{R}_+) = H^s(\mathbb{R}_+)$ *and* $H^{s,\delta;\lambda}_0(\overline{\mathbb{R}}_+) = H^s_0(\overline{\mathbb{R}}_+)$.

(e) *Let* $\overline{\lambda}(t)$ *be a smooth and strictly increasing function with* $\overline{\lambda}(t) = 1$ *for* $0 \leq t \leq 1$ *and* $\overline{\lambda}(t) = \lambda(t)$ *for* $t \geq 2$, *and set* $\overline{\Lambda}(t) = \int_0^t \overline{\lambda}(t')dt'$. *Then, for all* $s \geq 0$, $\delta \in \mathbb{R}$,

$$H^{s,\delta;\lambda}(\mathbb{R}_+) = \{\overline{\lambda}(t)^{\delta+1/2} w(\overline{\Lambda}(t)) \colon w \in H^s(\mathbb{R}_+)\},$$

$$H^{s,\delta;\lambda}_0(\overline{\mathbb{R}}_+) = \{\overline{\lambda}(t)^{\delta+1/2} w(\overline{\Lambda}(t)) \colon w \in H^s_0(\overline{\mathbb{R}}_+)\}.$$

2.2   The Spaces $H^{s,\delta;\lambda}(\mathbb{R}_+ \times \mathbb{R}^n)$

Recall that the norm of the usual Sobolev space $H^s(\mathbb{R}^{1+n})$ satisfies

$$\|u(t,x)\|^2_{H^s(\mathbb{R}\times\mathbb{R}^n)} = \int_{(\tau,\xi)} \langle(\tau,\xi)\rangle^{2s} \left|F_{(t,x)\to(\tau,\xi)}u(\tau,\xi)\right|^2 \, d\tau \, d\xi$$

$$= \int_{\mathbb{R}^n_\xi} \langle\xi\rangle^{2s} \left\|\kappa_{\langle\xi\rangle}^{-1}\hat{u}(.,\xi)\right\|^2_{H^s(\mathbb{R}_t)} \, d\xi,$$

where $\hat{u}(t,\xi) = F_{x\to\xi}u(t,\xi)$ denotes the partial Fourier transform with respect to x, and $\kappa_\nu : H^s(\mathbb{R}_t) \to H^s(\mathbb{R}_t)$ is the isomorphism defined by

$$\kappa_\nu w(t) = \nu^{1/2}w(\nu t), \quad t \in \mathbb{R}, \quad \nu > 0. \tag{2.1}$$

This leads us to the following definition.

Definition 2.4. Let E be a Hilbert space and $\{\kappa_\nu\}_{\nu>0}$ be a strongly continuous group of isomorphisms acting on E with $\kappa_\nu\kappa_{\nu'} = \kappa_{\nu\nu'}$ for ν, $\nu' > 0$ and $\kappa_1 = \mathrm{id}_E$.
For $s \in \mathbb{R}$, the abstract edge Sobolev space $\mathcal{W}^s(\mathbb{R}^n; (E, \{\kappa_\nu\}_{\nu>0}))$ consists of all $u \in S'(\mathbb{R}^n; E)$ such that $\hat{u} \in L^2_{\mathrm{loc}}(\mathbb{R}^n; E)$ and

$$\|u\|^2_{\mathcal{W}^s(\mathbb{R}^n;(E,\{\kappa_\nu\}_{\nu>0}))} = \int_{\mathbb{R}^n} \langle\xi\rangle^{2s} \left\|\kappa_{\langle\xi\rangle}^{-1}\hat{u}(\xi)\right\|^2_E \, d\xi < \infty.$$

Example 2.5. For $s \geq 0$, and with κ_ν from (2.1),

$$H^s(\mathbb{R}_+ \times \mathbb{R}^n) = \mathcal{W}^s(\mathbb{R}^n; (H^s(\mathbb{R}_+), \{\kappa_\nu\}_{\nu>0})),$$
$$H^s_0(\overline{\mathbb{R}}_+ \times \mathbb{R}^n) = \mathcal{W}^s(\mathbb{R}^n; (H^s_0(\overline{\mathbb{R}}_+), \{\kappa_\nu\}_{\nu>0})).$$

Now we are in a position to define the edge Sobolev spaces $H^{s,\delta;\lambda}(\mathbb{R}_+ \times \mathbb{R}^n)$.

Definition 2.6. Let $s \geq 0$, $\delta \in \mathbb{R}$. Then we define the group $\{\kappa_\nu^{(\delta)}\}_{\nu>0}$ by

$$\kappa_\nu^{(\delta)}w(t) = \nu^{\beta/2 - \beta\delta l_*}w(\nu^\beta t), \quad \nu > 0,$$

where $\beta = 1/(l_* + 1)$, and set

$$H^{s,\delta;\lambda}(\mathbb{R}_+ \times \mathbb{R}^n) = \mathcal{W}^s(\mathbb{R}^n; (H^{s,\delta;\lambda}(\mathbb{R}_+), \{\kappa_\nu^{(\delta)}\}_{\nu>0})),$$
$$H^{s,\delta;\lambda}_0(\overline{\mathbb{R}}_+ \times \mathbb{R}^n) = \mathcal{W}^s(\mathbb{R}^n; (H^{s,\delta;\lambda}_0(\overline{\mathbb{R}}_+), \{\kappa_\nu^{(\delta)}\}_{\nu>0})).$$

The following results are, in part, direct consequences of this definition and Proposition 2.3.

Proposition 2.7. (a) $S(\overline{\mathbf{R}}_+ \times \mathbf{R}^n)$ *is dense in* $H^{s,\delta;\lambda}(\mathbf{R}_+ \times \mathbf{R}^n)$. *If* $s \notin 1/2 + \mathbf{N}$, *then* $H_0^{s,\delta;\lambda}(\overline{\mathbf{R}}_+ \times \mathbf{R}^n)$ *is the closure of* $C_{\mathrm{comp}}^\infty(\mathbf{R}_+ \times \mathbf{R}^n)$ *in* $H^{s,\delta;\lambda}(\mathbf{R}_+ \times \mathbf{R}^n)$.

(b) *For every fixed* $\delta \in \mathbf{R}$, $\{H^{s,\delta;\lambda}(\mathbf{R}_+ \times \mathbf{R}^n): s \geq 0\}$ *forms an interpolation scale with respect to the complex interpolation method.*

(c) *If* $l_* = 0$, *then* $H^{s,\delta;\lambda}(\mathbf{R}_+ \times \mathbf{R}^n) = H^s(\mathbf{R}_+ \times \mathbf{R}^n)$ *and* $H_0^{s,\delta;\lambda}(\overline{\mathbf{R}}_+ \times \mathbf{R}^n) = H_0^s(\overline{\mathbf{R}}_+ \times \mathbf{R}^n)$.

(d) *For each* $a > 0$,

$$H^{s,\delta;\lambda}(\mathbf{R}_+ \times \mathbf{R}^n)\big|_{(a,\infty)\times\mathbf{R}^n}$$
$$= \{\lambda(t)^{1/2+\delta}v(\Lambda(t),x): v \in H^s(\mathbf{R}_+ \times \mathbf{R}^n)\}\big|_{(a,\infty)\times\mathbf{R}^n},$$

where $\big|_{(a,\infty)\times\mathbf{R}^n}$ *means restriction of functions* $u = u(t,x)$ *from the corresponding function space to* $(a,\infty) \times \mathbf{R}^n$. *In particular,*

$$H_{\mathrm{comp}}^s(\mathbf{R}_+ \times \mathbf{R}^n) \subset H^{s,\delta;\lambda}(\mathbf{R}_+ \times \mathbf{R}^n) \subset H_{\mathrm{loc}}^s(\mathbf{R}_+ \times \mathbf{R}^n)$$

with continuous embeddings.

The spaces $H^{s,\delta;\lambda}(\mathbf{R}_+ \times \mathbf{R}^n)$ admit traces at $t = 0$ in the following sense.

Proposition 2.8. *Let* $s \geq 0$, $\delta \in \mathbf{R}$. *Then, for each* $j \in \mathbf{N}$, $j < s - 1/2$, *the map* $S(\overline{\mathbf{R}}_+ \times \mathbf{R}^n) \to S(\mathbf{R}^n)$, $u \mapsto (\partial_t^j u)(0,x)$, *extends by continuity to a map*

$$\tau_j: H^{s,\delta;\lambda}(\mathbf{R}_+ \times \mathbf{R}^n) \to H^{s-\beta j+\beta\delta l_* - \beta/2}(\mathbf{R}^n).$$

Furthermore, the map

$$H^{s,\delta;\lambda}(\mathbf{R}_+ \times \mathbf{R}^n) \to \prod_{j<s-1/2} H^{s-\beta j+\beta\delta l_* - \beta/2}(\mathbf{R}^n), \quad u \mapsto \{\tau_j u\}_{j<s-1/2}$$

is surjective.

Sketch of proof. By interpolation, we may assume that $s \notin 1/2 + \mathbf{N}$. Then

$$H^{s,\delta;\lambda}(\mathbf{R}_+) = \left\{ \sum_{j<s-1/2} \omega(t)\frac{t^j}{j!} d_j: d_j \in \mathbf{C} \; \forall j \right\} \oplus H_0^{s,\delta;\lambda}(\overline{\mathbf{R}}_+),$$

where $\omega \in C^\infty(\overline{\mathbf{R}}_+)$ is a cut-off function close to $t = 0$. Applying the functor $\mathcal{W}^s(\mathbf{R}^n; (\cdot, \{\kappa_\nu^{(\delta)}\}_{\nu>0}))$ completes the proof. $\qquad\square$

Similar arguments lead to the following results.

Proposition 2.9. *For $s \geq 0$, $\delta \in \mathbb{R}$, we have continuity of the following maps:*
(a) $\partial_t \colon H^{s+1,\delta;\lambda}(\mathbb{R}_+ \times \mathbb{R}^n) \to H^{s,\delta+1;\lambda}(\mathbb{R}_+ \times \mathbb{R}^n)$;
(b) $t^l \colon H^{s,\delta;\lambda}(\mathbb{R}_+ \times \mathbb{R}^n) \to H^{s,\delta+l/l_*;\lambda}(\mathbb{R}_+ \times \mathbb{R}^n)$ *for $l = 0,1,\ldots,l_*$;*
(c) $\partial_{x_j} \colon H^{s+1,\delta;\lambda}(\mathbb{R}_+ \times \mathbb{R}^n) \to H^{s,\delta;\lambda}(\mathbb{R}_+ \times \mathbb{R}^n)$ *for $1 \leq j \leq n$;*
(d) $\varphi \colon H^{s,\delta;\lambda}(\mathbb{R}_+ \times \mathbb{R}^n) \to H^{s,\delta;\lambda}(\mathbb{R}_+ \times \mathbb{R}^n)$ *for each $\varphi = \varphi(t) \in \mathcal{S}(\overline{\mathbb{R}}_+)$.*
Here t^l means the operator of multiplication by t^l. Similarly for φ. In particular, the differential operator L from (1.3) is continuous from $H^{s+2,\delta;\lambda}((0,T_0) \times \mathbb{R}^n)$ to $H^{s,\delta+2;\lambda}((0,T_0) \times \mathbb{R}^n)$, where the space $H^{s,\delta;\lambda}((0,T_0) \times \mathbb{R}^n)$ for $s \geq 0$, $\delta \in \mathbb{R}$ is defined in (2.2) below.

2.3   The Spaces $H^{s,\delta;\lambda}((0,T) \times \mathbb{R}^n)$

For $T > 0$, we set

$$H^{s,\delta;\lambda}((0,T) \times \mathbb{R}^n) = H^{s,\delta;\lambda}(\mathbb{R}_+ \times \mathbb{R}^n)\big|_{(0,T)\times\mathbb{R}^n} \tag{2.2}$$

and equip this space with its infimum norm. There is an alternative description of this space provided that $s \in \mathbb{N}$.

Lemma 2.10. *Let $s \in \mathbb{N}$, $\delta \in \mathbb{R}$ and $T > 0$. Then the infimum norm of the space $H^{s,\delta;\lambda}((0,T) \times \mathbb{R}^n)$ is equivalent to the norm $\|.\|_{s,\delta;T}$, where*

$$\|u\|_{s,\delta;T}^2 = \sum_{l=0}^{s} T^{2l-1} \int_0^T \int_{\mathbb{R}_\xi^n} \vartheta_l(t,\xi)^2 \left|\partial_t^l \hat{u}(t,\xi)\right|^2 d\xi\, dt,$$

$$\vartheta_l(t,\xi) = \begin{cases} \langle\xi\rangle^{s-l}\lambda(t_\xi)^{-\delta-l} & : 0 \leq t \leq t_\xi, \\ \langle\xi\rangle^{s-l}\lambda(t)^{-\delta-l} & : t_\xi \leq t \leq T. \end{cases}$$

Here we have introduced the notation $t_\xi = \langle\xi\rangle^{-\beta}$, $\beta = 1/(l_* + 1)$.

For a proof of this and the following results, see [9].

Lemma 2.11. *For s, $s' \geq 0$, δ, $\delta' \in \mathbb{R}$, and $T > 0$,*

$$H^{s,\delta;\lambda}((0,T) \times \mathbb{R}^n) \subseteq H^{s',\delta';\lambda}((0,T) \times \mathbb{R}^n)$$

if and only if

$$s \geq s', \quad s + \beta\delta l_* \geq s' + \beta\delta' l_*. \tag{2.3}$$

The two conditions in (2.3) are related to the fact that the elements of the edge Sobolev spaces have different smoothness for $t > 0$ and $t = 0$, respectively.

The following two results provide a criterion when the superposition operators defined by the right–hand sides of the hyperbolic equations in (1.1) and (1.2) map an edge Sobolev space into itself.

Proposition 2.12. *Assume that $s + \delta \geq 0$. We suppose that $s \in \mathbb{N}$ and $\min\{s, s + \beta\delta l_*\} > (n + 2)/2$. Then $H^{s,\delta;\lambda}((0, T) \times \mathbb{R}^n)$ is an algebra under pointwise multiplication for any $0 < T \leq T_0$. In other words, we have*

$$\|uv\|_{s,\delta;T} \leq C \|u\|_{s,\delta;T} \|v\|_{s,\delta;T}$$

for $u, v \in H^{s,\delta;\lambda}((0, T) \times \mathbb{R}^n)$. Moreover, the constant C is independent of $0 < T \leq T_0$.

The proof consists in a direct, but quite long calculation using the representation of the norm from Lemma 2.10. Then, an interpolation argument (see [2]) gives us the following result:

Corollary 2.13. *Let $f = f(u)$ be an entire function with $f(0) = 0$, i.e., $f(u) = \sum_{j=1}^{\infty} f_j u^j$ for all $u \in \mathbb{R}$. Assume that $\lfloor s \rfloor + \delta \geq 0$ and $\min\{\lfloor s \rfloor, \lfloor s \rfloor + \beta\delta l_*\} > (n + 2)/2$. Then there is, for each $R > 0$, a constant $C_1(R)$ with the property that*

$$\|f(u)\|_{s,\delta;T} \leq C_1(R) \|u\|_{s,\delta;T},$$
$$\|f(u) - f(v)\|_{s,\delta;T} \leq C_1(R) \|u - v\|_{s,\delta;T}$$

provided that $u, v \in H^{s,\delta;\lambda}((0, T) \times \mathbb{R}^n)$ and $\|u\|_{s,\delta;T} \leq R$, $\|v\|_{s,\delta;T} \leq R$.

3   Linear and Semilinear Cauchy Problems

Our considerations start with the linear Cauchy problem

$$Lw(t, x) = f(t, x), \qquad (\partial_t^j w)(0, x) = w_j(x), \quad j = 0, 1. \tag{3.1}$$

We define

$$a(t, \xi) = \sum_{i,j=1}^{n} a_{ij}(t) \frac{\xi_i \xi_j}{|\xi|^2}, \quad b(t, \xi) = -\sum_{j=1}^{n} b_j(t) \frac{\xi_j}{|\xi|},$$

$$c(t, \xi) = \sum_{j=1}^{n} c_j(t) \frac{\xi_j}{|\xi|}.$$

Further we introduce the number

$$Q_0 = -\frac{1}{2} + \sup_{\xi} \frac{|b(0,\xi) + c(0,\xi)|}{2\sqrt{c(0,\xi)^2 + a(0,\xi)}}, \tag{3.2}$$

and fix $A_0 = Q_0 l_*/(l_* + 1) = \beta Q_0 l_*$.

Theorem 3.1. *Let s, $Q \in \mathbb{R}$, $s \geq 1$, $Q \geq Q_0$. Further let $w_0 \in H^{s+A}(\mathbb{R}^n)$, $w_1 \in H^{s+A-\beta}(\mathbb{R}^n)$, and $f \in H^{s-1,Q+1;\lambda}((0,T) \times \mathbb{R}^n)$, where $A = \beta Q l_*$. Then there is a solution $w \in H^{s,Q;\lambda}((0,T) \times \mathbb{R}^n)$ to (3.1). Moreover, the solution w is unique in the space $H^{s,Q_0;\lambda}((0,T) \times \mathbb{R}^n)$.*

Remark 3.2. The parameter A_0 describes the loss of regularity. The explicit representations of the solutions for special model operators in [14] and [19] show that the statement of the Theorem becomes false if $A < A_0$.

By interpolation, it suffices to prove Theorem 3.1 when $s \in \mathbb{N}_+$. In this case, Theorem 3.1 will follow by standard functional–analytic arguments if the following *a priori* estimate is established.

Proposition 3.3. *For each $s \in \mathbb{N}_+$, $Q \geq Q_0$, there is a constant $C_0 = C_0(s,Q)$ with the property that*

$$\|w\|_{s,Q;T} \leq C_0 \left(\|w_0\|_{H^{s+A}(\mathbb{R}^n)} + \|w_1\|_{H^{s+A-\beta}(\mathbb{R}^n)} + T \|f\|_{s-1,Q+1;T} \right)$$

for all $0 < T \leq T_0$. The constant C_0 does not depend on T.

We only sketch the proof and refer the reader to [9] for the details. Let $g = g(t,\xi)$ be the temperate weight function

$$g(t,\xi) = \omega(\Lambda(t)\langle\xi\rangle)t_\xi^{-1} + (1 - \omega(\Lambda(t)\langle\xi\rangle))\lambda(t)|\xi|, \quad (t,\xi) \in [0,T_0] \times \mathbb{R}^n,$$

where $\omega \in C^\infty(\overline{\mathbb{R}}_+)$ has support in $[0,2]$, satisfies $\omega(t) = 1$ for $0 \leq t \leq 1$, and takes values in $[0,1]$. Then we introduce the vector $W(t,\xi) = {}^t(g(t,\xi)\hat{w}(t,\xi), D_t\hat{w}(t,\xi))$ and obtain the first–order system

$$D_t W(t,\xi) = A(t,\xi)W(t,\xi) + F(t,\xi), \tag{3.3}$$

$$A(t,\xi) = \begin{pmatrix} \frac{D_t g(t,\xi)}{g(t,\xi)} & g(t,\xi) \\ \frac{\lambda(t)^2|\xi|^2 a(t,\xi) - i\lambda'(t)|\xi|b(t,\xi)}{g(t,\xi)} & -2c(t,\xi)\lambda(t)|\xi| + ic_0(t) \end{pmatrix},$$

$$F(t,\xi) = {}^t(0, -\hat{f}(t,\xi)).$$

It is clear that this first–order O.D.E system has a unique solution W for each fixed $\xi \in \mathbb{R}^n$. Point–wise estimates of $W(t, \xi)$ and its time derivatives will allow to establish the *a priori* estimate from Proposition 3.3.

If $X(t, t', \xi)$ denotes the fundamental matrix, i.e.,

$$D_t X(t, t', \xi) = A(t, \xi) X(t, t', \xi), \quad X(t, t, \xi) = I, \quad 0 \le t', t \le T_0, \quad (3.4)$$

then $W(t, \xi) = X(t, t', \xi) W(t', \xi) + i \int_{t'}^t X(t, t'', \xi) F(t'', \xi) dt''$. This immediately gives estimates of $|W(t, \xi)|$ if estimates of $X(t, t', \xi)$ have been found. The following lemma is the crucial tool.

Lemma 3.4. *There is a constant $C > 0$ such that*

$$\|X(t, t', \xi)\| \le C \left(\frac{g(t, \xi)}{g(t', \xi)} \right)^{Q_0 + 1}, \quad 0 \le t' \le t \le T_0,$$

holds for all $\xi \in \mathbb{R}^n$, where Q_0 is given by (3.2).

Proof. This has been proved in [9] for the case of $t_\xi \le t' \le t \le T_0$. From this, $\|A(t, \xi)\| \le C g(t, \xi)$, and $\|X(t, t', \xi)\| \le \exp(\int_{t'}^t \|A(t'', \xi)\| \, dt'')$ we derive the desired inequality for arbitrary $0 \le t' \le t \le T_0$. $\qquad \square$

The derivatives $D_t^l W(t, \xi)$ can be estimated by similar arguments and induction on l. Multiplying the resulting point–wise estimates of $|\partial_t^l \hat{w}(t, \xi)|$ with appropriate factors depending on ξ, and integrating the resulting expressions with respect to (t, ξ) completes the proof of Proposition 3.3.

This *a priori* estimate, Corollary 2.13, and the usual iteration procedure give us the following two theorems.

Theorem 3.5. *Let $s \in \mathbb{N}$ and assume that $\min\{s, s + \beta Q_0 l_*\} > (n + 2)/2$, where Q_0 be the number from (3.2). Suppose that $f = f(u)$ is an entire function with $f(0) = 0$. Let $Q \ge Q_0$ and $A = \beta Q l_*$. Then, for $u_0 \in H^{s+A}(\mathbb{R}^n)$, $u_1 \in H^{s+A-\beta}(\mathbb{R}^n)$, there is a number $T > 0$ with the property that a solution $u \in H^{s, Q; \lambda}((0, T) \times \mathbb{R}^n)$ to the Cauchy problem (1.1) exists. This solution u is unique in the space $H^{s, Q_0; \lambda}((0, T) \times \mathbb{R}^n)$.*

Theorem 3.6. *Let $s \in \mathbb{N}$ and assume that $s - 1 > (n + 2)/2$. Suppose that $f = f(u, v, v_1, \ldots, v_n)$ is entire with $f(0, \ldots, 0) = 0$. Let $Q \ge Q_0$ and $A = \beta Q l_*$. Then, for $u_0 \in H^{s+A}(\mathbb{R}^n)$, $u_1 \in H^{s+A-\beta}(\mathbb{R}^n)$, there is a number $T > 0$ with the property that a solution $u \in H^{s, Q; \lambda}((0, T) \times \mathbb{R}^n)$ to the Cauchy problem (1.2) exists. This solution u is unique in the space $H^{s, Q_0; \lambda}((0, T) \times \mathbb{R}^n)$.*

Eventually, we state a result concerning the propagation of mild singularities.

Theorem 3.7. *Let s satisfy the assumptions of Theorem 3.5. Assume $u_0 \in H^{s+\beta Q_0 l_*}(\mathbb{R}^n)$, $u_1 \in H^{s+\beta Q_0 l_* - \beta}(\mathbb{R}^n)$, where Q_0 is given by (3.2). Let v be the solution to*

$$Lv = 0, \quad (\partial_t^j v)(0, x) = u_j(x), \quad j = 0, 1. \tag{3.5}$$

Then the solutions $u, v \in H^{s, Q_0; \lambda}((0, T) \times \mathbb{R}^n)$ to (1.1) and (3.5) satisfy

$$u - v \in H^{s+\beta, Q_0; \lambda}((0, T) \times \mathbb{R}^n).$$

Proof. Corollary 2.13 implies $f(u) \in H^{s, Q_0; \lambda}((0, T) \times \mathbb{R}^n)$. From Lemma 2.11 we deduce that $f(u) \in H^{s-1+\beta, Q_0+1; \lambda}((0, T) \times \mathbb{R}^n)$. The function $w(t, x) = (u - v)(t, x)$ solves $Lw = f(u)$ and has vanishing initial data. An application of Theorem 3.1 concludes the proof. □

Example 3.8. Consider Qi Min–You's operator L from (1.4). Then $l_* = 1$, $\beta = 1/2$, and $Q_0 = 2m$. Theorems 3.1, 3.5, and 3.7 state that the solutions u, v to (1.1), (3.5) satisfy

$$u, \ v \in H^{s, 2m; \lambda}((0, T) \times \mathbb{R}), \qquad u - v \in H^{s+1/2, 2m; \lambda}((0, T) \times \mathbb{R})$$

if $u_0 \in H^{s+m}(\mathbb{R})$, $u_1 \in H^{s+m-1/2}(\mathbb{R})$. Proposition 2.7 then implies

$$u, \ v \in H_{\text{loc}}^s((0, T) \times \mathbb{R}), \qquad u - v \in H_{\text{loc}}^{s+1/2}((0, T) \times \mathbb{R}).$$

We find that the strongest singularities of u coincide with the singularities of v. The latter can be looked up in (1.5) in case $u_1 \equiv 0$.

4   Branching Phenomena for Solutions to Semilinear Equations

In this section, we consider the Cauchy problems

$$Lu = f(u), \quad (\partial_t^j u)(-T_0, x) = \varepsilon w_j(x), \quad j = 0, 1, \tag{4.1}$$

$$Lv = 0, \quad (\partial_t^j v)(-T_0, x) = \varepsilon w_j(x), \quad j = 0, 1, \tag{4.2}$$

with L from (1.3), and we are interested in branching phenomena for singularities of the solution u. Our main result is Theorem 4.5.

We know, e.g., from the example of Qi Min–You that we have to expect a loss of regularity when we pass from the Cauchy data at $\{t = 0\}$ to the solution at $\{t \neq 0\}$. However, we *also* will observe a loss of smoothness if we prescribe Cauchy data at, say, $t = -T_0$ and look at the solution for $t = 0$.

This can be seen as follows. For simplicity, we only consider the operator of Taniguchi–Tozaki,

$$L = \partial_t^2 - t^{2l_*}\partial_x^2 - bl_*t^{l_*-1}\partial_x, \quad x \in \mathbb{R}, \quad b \in \mathbb{R} \tag{4.3}$$

In [19], it was shown that the Fourier transform of the solution $v(t,x)$ to

$$Lv = 0, \quad (\partial_t^j v)(0,x) = u_j(x), \quad j = 0,1,$$

can be written in the form

$$\hat{v}(t,\xi) = \exp(-i\Lambda(t)\xi)\,{}_1F_1\left(\beta(1+b)l_*/2, \beta l_*, 2i\Lambda(t)\xi\right)\hat{u}_0(\xi)$$
$$+ t\exp(-i\Lambda(t)\xi)\,{}_1F_1\left(\beta(1+b)l_*/2 + \beta, \beta(l_*+2), 2i\Lambda(t)\xi\right)\hat{u}_1(\xi),$$

where $\Lambda(t) = \int_0^t \lambda(t')\,dt'$, and ${}_1F_1(\alpha,\gamma,z)$ is the confluent hypergeometric function satisfying

$$_1F_1(\alpha,\gamma,z) = \frac{\Gamma(\gamma)}{\Gamma(\gamma-\alpha)}e^{\pm i\pi\alpha}z^{-\alpha}(1+\mathcal{O}(|z|^{-1}))$$
$$+ \frac{\Gamma(\gamma)}{\Gamma(\alpha)}e^z z^{\hat{\alpha}-\gamma}(1+\mathcal{O}(|z|^{-1}))$$

for $|z| \to \infty$. This leads us to

$$\hat{v}(t,\xi) = (\exp(-i\Lambda(t)\xi)\mathcal{O}(|\xi|^{\alpha_0^-}) + \exp(i\Lambda(t)\xi)\mathcal{O}(|\xi|^{\alpha_0^+}))\hat{u}_0(\xi) \tag{4.4}$$
$$+ (\exp(-i\Lambda(t)\xi)\mathcal{O}(|\xi|^{\alpha_1^-}) + \exp(i\Lambda(t)\xi)\mathcal{O}(|\xi|^{\alpha_1^+}))\hat{u}_1(\xi),$$

for $|\xi| \to \infty$ provided that $t \neq 0$ has been fixed. Here we have set

$$\alpha_0^- = -\beta(1+b)l_*/2, \quad \alpha_0^+ = -\beta(1-b)l_*/2,$$
$$\alpha_1^- = \alpha_0^- - \beta, \qquad \alpha_1^+ = \alpha_0^+ - \beta.$$

In general, one of the exponents α_0^-, α_0^+ and one of the exponents α_1^-, α_1^+ is positive, the other one is negative. We observe a loss of $\max\{\alpha_0^-, \alpha_0^+\}$ derivatives when we start from $t = 0$, and a loss of $\max\{-\alpha_0^-, -\alpha_0^+\}$ derivatives when we arrive at $t = 0$. Therefore, we have to expect both these phenomena when we prescribe Cauchy data at $t = -T_0$ and cross the line $t = 0$. This leads us to the following definition.

Definition 4.1. Let $s \geq 0$, $\delta \in \mathbb{R}$. We say that $u \in H^{s,\delta;\lambda}((-T,T) \times \mathbb{R}^n)$ if $u(t,x) \in H^{s,\delta;\lambda}((0,T) \times \mathbb{R}^n)$, $u(-t,x) \in H^{s,\delta;\lambda}((0,T) \times \mathbb{R}^n)$, and $u(t,x) - u(-t,x) \in H_0^{s,\delta;\lambda}((0,T) \times \mathbb{R}^n)$.

Let $s_-, s_+ \geq 0$, $\delta_-, \delta_+ \in \mathbb{R}$ and suppose that

$$s_- + \beta\delta_- l_* = s_+ + \beta\delta_+ l_*, \quad s_+ \leq s_-.$$

We say that $u \in H^{s-,s+,\delta-,\delta+;\lambda}((-T,T) \times \mathbb{R}^n)$ if $u \in H^{s+,\delta+;\lambda}((-T,T) \times \mathbb{R}^n)$ and $u(-t,x) \in H^{s-,\delta-;\lambda}((0,T) \times \mathbb{R}^n)$. We define the norm by

$$\|u(t,x)\|_{H^{s-,s+,\delta-,\delta+;\lambda}((-T,T)\times\mathbb{R}^n)}$$
$$= \|u(t,x)\|_{H^{s+,\delta+;\lambda}((0,T)\times\mathbb{R}^n)} + \|u(-t,x)\|_{H^{s-,\delta-;\lambda}((0,T)\times\mathbb{R}^n)}.$$

This choice of the norm is possible, since $H^{s-,\delta-;\lambda}((0,T) \times \mathbb{R}^n) \subset H^{s+,\delta+;\lambda}((0,T) \times \mathbb{R}^n)$, compare Lemma 2.11.

Let us consider the equation $Lv = 0$, again, where the operator L is from (4.3), and suppose that the initial data at $t = -T_0$ satisfy $w_0 \in H^{s-}(\mathbb{R})$, $w_1 \in H^{s--1}(\mathbb{R})$. We set $\delta_+ = Q_0 = (|b|-1)/2$, $\delta_- = -1 - Q_0 = (-|b|+1)/2$, $s_+ = s_- + \beta\delta_- l_* - \beta\delta_+ l_*$. The identity (4.4) leads, after some calculation, to a representation of $\hat{v}(t,\xi)$ in terms of $\hat{w}_0(\xi)$ and $\hat{w}_1(\xi)$ which shows that the solution v belongs to $H^{s-,s+,\delta-,\delta+;\lambda}((-T_0,T_0) \times \mathbb{R})$. Moreover,

$$v(0,x) \in H^{s-+\beta\delta-l_*}(\mathbb{R}), \quad (\partial_t v)(0,x) \in H^{s-+\beta\delta-l_*-\beta}(\mathbb{R}),$$

and these statements about the smoothness are best possible.

For the general operator L from (1.3), we can prove the following well-posedness result:

Proposition 4.2. *Let L be the operator from (1.3), Q_0 be the number defined in (3.2), and set*

$$Q_+ = Q_0, \quad Q_- = -1 - Q_0, \quad s_+ = s_- + \beta Q_- l_* - \beta Q_+ l_*,$$

where $s_-, s_+ \geq 1$. Suppose that

$$w_0(x) \in H^{s-}(\mathbb{R}^n), \quad w_1(x) \in H^{s--1}(\mathbb{R}^n),$$
$$f(t,x) \in H^{s--1,s+-1,Q-+1,Q++1;\lambda}((-T_0,T_0) \times \mathbb{R}^n).$$

Then there is a unique solution $w \in H^{s-,s+,Q-,Q+;\lambda}((-T_0,T_0) \times \mathbb{R}^n)$ to

$$Lw = f(t,x), \quad (\partial_t^j w)(-T_0,x) = w_j(x), \quad j = 0,1.$$

Moreover, $(\partial_t^j w)(0,x) \in H^{s-+\beta Q-l_*-\beta j}(\mathbb{R}^n)$ for $j = 0,1$, and

$$\|w(0,x)\|_{H^{s-+\beta Q-l_*}(\mathbb{R}^n)} + \|(\partial_t w)(0,x)\|_{H^{s-+\beta Q-l_*-\beta}(\mathbb{R}^n)} \tag{4.5}$$
$$\leq C(\|w_0\|_{H^{s-}(\mathbb{R}^n)} + \|w_1\|_{H^{s--1}(\mathbb{R}^n)} + \|f(-t,x)\|_{H^{s--1,Q-+1;\lambda}((0,T_0)\times\mathbb{R}^n)}),$$
$$\|w(t,x)\|_{H^{s-,s+,Q-,Q+;\lambda}((-T_0,T_0)\times\mathbb{R}^n)} \leq C(\|w_0\|_{H^{s-}(\mathbb{R}^n)} + \|w_1\|_{H^{s--1}(\mathbb{R}^n)})$$
$$+ C\|f(t,x)\|_{H^{s--1,s+-1,Q-+1,Q++1;\lambda}((-T_0,T_0)\times\mathbb{R}^n)}.$$

The key of the proof is an estimate of the fundamental matrix.

Lemma 4.3. *Let $X(t, t', \xi)$ be the solution to (3.4). Then it holds*

$$\|X(t, t', \xi)\| \leq C \left(\frac{g(t, \xi)}{g(t', \xi)} \right)^{-Q_0}, \quad 0 \leq t \leq t' \leq T_0.$$

Proof. From $X(t, t', \xi) = X(t', t, \xi)^{-1}$ and Lemma 3.4 we deduce that

$$\|X(t, t', \xi)\| \leq C \left(\frac{g(t', \xi)}{g(t, \xi)} \right)^{Q_0+1} |\det X(t', t, \xi)|^{-1}, \quad 0 \leq t \leq t' \leq T_0.$$

Since $\partial_t X = iAX$, we obtain

$$\partial_t \det X(t, t', \xi) = \text{trace}(iA(t, \xi)) \det X(t, t', \xi),$$

$$\text{trace}(iA(t, \xi)) = \frac{\partial_t g(t, \xi)}{g(t, \xi)} - 2ic(t, \xi)\lambda(t)|\xi| - c_0(t),$$

$$|\det X(t, t', \xi)| = \exp \left(\int_{t'}^t \frac{\partial_\tau g(\tau, \xi)}{g(\tau, \xi)} - c_0(\tau) \, d\tau \right), \quad 0 \leq t, t' \leq T_0.$$

This completes the proof.   □

Inverting the time direction, we will obtain an estimate of the fundamental matrix for negative times from this lemma.

Proof of Proposition 4.2. The existence and uniqueness of w is clear since its Fourier transform solves an O.D.E. with parameter ξ. It remains to discuss the smoothness of w. By interpolation, we may assume $s_- \in \mathbb{N}$. Then we can make use of Lemma 2.10 to represent the norms of $w|_{(-T_0,0)\times\mathbb{R}^n}$ and $f|_{(-T_0,0)\times\mathbb{R}^n}$. We define $g(t, \xi) = g(-t, \xi)$ for $(t, \xi) \in (-T_0, 0) \times \mathbb{R}^n$ and are going to show (4.5) and

$$\sum_{l=0}^{s_-} \int_{-T_0}^0 \int_{\mathbb{R}_\xi^n} \langle \xi \rangle^{2(s_-+Q-)} g(t, \xi)^{-2(l+Q-)} |\partial_t^l \hat{w}(t, \xi)|^2 \, d\xi \, dt \qquad (4.6)$$

$$\leq C(\|w_0\|_{H^{s_-}(\mathbb{R}^n)}^2 + \|w_1\|_{H^{s_--1}(\mathbb{R}^n)}^2)$$

$$+ C \sum_{l=0}^{s_--1} \int_{-T_0}^0 \int_{\mathbb{R}_\xi^n} \langle \xi \rangle^{2(s_-+Q-)} g(t, \xi)^{-2(l+Q-+1)} |\partial_t^l \hat{f}(t, \xi)|^2 \, d\xi \, dt,$$

i.e., the *a priori* estimate for $t \leq 0$. Proposition 3.3 then gives the *a priori* estimate for $t \geq 0$, which completes the proof.

For $(t, \xi) \in (-T_0, 0) \times \mathbb{R}^n$, we set $W(t, \xi) = {}^t(g(t, \xi)\hat{w}(t, \xi), D_t\hat{w}(t, \xi))$ as above, and conclude that

$$W(t, \xi) = X(t, -T_0, \xi)W(-T_0, \xi) + i \int_{-T_0}^t X(t, t', \xi)F(t', \xi)\, dt',$$

for $-T_0 \le t \le 0$, where $X(t, t', \xi)$ solves (3.4). Inverting the time direction in (3.3), we conclude from Lemma 4.3 that

$$\|X(t, t', \xi)\| \le C \left(\frac{g(t, \xi)}{g(t', \xi)} \right)^{-Q_0}, \quad -T_0 \le t' \le t \le 0.$$

Consequently,

$$g(t, \xi)^{Q_0}|W(t, \xi)| \tag{4.7}$$
$$\le Cg(-T_0, \xi)^{Q_0}|W(-T_0, \xi)| + C \int_{-T_0}^t g(t', \xi)^{Q_0}|F(t', \xi)|\, dt'.$$

Taking squares, setting $t = 0$, and using $g(-T_0, \xi) \sim \langle \xi \rangle$ yields

$$\langle \xi \rangle^{-2\beta Q_0 l_*}|W(0, \xi)|^2$$
$$\le C|W(-T_0, \xi)|^2 + C \int_{-T_0}^0 \langle \xi \rangle^{-2Q_0}g(t', \xi)^{2Q_0}|F(t', \xi)|^2\, dt'.$$

Multiplying with $\langle \xi \rangle^{2(s_- - 1)}$ and integrating over $\mathbb{R}_\xi^n$ gives (4.5). Clearly,

$$D_t^l W(t, \xi) = \sum_{m=0}^{l-1} \binom{l-1}{m} (D_t^m A(t, \xi))(D_t^{l-1-m}W(t, \xi)) + D_t^{l-1}F(t, \xi),$$

for every $l \ge 1$. From $\|D_t^m A(t, \xi)\| \le Cg(t, \xi)^{m+1}$ and (4.7) we then get

$$\sum_{l=0}^{s_- - 1} g(t, \xi)^{-l}|D_t^l W(t, \xi)| \le C|W(t, \xi)| + \sum_{l=0}^{s_- - 2} g(t, \xi)^{-l-1}|D_t^l F(t, \xi)|,$$

$$\sum_{l=0}^{s_- - 1} \langle \xi \rangle^{s_- - 1 - Q_0}g(t, \xi)^{Q_0 - l}|D_t^l W(t, \xi)|$$

$$\le C\langle \xi \rangle^{s_- - 1 - Q_0}g(-T_0, \xi)^{Q_0}|W(-T_0, \xi)|$$

$$+ C \int_{-T_0}^t \langle \xi \rangle^{s_- - 1 - Q_0}g(t', \xi)^{Q_0}|F(t', \xi)|\, dt'$$

$$+ \sum_{l=0}^{s_- - 2} \langle \xi \rangle^{s_- - 1 - Q_0}g(t, \xi)^{Q_0 - l - 1}|D_t^l F(t, \xi)|.$$

Squaring this relation and integration over $(-T_0, 0) \times \mathbb{R}^n$ gives

$$\left\| F_{\xi \to x}^{-1} W(-t, \xi) \right\|_{H^{s_- - 1, Q_- + 1; \lambda}((0, T_0) \times \mathbb{R}^n)}^2$$
$$\leq C(\|w_0\|_{H^{s_-}(\mathbb{R}^n)}^2 + \|w_1\|_{H^{s_- - 1}(\mathbb{R}^n)}^2)$$
$$+ C \|f(-t, x)\|_{H^{s_- - 1, Q_- + 1; \lambda}((0, T_0) \times \mathbb{R}^n)}^2,$$

which implies (4.6). The proof is complete.   □

Remark 4.4. Due to (3.2), $Q_0 \geq -1/2$, which is equivalent to $s_+ \leq s_-$. If $s_+ = s_-$, no loss of regularity occurs when we cross the line of degeneracy. The case of an hyperbolic operator with this property and countably many points of degeneracy (or singularity) accumulating at $t = 0$ has been discussed in [20].

The next theorem relates branching phenomena for the semilinear problem (4.1) with branching phenomena for the linear reference problem (4.2). This relation between a semilinear Cauchy problem and an associated linear reference problem has already been discussed in Example 3.8.

Theorem 4.5. *Let L be the operator from (1.3), Q_0 be the number from (3.2), and suppose that*

$$\min\{\lfloor s_\pm \rfloor, \lfloor s_\pm \rfloor + \beta Q_\pm l_*\} > \frac{n+2}{2}, \quad \lfloor s_\pm \rfloor + Q_\pm \geq 0, \quad s_\pm \geq 1,$$

where $Q_+ = Q_0$, $Q_- = -1 - Q_0$, and $s_+ = s_- + \beta Q_- l_* - \beta Q_+ l_*$.
Assume that $w_0 \in H^{s_-}(\mathbb{R}^n)$, $w_1 \in H^{s_- - 1}(\mathbb{R}^n)$, and that $f = f(u)$ is an entire function with $f(0) = f'(0) = 0$.
Then there is an $\varepsilon_0 > 0$ such that for every $0 < \varepsilon \leq \varepsilon_0$ there are unique solutions $u, v \in H^{s_-, s_+, Q_-, Q_+; \lambda}((-T_0, T_0) \times \mathbb{R}^n)$ to (4.1) and (4.2), respectively, which, in addition, satisfy

$$u - v \in H^{s_- + \beta, s_+ + \beta, Q_-, Q_+; \lambda}((-T_0, T_0) \times \mathbb{R}^n). \tag{4.8}$$

Proof. For the sake of simplicity, let us denote the Banach space $H^{s_-, s_+, Q_-, Q_+; \lambda}((-T_0, T_0) \times \mathbb{R}^n)$ by B. The conditions on $s_\pm$ and $Q_\pm$ imply that B is an algebra, see Corollary 2.13. Since $f(0) = f'(0) = 0$, we can conclude that

$$\|f(u)\|_{H^{s_- - 1, s_+ - 1, Q_- + 1, Q_+ + 1; \lambda}((-T_0, T_0) \times \mathbb{R}^n)} \leq C \|f(u)\|_B \leq C(R) \|u\|_B^2,$$
$$\|f(u) - f(v)\|_B \leq C(R)R \|u - v\|_B,$$

$\qquad$ for all u, v with $\|u\|_B \leq R, \quad \|v\|_B \leq R$.

If we choose R small enough, and then choose ε small enough, the usual iteration approach works, leading to a solution u to (4.1) with $\|u\|_B \leq R$. From Lemma 2.11 we then deduce that

$$f(u) \in B \subset H^{s_- - 1 + \beta, s_+ - 1 + \beta, Q_- + 1, Q_+ + 1; \lambda}((-T_0, T_0) \times \mathbb{R}^n).$$

The difference $u - v$ solves $L(u - v) = f(u)$ and has vanishing Cauchy data for $t = -T_0$. Then Proposition 4.2 yields (4.8). $\square$

References

[1] K. Amano and G. Nakamura. Branching of singularities for degenerate hyperbolic operators. *Publ. Res. Inst. Math. Sci.*, 20(2):225–275, 1984.

[2] J. Bergh and J. Löfström. *Interpolation Spaces. An Introduction*, Grundlehren Math. Wiss., Vol. 223, Springer, Berlin, 1976.

[3] F. Colombini, E. Jannelli, and S. Spagnolo. Well–posedness in the Gevrey classes of the Cauchy problem for a non–strictly hyperbolic equation with coefficients depending on time. *Ann. Scuola Norm. Sup. Pisa Cl. Sci. (4)*, 10:291–312, 1983.

[4] F. Colombini and S. Spagnolo. An example of a weakly hyperbolic Cauchy problem not well posed in C^∞. *Acta Math.*, 148:243–253, 1982.

[5] P. D'Ancona and M. Di Flaviano. On quasilinear hyperbolic equations with degenerate principal part. *Tsukuba J. Math.*, 22(3):559–574, 1998.

[6] M. Dreher. Weakly hyperbolic equations, Sobolev spaces of variable order, and propagation of singularities. *submitted*.

[7] M. Dreher. *Local solutions to quasilinear weakly hyperbolic differential equations*. Ph.D. thesis, Technische Universität Bergakademie Freiberg, Freiberg, 1999.

[8] M. Dreher and M. Reissig. Propagation of mild singularities for semilinear weakly hyperbolic differential equations. *J. Analyse Math.*, 82, 2000.

[9] M. Dreher and I. Witt. Edge Sobolev spaces and weakly hyperbolic equations. *submitted*.

[10] Y. Egorov and B.-W. Schulze. *Pseudo-Differential Operators, Singularities, Applications*, volume 93 of *Oper. Theory Adv. Appl.* Birkhäuser, Basel, 1997.

[11] V. Ivrii and V. Petkov. Necessary conditions for the Cauchy problem for non–strictly hyperbolic equations to be well–posed. *Russian Math. Surveys*, 29(5):1–70, 1974.

[12] K. Kajitani and K. Yagdjian. Quasilinear hyperbolic operators with the characteristics of variable multiplicity. *Tsukuba J. Math.*, 22(1):49–85, 1998.

[13] O. Oleinik. On the Cauchy problem for weakly hyperbolic equations. *Comm. Pure Appl. Math.*, 23:569–586, 1970.

[14] M.-Y. Qi. On the Cauchy problem for a class of hyperbolic equations with initial data on the parabolic degenerating line. *Acta Math. Sinica*, 8:521–529, 1958.

[15] M. Reissig. Weakly hyperbolic equations with time degeneracy in Sobolev spaces. *Abstract Appl. Anal.*, 2(3,4):239–256, 1997.

[16] M. Reissig and K. Yagdjian. Weakly hyperbolic equations with fast oscillating coefficients. *Osaka J. Math.*, 36(2):437–464, 1999.

[17] B.-W. Schulze. *Boundary Value Problems and Singular Pseudodifferential Operators*. Wiley Ser. Pure Appl. Math. J. Wiley, Chichester, 1998.

[18] K. Shinkai. Stokes multipliers and a weakly hyperbolic operator. *Comm. Partial Differential Equations*, 16(4,5):667–682, 1991.

[19] K. Taniguchi and Y. Tozaki. A hyperbolic equation with double characteristics which has a solution with branching singularities. *Math. Japon.*, 25(3):279–300, 1980.

[20] T. Yamazaki. Unique existence of evolution equations of hyperbolic type with countably many singular or degenerate points. *J. Differential Equations*, 77:38–72, 1989.

Sur les ondes superficieles de l'eau et le dévelopement de Friedrichs dans le système de coordonnées de Lagrange

TADAYOSHI KANO   Département de Mathématiques, Université d'Osaka, Japon

SAE MIKI   Département de Mathématiques, Université d'Osaka, Japon

1. LES ONDES SUPERFICIELLES DE L'EAU.

Les ondes de surface de l'eau ont tendance à s 'écraser sur elles-mêmes et c'est ainsi non seulement sur la plage mais aussi en pleine mer. En effet, G.B.Airy a donné (non pas demontré mathématiquement) en 1848 [1] ses fameuses équations des ondes en eau peu profonde comme la seconde approximation pour les ondes de surface de l'eau d'ampleur finie après la première approximation de Lagrange dans son livre [14] par l'équation linéaire des ondes pour les oscillations infinitésimales :

$$(1.1) \qquad u_t + uu_x + g\gamma_x = 0$$

$$(1.2) \qquad \gamma_t + u\gamma_x + \gamma u_x = 0, \quad \gamma > 0, \quad (\text{Airy}, 1848),$$

où u est la vitesse et $y - \gamma(t,x) = 0$ représente le profil des ondes de surface de l'eau. Elles s 'écrasent, en effet, dans un temps fini; soient :

$$(1.3) \qquad P = u + 2(g\gamma)^{1/2}, \qquad Q = u - 2(g\gamma)^{1/2},$$

199

alors elles satisfont aux équations

$$(1.4) \qquad P_t + ((g\gamma)^{1/2} + u)P_x = 0, \quad Q_t - ((g\gamma)^{1/2} - u)Q_x = 0,$$

représentant des mouvements de surface de l'eau avec les vitesses dépendantes d'apmleur des ondes, c'est-à-dire que la crête s'avance plus vite que le creux et ainsi la "tête" dépasse "le corps" pour s'écraser sur elle-même.

La découverte de l'onde solitaire de surface de l'eau par J.Scott Russell en 1834[27] est donc quelque chose de très singulier et par la suite, un nombre considérable de mathématiciens éminents de 19e siècle se sont intéressés aux ondes superficielles de l'eau tels que, entre autres, G.G.Stokes, Lord Rayleigh...

Airy s'est violemment opposé à l'idée de l'onde solitaire de Scott Russell, alors que G.G.Stokes donnait une seconde approximation de vitesse de propagation indépendante d'ampleur pour les ondes d'ampleur finie en 1848 [30]. L'idée de l'onde solitaire est considérablement renforcée par la découverte de l'équation de Boussinesq [2]:

$$(1.5) \qquad u_{tt} - u_{xx} - (3u^2 + (1/3)u_{xx})_{xx} = 0,$$

et celle de Korteweg - de Vries[13]:

$$(1.6) \qquad u_t + u_x + 3uu_x + (1/3)u_{xxx} = 0,$$

qui admettent les ondes solitaires comme des solutions particulières.

Mais il est très difficile de voir la structure des ondes superficielles de l'eau à travers des équations "approchées" ainsi déduites, même après la démonstration mathématique de l'existence des solutions analytiques périodiques d'ondes de surface de l'eau par Levi-Civita en 1925[17] et ensuite par Struik également en 1925[31], et celle d'ondes solitaires par M.Lavrentiev en 1947[16] et Friedrichs-Hyers en 1954[6].

Or Friedrichs a proposé (ou plutôt découvert) en 1948[5] une procédure systématique pour obtenir les équations d'Airy à partir d'équations d'Euler originales pour les ondes superficielles de l'eau en considérant le problème nondimensionnel pour le potentiel Φ et le profil de surface $y = \Gamma(t,x)$ pour un écoulement bidimensionnel:

(1.7) $\delta^2\Phi_{xx} + \Phi_{yy} = 0$ dans $\Omega = \{(x,y): x \in \mathbf{R},\ 0 < y < \Gamma(t,x)\}$,

(1.8) $\Phi_y = 0,\ x \in \mathbf{R},\ y = 0$,

(1.9) $\delta^2(\Phi_t + (1/2)\Phi_x^2 + y) + (1/2)\Phi_y^2 = 0,\ x \in \mathbf{R},\ y = \Gamma(t,x)$,

(1.10) $\delta^2(\Gamma_t + \Gamma_x\Phi_x) - \Phi_y = 0$, $x \in \mathbf{R},\ y = \Gamma(t,x)$.

En supposant un développement de solutions et d'équations en séries entières par rapport à un paramètre non-dimensionnel δ qui intervient d'une manière naturelle dans la non-dimensionalisation du problème représentant le ratio de la profondeur moyenne h de l'eau à la longuer des ondes λ: $\delta = = h/\lambda$. Friedrichs a montré que les premiers termes de développement satisfaisaient aux équations d'Airy. Il s'est particulièrement intéressé à cette procédure, car les équations d'Airy sont exactement les mêmes que celles pour le mouvement de gaz polytropique [3].

Lui-même n'a pas démontré la légitimité mathématique de son développement, et c'est L.V.Ovsjannikov qui a donné une justification mathématique pour les premiers termes lorsque les ondes étaient spatialement périodiques [25]. Nous avons démontré que les solutions analytiques du problème nondimensionnel existent et sont indéfiniment différentiables par rapport à ce paramètre nondimensionnel δ et que par la suite le développement de Friedrichs est mathématiquement légitime pour les solutions analytiques[9]. On a par conséquent donné une justification mathématique pour les équations d'Airy également pour les solutions analytiques qui ne sont pas nécessairement périodiques spatialement [8].

Le développement de Friedrichs nous a permis de donner une justification mathématique pour l'équation de Boussinesq et celle de Korteweg-de Vries comme des équations approchées découvrant la structure des ondes "longues"[10]; elles satisfont aux relations telles que les quantités ε et δ^2 définies par

$$\varepsilon = \text{(l'ampleur } \alpha\text{)/(la profondeur } \lambda\text{)},$$

(#)

$$\delta^2 = \text{((la profondeur h)/(la longuer } \lambda\text{))}^2$$

sont de même ordre comme infinitésimaux lorsque l'on fait tendre λ vers l'infini et α vers zéro, alors que ε reste fini, δ tendant vers zéro pour les ondes en eau peu profonde d'Airy, (voir également F.Ursell [32]). Notons enfin que nous avons donné également une justification mathématique pour l'équation de Kadomtsev-Petviashvili comme une équation approchée pour les ondes longues caractérisées en haut dans l'écoulement tridimensionnel [12].

Dans cet article, on donnera exactement la même théorie que celle mentionnée plus haut, cette fois-ci dans le système de coordonnées de Lagrange. Les démonstrations détaillées seront publiées ultérieurement sous le titre: Ondes lagrangiennes de surface de l'eau I, II.

2. LE PROBLEME DANS LE SYSTEME DE COORDONNEES DE LAGRANGE .

Le problème est régi par les équations suivantes:

les équations dynamiques

(2.1) $$x_{tt}x_\xi + (y_{tt} + g)y_\xi + p_\xi /\rho = 0$$

(2.2) $$x_{tt}x_\eta + (y_{tt} + g)y_\eta + p_\eta /\rho = 0,$$

et l' équation de continuité

(2.3) $$x_\xi y_\eta - x_\eta y_\xi = 1,$$

dans

(2.4) $$\Omega_0 = \{(\xi,\eta): 0 < \eta < h(\xi)\} \quad,$$

avec les données initiales:

(2.5) $$x(0,\xi,\eta) = \xi, \quad y(0,\xi,\eta) = \eta,$$

(2.6) $$x_t(0,\xi,\eta) = x^1(\xi,\eta), \quad y_t(0,\xi,\eta) = y^1(\xi,\eta);$$

où $p = p(t,\xi,\eta)$ représente la pression, ρ la densité comme une constante et g la pesanteur.

On s'explique, avant d'aborder le problème, pourquoi on essaie de refaire les mêmes études que nous avons déjà faites dans les années 80 dans le système de coordonnées de Lagrange cette fois-ci. Dans le système de coordonnées d'Euler, il s'agissait du problème aux limites pour l'équation elliptique (essentiellement l'équation de Laplace) avec les conditions aux bords définies par les équations (1.9) - (1.10). Et tout le problème était concentré dans la façon d'écrire Φ_y sur la frontière libre (ondes superficielle de l'eau) par les valeurs de Φ_x sur la frontière pour obtenir un système complet d'équations sur la frontière libre. La non-linéarité est concentrée sur cet opérateur Dirichlet-Neumann de fonction du potentiel sur la frontière libre. Pour une solution analytique, nous l'avons explicité suivant Ovsjannikov par le noyau de Cauchy pour l'écoulement bidimensionnel et obtenu des estimations a priori uniformes par rapport au paramètre δ pour Φ_z par les valeurs de $\{\Phi_x,\Phi_y\}$ sur la frontière libre dans le cas de l'écoulement tridimensionnel.

On considère, dans cet exposé, les ondes superficielles de l'eau en écoulement bidimensionnel dans le système de coordonnées de Lagrange où la couche initiale est une bande, c'est-à-dire (2.4) avec $h(\xi) = h$: constante:

(2.4)' $$\Omega_0 = \{(\xi,\eta): \xi \in \mathbf{R}, 0 < \eta < h : \text{constante}\} \quad.$$

Le cas plus général où $h = h(\xi)$ est une fonction de ξ peut être ramené au cas présent si les ondes ne sont pas trop grandes, c'est-à-dire que d'après l'énergie $E(f)$ que l'on définira ci-dessous $h = h(\xi)$ est petite dans le sens qu'on ait $E(h(\xi)-h) + E(D_\xi(h(\xi)-h))$, $D_\xi = \partial / \partial \xi$, est petite.

On va donc considérer l'évolution temporaire de Ω_0 dans

$$(2.7) \qquad \Omega_t = \{(x,y): x \in \mathbf{R}, 0 < y < y(t,\xi, h)\}$$

d'après les équations (2.1),(2.2) et (2.3). On étendra d'abord le problème sur une bande

$$(2.4)'' \qquad \Omega = \{(\xi,\eta): \xi \in \mathbf{R}, 0 < |\eta| < h : \text{constante}\}$$

en définissant les inconnues pour $\eta < 0$ par

$$(2.8) \qquad \begin{aligned} x(t,\xi,\eta) &= x(t,\xi, -\eta) \\ y(t,\xi,\eta) &= - y(t,\xi, -\eta) \\ p(t,\xi,\eta) &= p(t,\xi, -\eta). \end{aligned}$$

Le problème nondimensionnel qu'on va résoudre ici s'écrit comme suit pour les quantités nondimensionnelles, avec prime, définies par

$$(*) \qquad (t,\xi,\eta) = ((\lambda/c)t', \lambda\xi', h\eta'), \quad (x,y,p) = (\lambda x', hy', \rho ghp'),$$

où g, la pesanteur et $c=(gh)^{1/2}$, la vitesse sonore; on écrit les équations en laissant tomber le prime de quantités nondimensionnelles:

$$(2.9) \qquad x_{tt}x_\xi + (\delta^2 y_{tt} + 1)y_\xi + p_\xi = 0$$

$$(2.10) \qquad x_{tt}x_\eta + (\delta^2 y_{tt} + 1)y_\eta + p_\eta = 0,$$

l' équation de continuité:

$$(2.11) \qquad x_\xi y_\eta - x_\eta y_\xi = 1,$$

dans

$$(2.12) \qquad \Omega_1 = \{(\xi,\eta): 0 < |\eta| < 1\} \quad,$$

avec les données initiales

$$(2.13) \qquad x(0,\xi,\eta) = \xi, \ \ y(0,\xi,\eta) = \eta,$$

$$(2.14) \qquad x_t(0,\xi,\eta) = x^1(\xi,\eta), \ \ y_t(0,\xi,\eta) = y^1(\xi,\eta);$$

et les conditions aux bords

$$(2.15) \qquad p(t,\xi,1) = 0, \ \ y(t,\xi,0) = 0,$$

et enfin l'équation de surface sur $\eta = 1$:

$$(2.16) \qquad x_{tt}(1)x_\xi(1) + (\delta^2 y_{tt}(1) + 1)y_\xi(1) = 0$$

avec $x_{tt}(1) = x_{tt}(t,\xi,1)$ etc.

Pour obtenir les solutions analytiques du problème (2.9)-(2.16) pour le système d'équations elliptiques nonlinéaires (2.9)-(2.11), on le transforme suivant M.Shinbrot [28] au problème elliptique linéaire non-homogène pour les coefficients de développement en séries formelles de Taylor de supposées solutions. Soient en effet:

$$(2.17)^1 \qquad x = \xi + \sum_{1 \leq n < \infty} (x^n(\xi,\eta;\delta) / n!) \, t^n,$$

$$(2.17)^2 \qquad y = \eta + \sum_{1 \leq n < \infty} (y^n(\xi,\eta;\delta) / n!) \, t^n,$$

$(2.17)^3$ $\qquad p = -(\eta-1) + \sum_{0 \leq n < \infty} (p^n(\xi,\eta;\delta) / n!)\, t^n\,,$

alors on obtient les équations suivantes pour les coefficients de Taylor, d'ordre n, $(x_n, y_n, p_n)(\xi,\eta;\delta)$:

$(2.18)\quad X^n + (1/\delta)\nabla(\delta y^{n-2}) + \nabla p^{n-2} + \Phi^{n-1} = 0,\; n\geq 3;\quad X^2 + \nabla p^0 = 0,$

$(2.19)\qquad \nabla\cdot X^n + w^{n-1} = 0,\; n\geq 2;\quad D_\xi x^1 + D_\eta(\delta y^1) = 0;$

avec les conditions aux bords:

$(2.20)\qquad p^n(\xi,1;\delta) = 0,\; n\geq 0;\; \delta y^n(\xi,0;\delta) = 0,\; n\geq 1;$

et enfin l'équation de la surface de l'eau:

$(2.21)\quad x^n(1) + (1/\delta)D_\xi(\delta y^{n-2}(1)) + \phi^{n-1}(1) = 0,\; n\geq 3;\quad x^2(1) = 0\,,$

où

$$X^n = (x^n, \delta y^n),\quad \nabla = (D_\xi, D_\eta) \equiv (\partial/\partial\xi, (1/\delta)\partial/\partial\eta),$$

$$\Phi^{n-1} = (\phi^{n-1}, \psi^{n-1}) =$$

$$(2.22)\quad = \sum_{2\leq v \leq n-1} {}_{n-2}C_{v-2}\, (x^v \nabla x^{n-v} + (\delta y^v)(\nabla \delta y^{n-v})),\; n\geq 3,$$

$$\Phi^1 = 0,$$

et

$$(2.23) \quad w^{n-1} = \sum_{1 \leq v \leq n-1} {}_nC_v(D_\xi x^v D_\eta(\delta y^{n-v}) - D_\eta x^v D_\xi(\delta y^{n-v})), \quad n \geq 2.$$

En définissant une certaine énergie $E(X^n)$ pour dérivées par rapport au temps et $E(D^mX^n)$, avec $D^m = \sum_{0 \leq \mu \leq m} D_\xi^{m-\mu}D_\eta^\mu$, pour dérivées par rapport aux variables spatiales, on obtient (voir § 3) une série d'estimations a priori telles que:

PROPOSITION 2.1 Il existe une constante C positive indépendante de (n,δ) telle que l'on ait:

$$E(X^n) \leq C\left[E(D^2X^{n-2}) + E(DX^{n-2}) + \right.$$

$$+ \sum_{2 \leq v \leq n-1} {}_{n-2}C_{v-2} E(X^v)(E(X^{n-v}) + E(DX^{n-v})) +$$

$$(2.24) \quad + \sum_{1 \leq v \leq n-1} {}_nC_v E(X^v)(E(X^{n-v}) + E(DX^{n-v})) +$$

$$+ \sum_{1 \leq v \leq n-3} {}_nC_v \{E(DX^v)E(DX^{n-v-2}) + E(X^v)(E(DX^{n-v-2}) +$$

$$\left. + E(D^2X^{n-v-2}))\} \right].$$

PROPOSITION 2.2 Il existe une constante C positive indépendante de (n,δ) telle que l'on ait:

$$E(D^mX^n) \leq C\left[E(D^{m+2}X^{n-2}) + E(D^{m+1}X^{n-2}) + \right.$$

$$+ \sum_{2 \leq v \leq n-1} {}_{n-2}C_{v-2} \{\sum_{0 \leq \mu \leq m} {}_mC_\mu E(D^\mu X^v)E(D^{m-\mu}X^{n-v}) +$$

$$(2.25) \qquad + \sum_{0 \leq \mu \leq m+1} {}_{m+1}C_{\mu} \, E(D^{\mu}X^{\nu})E(D^{m+1-\mu}X^{n-\nu}) \} \ +$$

$$+ \sum_{1 \leq \nu \leq n-1} {}_{n}C_{\nu} \ \{ \sum_{0 \leq \mu \leq m} {}_{m}C_{\mu} \, E(D^{\mu}X^{\nu})E(D^{m-\mu}X^{n-\nu}) +$$

$$+ \sum_{0 \leq \mu \leq m+1} {}_{m+1}C_{\mu} E(D^{\mu}X^{\nu})E(D^{m+1-\mu}X^{n-\nu}) +$$

$$+ \sum_{1 \leq \nu \leq n-3} {}_{n-2}C_{\nu} \ \{ \sum_{0 \leq \mu \leq m+2} {}_{m+2}C_{\mu} E(D^{\mu}X^{\nu})E(D^{m+2-\mu}X^{n-\nu-2}) +$$

$$+ \sum_{0 \leq \mu \leq m+1} {}_{m+1}C_{\mu} E(D^{\mu}X^{\nu})E(D^{m+1-\mu}X^{n-\nu-2}) \} \quad].$$

Pour obtenir ces estimations a priori uniformes par rapport au paramètre δ, supposons que le mouvement de l'eau soit initialement irrotationnel: en effet, en supprimant la pression p de (2.9)-(2.10) et en intégrant une fois par rapport au temps, on obtient pour la vitesse $(u, \delta v) = = (x_t, \delta y_t)$, l'équation suivante:

$$(2.26) \qquad x_{\xi} u_{\eta} + (\delta y_{\xi})(\delta v_{\eta}) - x_{\eta} u_{\xi} - (\delta y_{\eta})(\delta v_{\xi}) = 0,$$

qui donne

$$D_{\eta} x^1 - D_{\xi}(\delta y^1) = 0,$$

$$(2.27)$$

$$D_{\eta} x^n - D_{\xi}(\delta y^n) + D_{\eta} \phi^{n-1} - D_{\xi} \psi^{n-1} = 0, \quad n \geq 2.$$

D'où, l'on obtient, compte tenu de l'équation (2.19), les équations elliptiques linéaires non-homogènes pour $(x^n, \delta y^n)$, en supprimant p^n :

(2.28) $D_\xi^2 x^n + D_\eta^2 x^n = f^{n-1} \equiv -D_\eta(D_\xi\phi^{n-1} - D_\xi\psi^{n-1}) - D_\xi w^{n-1},$

(2.29) $D_\xi^2(\delta y^n) + D_\eta^2(\delta y^n) = g^{n-1} \equiv D_\xi(D_\eta\phi^{n-1} - D_\xi\psi^{n-1}) - D_\eta w^{n-1},$

$$n \geqq 2.$$

Il est à noter ainsi que les données initiales $(x^1, \delta y^1)$ doivent satisfaire aux équations

(2.30) $D_\xi^2 x^1 + D_\eta^2 x^1 = 0, \ \ D_\xi^2(\delta y^1) + D_\eta^2(\delta y^1) = 0,$

ce qui nous permet d'obtenir les estimations a priori uniformes par rapport à δ par récurrence. On donnera les démonstrations dans le paragraphe suivant.

On va ensuite démontrer dans §5 la convergence de série $\sum_n (E(X^n)/n!)t^n$ qui est la série majorante pour les développements de Taylor $(2.17)^{1,2,3}$. Elle revient à la démonstration de la sommabilité de série double d'estimations a priori des seconds membres dans les propositions 2.1-.2.2. Donc toute la difficulté pour obtenir des estimations a priori uniformes par rapport au paramètre δ pour une solution d'équation elliptique nonlinéaire a été déplacée vers la démonstration de la sommabilité de série double d'estimations a priori.

3. LES ESTIMATIONS A PRIORI. PREUVES DES PROPOSITIONS 2.1-2.2.

Résumons d'abord ce qu'on a dit dans le précédent paragraphe: on voit que les vitesses $(u, \delta v) = (x_t, \delta y_t)$ satisfont aux équations de continuité (2.11) et de sans-vorticité (2.26). L'équation (2.26) et une différentiation de (2.11) par rapport au temps nous donnent le système elliptique suivant d'équations nonlinéaires :

$$x_\xi u_\eta + (\delta y_\xi)(\delta v_\eta) - x_\eta u_\xi - (\delta y_\eta)(\delta v_\xi) = 0,$$

(3.1)

$$(\delta y_\xi)u_\eta - x_\xi(\delta v_\eta) - (\delta y_\eta)u_\xi + x_\eta(\delta v_\xi) = 0.$$

Et ainsi, les coefficients de séries formelles de Taylor de solution $(x,\delta y)$ satisfont aux équations (2.28) - (2.29).

On définit des énergies $E(DX^n)$ et $E(D^m X^n)$, $m \geq 1$, pour $n \geq 1$ par

DEFINITION 3.1 $E(\phi)$ pour $\phi \in C^\infty(\Omega_1) \Leftrightarrow$

$\Leftrightarrow$ $E(\phi) \equiv \||\phi\|| + \||D_\xi\phi\|| + \||D_\eta\phi\|| + \|\phi(1)\| + \|D_\xi\phi(1)\| + \|(D_\eta\phi)(1)\|,$

où

$$\||\phi\|| = \|\phi\|_0 + \|D\phi\|_0 + \|D^2\phi\|_0 \; ; \; \|\phi(1)\| = \|\phi(1)\|_0 + \|D\phi(1)\|_0 + \|D^2\phi(1)\|_0$$

avec $D = D_\xi$ ou D_η et

$$\|\phi\|_0^2 = \int_{-1}^{1}\int_{-\infty}^{\infty} |\phi(\xi,\eta)|^2\, d\xi d\eta; \quad \|\phi(1)\|_0 = \int_{-1}^{1}\int_{-\infty}^{\infty} |\phi(\xi,1)|^2\, d\xi;$$

Et pour $m \geq 1$, pour $n \geq 1$:

$$E(D^m\phi) \quad \text{pour} \quad \phi \in C^\infty(\Omega_1) \Leftrightarrow$$

$\Leftrightarrow$ $E(D^m\phi) \equiv \||D^m\phi\|| + \||D_\xi D^m\phi\|| + \||D_\eta D^m\phi\|| +$

$$+ \|D^m \phi(1)\| + \|D_\xi D^m \phi(1)\| + \|(D_\eta D^m \phi)(1)\|,$$

où

$$\|D^m \phi\| = \Sigma_{0 \leq \mu \leq m} \; \theta^\mu \|D_\xi^{m-\mu} D_\eta^{\;\mu} \phi\| \; ;$$

$$\|D^m \phi(1)\| = \Sigma_{0 \leq \mu \leq m} \; \theta^\mu \|(D_\xi^{m-\mu} D_\eta^{\;\mu}) \phi(1)\|, \; 0 \leq \theta \leq 1.$$

D'après la transformée de Fourier, l'équation (2.28) se ramène à l'équation intégrale suivante pour la transformée de Fourier $x^{n^\wedge}(\eta)$ avec la fonction de Green:

$$x^{n^\wedge}(\eta) = [ch(2\pi\delta k\eta)/ch(2\pi\delta k)]x^{n^\wedge}(1) +$$

$$+ \int_{-1}^{\eta} \delta^2/[(2\pi\delta k)sh(4\pi\delta k)][sh(2\pi\delta k(\eta-1))sh(2\pi\delta k(\sigma+1))]f^{n-1^\wedge}(\sigma)d\sigma +$$

$$+ \int_{\eta}^{1} \delta^2/[(2\pi\delta k)sh(4\pi\delta k)][sh(2\pi\delta k(\eta+1))sh(2\pi\delta k(\sigma-1))]f^{n-1^\wedge}(\sigma)d\sigma =$$

$$=[ch(S\eta)/ch(S)]x^{n^\wedge}(1) +$$

$$+ \delta^2/[Ssh(2S)]\Big[\int_{-1}^{\eta} [sh(S(\eta-1))sh(S(\sigma+1))]f^{n-1^\wedge}(\sigma)d\sigma +$$

$$+ \int_{\eta}^{1} [sh(S(\eta+1))sh(S(\sigma-1))]f^{n-1^\wedge}(\sigma)d\sigma \Big] =$$

$$= [ch(S\eta)/ch(S)]x^{n^\wedge}(1) -$$

$$-S[\delta^2/sh(2S)][\int_{-1}^{\eta} sh(S(\eta-1))sh(S(\sigma+1))\phi^{n-1^\wedge}(\sigma)d\sigma+$$

$$+ \int_{\eta}^{1} sh(S(\eta+1))sh(S(\sigma-1))\phi^{n-1^\wedge}(\sigma)d\sigma] -$$

$$-iS[\delta^2/sh(2S)][\int_{-1}^{\eta} sh(S(\eta-1))ch(S(\sigma+1))\psi^{n-1^\wedge}(\sigma)d\sigma +$$

$$+ \int_{\eta}^{1} sh(S(\eta+1))ch(S(\sigma-1))\psi^{n-1^\wedge}(\sigma)d\sigma] -$$

$$- i[\delta/sh(2S)][\int_{-1}^{\eta} sh(S(\eta-1))sh(S(\sigma+1))w^{n-1^\wedge}(\sigma)d\sigma +$$

$$+ \int_{\eta}^{1} sh(S(\eta+1))sh(S(\sigma-1))w^{n-1^\wedge}(\sigma)d\sigma] +$$

$$- \phi^{n-1^\wedge}(\eta) - [sh(S\eta)/shS]\phi^{n-1^\wedge}(1); \quad \text{où} \quad S = 2\pi\delta k, \; i = (-1)^{1/2}.$$

De même pour δy^n, d'où les lemmes 2.1-2.2 d'après un calcul ordinaire.

4. LES ESTIMATIONS A PRIORI POUR LA PRESSION.

Les coefficients p^n sont définis par X^n puisqu'ils satisfont aux équations (2.18):

$$(2.18) \quad X^n + (1/\delta)\nabla(\delta y^{n-2}) + \nabla p^{n-2} + \Phi^{n-1} = 0, \; n \geq 3; \quad X^2 + \nabla p^0 = 0,$$

c'est-à-dire:

$$D_\xi{}^2 p^{n-2} + D_\eta{}^2 p^{n-2} = -D_\xi x^n - D_\eta(\delta y^n) - D_\xi \phi^{n-1} - D_\eta \psi^{n-1} -$$

$$(4.1) \qquad - (1/\delta)[D_\eta{}^2(\delta y^{n-2}) + D_\xi{}^2(\delta y^{n-2})], \; n \geq 3;$$

$$D_\xi{}^2 p^0 + D_\eta{}^2 p^0 = -D_\xi x^2 - D_\eta(\delta y^2) - D_\xi \phi^1 - D_\eta \psi^1.$$

D'où, l'on obtient la proposition suivante comme les propositions 2.1-2.2:

PROPOSITION 4.1   Il existe une constante positive indépendante de (n, δ) telle que l'on ait

$$E(\delta p^{n-2}) \leqq C[E(X^n) + E(X^{n-2}) + E(DX^{n-2}) +$$

$$+ \sum_{2 \leqq v \leqq n-1} {}_{n-2}C_{v-2} \, E(X^v)E(X^{n-v})],$$

$$E(D^m(\delta p^{n-2})) \leqq C[E(D^mX^n)+E(D^{m-1}X^n)+E(D^mX^{n-2})+$$

$$+\sum_{2 \leqq v \leqq n-1} {}_{n-2}C_{v-2} \, \{\sum_{0 \leqq \mu \leqq m-1} {}_{m-1}C_\mu E(D^\mu X^v)E(D^{m-\mu-1}X^{n-v})+$$

$$+ \sum_{0 \leqq \mu \leqq m} {}_m C_\mu \, E(D^\mu X^v)E(D^{m-\mu}X^{n-v}) \} \;].$$

5. LA CONVERGENCE DE LA SERIE $\sum_n (E(X^n)/n!)t^n$.

Nous suivrons essentiellement les idées de Shinbrot dans [28]. Cette majorante pour (2.17) est convergente pour un temps suffisamment petit pour les données initiales $X^1 = (x^1, \delta y^1)(\xi,\eta)$ uniformément analytiques définies par

DEFINITION 5.1 $X^1 = (x^1, \delta y^1)(\xi,\eta)$ sont uniformément analytiques sur la bande

$$\Omega_1 = \{(\xi,\eta): 0 < |\eta| \leqq 1\} \iff$$

$\iff$ Il existe les constantes C et R positives telles que l'on ait

$$\|D^mX^1\| \leqq CR^m m!, \quad m = 0,1,2,3\dots$$

$\iff$ étant donné un nombre q positif, il existe les constantes c et r positives telles que l'on ait

$$\|D^mX^1\| \leqq cr^m(m+1)^{-q}m!, \quad m = 0,1,2,3\dots.$$

On peut démontrer par récurrence le

THEOREME 5.2 Supposons que les données initiales $X^1 = (x^1, \delta y^1)(\xi, \eta)$ soient uniformément analytiques. Soit $q > 3/2$, il existe $\gamma > 0, r > 0$ et $\varepsilon > 0$, petite, telles que l'on ait:

(1) $$E(X^n) \leq \gamma \, n!(n+1)^{-2q}(r/\varepsilon)^{n-1}, \quad n \geq 1;$$

(2) $$E(D^m X^n) \leq \gamma \,(m+n-1)!(n+1)^{-q}(m+n+1)^{-q} \, r^m (r/\varepsilon)^{n-1}, \quad m \geq 1, n \geq 1.$$

Ceci montre l'existence de solution analytique pour notre problème des ondes surperficielles de l'eau en assurant la convergence de séries $(2.17)^{1,2,3}$. Soient en effet

$$\gamma \, n!(n+1)^{-2q}(r/\varepsilon)^{n-1} \equiv c_n, \quad n \geq 1;$$

$$\gamma \,(m+n-1)!(n+1)^{-q}(m+n+1)^{-q} r^m (r/\varepsilon)^{n-1} \equiv a_n b_{m+n},$$

avec

$$a_n = \gamma \,(n+1)^{-q}(1/\varepsilon)^{n-1}, \quad b_k = (k-1)!(k+1)^{-q} r^{k-1},$$

alors l'estimation dans la Proposition 2.1, par exemple, s'écrit comme:

$$E(DX^n) \leq c[\, a_{n-2} b_{n-1} + a_{n-2} b_n +$$

$$+ \sum_{2 \leq v \leq n-1} {}_{n-2}C_{v-2} \; c_v(c_{n-v} + a_{n-v} b_{n-v+1}) + \ldots],$$

on voit qu'il existe des constantes γ_1, γ_2, γ_3 ... etc telles que l'on ait les inégalités:

$$a_{n-2}b_{n-1} \leqq \gamma_1\varepsilon^2 c_n \, , \quad a_{n-2}b_n \leqq \gamma_2\varepsilon^2 c_n, ... \, , \text{ etc.,}$$

qui nous donnent enfin

$$E(DX^n) \leqq c\varepsilon(\gamma_1\varepsilon + \gamma_2\varepsilon + \gamma_3 + ...)c_n.$$

On a donc $E(DX^n) \leqq c_n$ si l'on choisit ε convenablement petit pour que $c\varepsilon(\gamma_1\varepsilon + \gamma_2\varepsilon + \gamma_3 + +....)$ soit plus petite que 1.

Notons comme dernière remarque que toutes les estimations a priori sont uniformes par rapport au paramètre δ, et que par conséquent notre solution $\{x, \delta y, \delta p\}$ analytique est indéfiniment différentiable par rapport à δ. D'où la légitimité mathématique du développement de Friedrichs pour les ondes de surface de l'eau.

REMARQUE 5.3 Plus haut, nous avons démontré qu'en fait notre solution $\{x, \delta y, \delta p\}$ satisfait aux équations comme les fonctions analytiques par rapport aux $(\xi, \delta\eta)$ qui sont indéfiniment différentiables par rapport à δ. Nous avons considéré en effet le développement de solution en séries de Taylor par rapport aux (D_ξ, D_η) avec $D_\eta = (1/\delta)\, \partial / \partial \eta$, ce qui revient à la dérivation par rapport à $\delta\eta$.

6. EQUATIONS APPROCHEES.

Dans ce dernier paragraphe, nous donnons l'équation des ondes en eau peu profonde et l'équation de Boussinesq via l'équation de Korteweg-de Vries comme des équations approchées dans le système de coordonnées de Lagrange.

(1) L'équation des ondes en eau peu profonde.

On a d'abord le

THEOREME 6.1   La solution uniformément analytique du problème (2.1) - (2.6) satisfait à l'équation $x_{tt}x_\xi - x_{\xi\xi} = O(\delta^2)$ sur la surface de l'eau $\eta = 1$, autrement dit, $X(t,\xi)$ définie par $x(t,\xi,1) = \xi + X(t,\xi)$ satisfait à $X_{tt} - (1+X_\xi)^{-1}X_{\xi\xi} = O(\delta^2)$ au sens d'énergie $E(X)$.

C'est-à-dire que l'équation hyperbolique quasilinéaire $u_{tt} - (1+u_\xi)^{-1}u_{\xi\xi} = 0$ est en fait l'équation d'Airy pour les solutions analytiques dans le système de coordonnées de Lagrange:

TTHEOREME 6.2   Le problème de Cauchy pour $u_{tt} - (1+u_\xi)^{-1}u_{\xi\xi} = 0$ avec les données initiales $u(0,\xi) = X(0,\xi) = 0$ et $u_t(0,\xi) = X_t(0,\xi) = x^1(0,\xi)$ admet une solution unique uniformément analytique satisfaisant à $E(u) \leqq c_0 < 1$ et qui approche $X(t,\xi)$ satisfaisant à $E(X(t,\xi)) \leqq c_0 < 1$ au sens que l'on ait:

$$E(X_t(t) - u_t(t)) = O(\delta^2), \ E(X_\xi(t) - u_\xi(t)) = O(\delta^2),$$

d'après la continuité de solutions par rapport aux seconds membres.

REMARQUE 6.3   F.Ursell "approche" $(1+X_\xi)^{-1}$ par $1 - X_\xi$ et il donne[32] "son" équation des ondes en eau peu profonde

$$u_{tt} - u_{\xi\xi} - (1/2)(u^2)_{\xi\xi} = 0$$

comme l'équation d'Airy.

En effet, en partant de l'équation (2.21) sur la surface, on obtient le théorème 6.1 comme la conséquence des deux propositions suivantes: d'abord

LEMME 6.4 $\delta y^{n-1} = O(\delta)$, $D_\xi(\delta y^{n-1}) = O(\delta)$, $D_\eta(\delta y^{n-1}) = O(\delta)$,

$$D_\xi(x^{n-1}) = O(\delta), \quad D_\eta(x^{n-1}) = O(\delta).$$

LEMME 6.5 $w^{n-1} = O(\delta^2)$, $n \geq 2$; $(D_\eta(\delta^2) - D_\xi \psi^{n-1}) = O(\delta^2)$, $n \geq 3$.

On montre ces deux lemmes par récurrence en appliquant l'un à l'autre alternativement. Ils impliquent la

PROPOSITION 6.6 $\sum_n x^n(1) + (\phi^{n-1})(1) = x_{tt} x_\xi + O(\delta^2)$.

On a ensuite

LEMME 6.7 $$\int_{-1}^{1} D_\xi w^{n-1} d\eta = O(\delta^2).$$

LEMME 6.8 $$(-1/\delta) D_\eta(x^{n-2})(1) =$$

$$= (A_\delta/\delta) D_\xi(x^{n-2})(1) + (1/\delta)(D_\eta \phi^{n-1} - D_\xi \psi^{n-1})(1) + O(\delta^2),$$

où A_δ/δ est l'opérateur pseudo-différentiel défini par l'image inverse de transformée de Fourier de $i(\text{th}(2\pi\delta s))/\delta$ par rapport à s. D'où la

PROPOSITION 6.9 $\quad 1/2 \displaystyle\int_{-1}^{1} D_\xi^2 x^{n-2} d\eta = D_\xi^2 x^{n-2}(1) + O(\delta^2).$

PREUVE $\quad$ On voit en effet que

$$1/2 \int_{-1}^{1} D_\xi^2 x^{n-2} d\eta = 1/2 \int_{-1}^{1} (-D_\eta^2 x^{n-2} + f^{n-3}) d\eta =$$

$$= -(1/\delta)D_\eta x^{n-2}(1) - (1/\delta)(D_\eta \phi^{n-3} - D_\xi \psi^{n-3})(1) - 1/2 \int_{-1}^{1} D_\xi^2 w^{n-3} d\eta ,$$

d'où la proposition étant donnés les lemmes 6.7 - 6.8.

(2) L'équation de Boussinesq et de Korteweg-de Vries pour les ondes longues.

Considérons maintenant le mouvement de déplacement moyen de surface de la surface de l'eau en repos. C'est-à-dire qu'on étudie lafonction $Y(t, \xi, \eta)$ définie par $y = \eta + Y$, ce qui revient à l'étude de la surface libre de l'eau $Y = y - h$, h étant la profondeur moyenne de l'eau en repos.

Soit, en outre, α l'ampleur moyenne des ondes superficielles de l'eau. Au moyent de variables non-dimensionnelles $(*)$ de § 2, on définit l'ampleur non-dimensionnelle Y' par

$$y = \eta + Y = hy' =$$
$$= h\eta' + Y = h(\eta' + Y/h) =$$
$$= h(\eta' + (\alpha Y')/h) = h(\eta' + \varepsilon Y'), \quad \varepsilon = \alpha/h.$$

D'où $y' = \eta' + \varepsilon Y' = \eta' + \delta^2 Y'$ en définissant les ondes longues (voir § 1) par le mouvement d'une surface $Y'(1)$ définie par

$$(\#\#) \qquad\qquad y'(1) = 1 + \delta^2 Y'(1),$$

où $y'(1) = y'(t,\xi,1)$ etc, et la condition (#) "simplifiée" des ondes longues : $\varepsilon = \delta^2$.

Correspondant à Y', on définit les autres quantités non-dimensionnelles X' et P' par

$$x = \lambda x' = \xi + \delta^2 X =$$
$$= \lambda(\xi' + \delta^2 X') \ ;$$
$$p = \rho g h p' =$$
$$= \rho g(h - \eta) + P =$$
$$= \rho g h(1 - \eta') + \rho g h \delta^2 P' =$$
$$= \rho g h((1 - \eta') + \delta^2 P')$$

d'où

$$x' = \xi' + \delta^2 X' \quad \text{et} \quad p' = (1 - \eta') + \delta^2 P'.$$

Elles satisfont aux équations suivantes en laissant tomber le signe prime:

les équations du mouvement,

(6.1) $X_{tt} + (1/\delta)D_\xi(\delta Y) + \delta^2 X_{tt}D_\xi X + \delta^2(\delta Y)_{tt}D_\xi(\delta Y) + D_\xi P = 0,$

(6.2) $(\delta Y)_{tt} + (1/\delta)D_\eta(\delta Y) + \delta^2 X_{tt}D_\eta X + \delta^2(\delta Y)_{tt}D_\eta(\delta Y) + D_\eta P = 0;$

l' équation de continuité,

(6.3) $D_\xi X + D_\eta Y + \delta^2(D_\xi X D_\eta Y - D_\eta X D_\xi Y) = 1,$

dans

(6.4) $\Omega_1 = \{(\xi,\eta): 0 < |\eta| < 1\}$.

Les séries formelles de Taylor sont maintenant:

(6.5)[1] $x = \xi + \delta^2 X = \xi + \sum_{1 \leq n < \infty} \delta^2 (X^n(\xi,\eta;\delta) / n!)\, t^n,$

(6.5)[2] $y = \eta + \delta^2 Y = \eta + \sum_{1 \leq n < \infty} \delta^2 (Y^n(\xi,\eta;\delta) / n!)\, t^n,$

(6.5)[3] $p = (1 - \eta) + \delta^2 P = -(\eta-1) + \sum_{0 \leq n < \infty} \delta^2 (P^n(\xi,\eta;\delta) / n!)\, t^n .$

D'où l'on obtient, en minuscules comme plus haut, les équations suivantes:

(6.6) $x^n + (1/\delta)D_\xi(\delta y^{n-2}) + \delta^2 \phi^{n-1} + D_\xi p^{n-2} = 0,\ n \geq 3;\ x^2 + D_\xi p^0 = 0,$

(6.7) $\delta y^n + (1/\delta)D_\eta(\delta y^{n-2}) + \delta^2 \psi^{n-1} + D_\eta p^{n-2} = 0,\ n \geq 3;\ \delta y^2 + D_\eta p^0 = 0,$

c'est-à-dire

$$D_\eta x^n - D_\xi(\delta y^n) + \delta^2(D_\eta \phi^{n-1} - D_\xi \psi^{n-1}) = 0,$$

avec $\phi^1 = \psi^1 = 0$, et aussi

(6.8) $D_\xi x^n + D_\eta(\delta y^n) + \delta^2 w^{n-1} = 0, n \geq 2;$ $D_\xi x^1 + D_\eta(\delta y^1) = 0,$

$$w^0 = 0;$$

avec les conditions aux bords:

(6.9) $p^n(\xi, 1; \delta) = 0, n \geq 0;$ $\delta y^n(\xi, 0; \delta) = 0, n \geq 1;$

Enfin l'équation de la surface de l'eau $\eta = 1$:

(6.10) $x^n(1) + (1/\delta)D_\xi(\delta y^{n-2}(1)) + \delta^2 \phi^{n-1}(1) = 0, n \geq 3;$ $x^2(1) = 0$;

avec

(6.11) $(1/\delta)D_\xi(\delta y^{n-2}(1)) = -(1/2)\int_{-1}^{1} D_\xi x^{n-2} d\eta - (1/2)\delta^2 \int_{-1}^{1} D_\xi w^{n-3} d\eta,$

c'est-à-dire

$$x^n(1) - (1/2) \int_{-1}^{1} D_\xi^2 x^{n-2} d\eta + \delta^2 \phi^{n-1}(1) - (1/2)\delta^2 \int_{-1}^{1} D_\xi w^{n-3} d\eta = 0.$$

De même que dans § 1, on a la

PROPOSITION 6.10

$$D_\xi y^{n-2}(1) = - D_\xi^2 x^{n-2} - (1/3)\delta^2 D_\xi^4 x^{n-2}(1) + O(\delta^4),$$

d'après le

LEMME 6.11 $\quad D_\xi(D_\eta \phi^{n-3} - D_\xi \psi^{n-3})(\eta) = O(\delta^2), \quad D_\xi w^{n-3}(\eta) = O(\delta^2).$

Cela implique alors le

THEOREME 6.12 $\quad$ Sur la surface de l'eau $\eta = 1$, on a

$$(6.12) \qquad x_{tt} + \delta^2 x_{tt} x_\xi - x_{\xi\xi} - (1/3)\delta^2 x_{\xi\xi\xi\xi} = O(\delta^4),$$

pour $\quad x(t,\xi) = X(t,\xi,1).$

COROLLAIRE 6.13   Le théorème revient à dire que $u = -x_\xi$ satisfait à

$$u_{tt} - u_{\xi\xi} - \delta^2((1/2)u^2 + (1/3)u_{\xi\xi})_{\xi\xi} = O(\delta^4).$$

En effet, on voit que

$$x_{\xi\xi}/(1 + \delta^2 x_\xi) - (1 - \delta^2 x_\xi)\, x_{\xi\xi} = \delta^4 x_\xi^2\, x_{\xi\xi}/(1 + \delta^2 x_\xi) + O(\delta^4),$$

d'où le corollaire. Mais il ne s'ensuit pas immédiatement une justification mathématique pour l'équation de Boussinesq comme une équation approchée des ondes longues superficielles de l'eau. En effet, l'équation de Boussinesq

$$U_{tt} - U_{\xi\xi} - \delta^2((1/2)U^2 + (1/3)U_{\xi\xi})_{\xi\xi} = 0$$

a une " mauvaise" relation de dispersion. Nous avons en fait le

THEOREME 6.14   Les ondes longues de surface de l'eau x sont approchées par les solutions $F(t, \xi)$ et $G(t, \xi)$ des équations de Korteweg - de Vries

$$F_t - F_\xi - \delta^2(FF_\xi + (1/3)F_{\xi\xi\xi}) = \delta^2 R^0,$$
$$G_t + G_\xi + \delta^2(GG_\xi + (1/3)G_{\xi\xi\xi}) = \delta^2 Q^0,$$

satisfaiant aux conditions initiales

$$F(0) = x_\xi(0, \xi) + x_t(0, \xi) \quad \text{et} \quad G(0) = x_\xi(0, \xi) - x_t(0, \xi)$$

avec

$$R^0 = -(u^0 v^0 + (1/2)(v^0)^2 + (1/3)v^0_{\xi\xi})_\xi, \quad Q^0 = -(u^0 v^0 - (1/2)(v^0)^2 + (1/3)v^0_{\xi\xi})_\xi,$$

où $u^0 = x^0_\xi$, $v^0 = x^0_t$, x^0 étant la première approximation

$$x^0_{tt} - x^0_{\xi\xi} = 0 \quad \text{sur } \eta = 1$$

pour (6.1) - (6.3), avec des données initiales

$$x^0(0, \xi) = x(0, \xi) \quad \text{et} \quad x^0_t(0, \xi) = x_t(0, \xi)$$

satisfaisant à

$$E((x^0_t - x_t)(t)) = O(\delta^2), \quad E((x^0_\xi - x_\xi)(t)) = O(\delta^2).$$

PREUVE   En effet, (6.12) s'écrit comme suit:

$$u_t - v_\xi = 0, \quad v_t - u_\xi - \delta^2(uu_\xi + (1/3)u_{\xi\xi\xi}) = O(\delta^4),$$

avec $u = x_\xi$, $v = x_t$ et des données initiales

$$u(0, \xi) = x_\xi(0, \xi) = X_\xi(0, \xi, 1), \quad v(0, \xi) = x_t(0, \xi) = X^1(\xi,1).$$

Alors, f = u+v, g = u−v   satisfait, respectivement, à l'équation suivante:

$$f_t - f_\xi - \delta^2(ff_\xi + (1/3f_{\xi\xi\xi}) = \delta^2 R + O(\delta^4),$$

$$g_t + g_\xi + \delta^2(gg_\xi + (1/3)g_{\xi\xi\xi}) = \delta^2 Q + O(\delta^4),$$

avec

$$R = -(uv + (1/2)v^2 + (1/3)v_{\xi\xi})_\xi, \quad Q = -(uv - (1/2)v^2 + (1/3)v_{\xi\xi})_\xi.$$

On a donc

$$E(f(t) - F(t)) = O(\delta^4) \quad \text{et} \quad E(g(t) - G(t)) = O(\delta^4),$$

d'après la dépendance continue de solutions des seconds membres, vu que $R(t) - R^0(t) = O(\delta^2)$, $Q(t) - Q^0(t) = O(\delta^2)$. D'où le théorème, puisque l'on a des équations

$$u(t) = (1/2)(f+g)(t) = (1/2)(F+G)(t) + O(\delta^4),$$

$$v(t) = (1/2)(f-g)(t) = (1/2)(F-G)(t) + O(\delta^4);$$

c'est-à-dire que l'on obtient

$$E(2x_\xi(t) - (F + G)(t)) = O(\delta^4) \quad \text{et} \quad E(2x_t(t) - (F - G)(t)) = O(\delta^4).$$

Pour clore, remarquons que l'on retrouve le même résultat que celui dans [10], page 405, la proposition 3.1: l'équation de Koreteweg-de Vries est justifiée mathématiquement comme l'équation approchée pour les ondes longues de surface de l'eau dont les données initiales satisfont aux conditions $E(f(0)) = O(1)$ et $E(g(0)) = O(\delta^2)$. C'est-à-dire que l'on a la

PROPOSITION 6.15   Supposons que les données initiales satisfassent aux conditions suivantes:

(##) $E(\mathbf{x}_\xi(0, \xi) + \mathbf{x}_t(0, \xi)) = O(1)$, $E(\mathbf{x}_\xi(0, \xi) - \mathbf{x}_t(0, \xi)) = O(\delta^2)$.

Alors les ondes longues de surface de l'eau sont approchées par les solutions des équations de Korteweg-de Vries suivantes:

$$\mathbf{F}_t - \mathbf{F}_\xi - \delta^2((3/4)\mathbf{F}\,\mathbf{F}_\xi + (1/2)\mathbf{F}_{\xi\xi\xi}) = 0 \ ,$$

$$\mathbf{G}_t + \mathbf{G}_\xi + \delta^2((3/4)\mathbf{G}\,\mathbf{G}_\xi + (1/2)\mathbf{G}_{\xi\xi\xi}) = \delta^2(-(3/4)\mathbf{F}\,\mathbf{F}_\xi + (1/6)\mathbf{F}_{\xi\xi\xi}) \ ,$$

satisfaisant aux conditions initiales $\mathbf{F}(0) = F(0)$ et $\mathbf{G}(0) = G(0)$, au sens que l'on ait

$$E(f(t) - \mathbf{F}(t)) = O(\delta^4) \quad \text{et} \quad E(g(t) - \mathbf{G}(t)) = O(\delta^4).$$

PREUVE   D'après les définitions   de   f   et   g, on voit que $\delta^2 R$ et $\delta^2 Q$ s'écrivent   comme

$$\delta^2 R = \delta^2(-(1/4)ff_\xi + (1/6)f_{\xi\xi\xi}) + \delta^2((3/4)gg_\xi - (1/4)(fg)_\xi - (1/6)g_{\xi\xi\xi}),$$

$$\delta^2 Q = \delta^2((1/4)gg_\xi - (1/6)g_{\xi\xi\xi}) + \delta^2(-(3/4)ff_\xi + (1/4)(fg)_\xi + (1/6)f_{\xi\xi\xi}).$$

Par conséquent, on obtient d'abord

$$f_t - f_\xi - \delta^2((3/4)ff_\xi + (1/2)f_{\xi\xi\xi}) = O(\delta^4),$$

puisque l'on a $g(t) = O(\delta^2)$ d'après la dépendance continue des solutions des données initiales et des seconds membres d'équation pour g. Ainsi $f(t) - \mathbf{F}(t)$ satisfait à l'équation linéarisée non-homogène de Korteweg-deVries avec un second membre d'ordre $O(\delta^4)$ et la donnée initiale nulle. D'où l'on obtient $f(t) = \mathbf{F}(t) + O(\delta^4)$. De même, on obtient $g(t) = \mathbf{G}(t) + O(\delta^4)$ puisque l'on a

$$g_t + g_\xi + \delta^2((3/4)gg_\xi + (1/2)g_{\xi\xi\xi}) = \delta^2(-(3/4)ff_\xi + (1/6)f_{\xi\xi\xi}) + O(\delta^4),$$

compte tenu du fait que $E(f(t) - \mathbf{F}(t)) = O(\delta^4)$ et $g(0) - \mathbf{G}(0) = 0$. D'où la proposition, puisque l'on a

$$u(t) = x_\xi(t, \xi) = (1/2)(f + g)(t, \xi) = (1/2)(\mathbf{F}(t, \xi) + \mathbf{G}(t, \xi)) + O(\delta^4),$$

$$v(t) = x_t(t, \xi) = (1/2)(f - g)(t, \xi) = (1/2)(\mathbf{F}(t, \xi) - \mathbf{G}(t, \xi)) + O(\delta^4).$$

Bibliographie

[1] G.B.Airy:Tides and wave, B.Fellowes, London, 1845, 241-396.

[2] J.Boussinesq:Théorie des ondes, J.Math.Pure.Appl., 2e série, t.17 (1872), 55-108.

[3] R.Courant-K.-O.Friedrichs:Supersonic flow and shock waves, Springer, 1948 and 1976.

[4] W.Craig:An existence theory for water waves and the Boussinesq and Korteweg-de Vries scaling limits, Comm.in PDE, 10(1985), 787-1003.

[5] K.-O.Friedrichs:On the derivation of the shallow water theory, Appenndix to: "The formation of breakers and bores " by J.J.Stoker in Comm.Pure Appl.Math., 1(1948), 1-87.

[6] K.-O.Friedrichs-D.H.Hyers:The existence of the solitary waves, Comm. Pure Appl. Math., 7(1954), 517-550.
[7]_____:Asymptotic phenomena in mathematical physics, Bull.AMS, 61(1955), 485-504.

[8] T.Kano-T. Nishida:Sur les ondes de surface de l'eau avec une justification mathématique des équations des ondes en eau peu profonde, J.Math.Kyoto Univ., 19(1979), 335-370.

[9]_____:Water waves and Friedrichs expansion., Lect.Note. Num.Appl.Anal., Kinokuniya-North Holland, 6(1983), 39-57.

[10] _____:A mathematical justification for Korteweg-de Vries equation and Boussinesq equation of water surface waves, Osaka J. Math., 23(1986), 389-413.

[11] T.Kano:Une théorie trois-dimensionnelle des ondes de surface de l'eau et le développement de Friedrichs, J.Math. Kyoto Univ., 26(1986), 101-155 et 157-175.

[12] T.Kano:L'équation de Kadomtsev-Petviashvili approchant des ondes longues de surface de l'eau en écoulement trois-dimensionnel., Studies Math.Appl., 18: "Patterns and Waves-Qualitative analysis of nonlinear differential equations ", North Holland, 1986, 431-444.

[13] D.J.Korteweg-G.de Vries:On the change of the form of long waves advancing in a rectangular canal and a new type of water surface waves, Phil. Magazine, 39(1895), 422-443.

[14] L.Lagrange:Mécanique analytique, t.2, Mme Ve Courcier, Paris, 1815.

[15] H.Lamb:Hydrodynamics,6th ed., Dover, 1932.

[16] M.Lavrentiev:On the theory of long waves, Recueil des Travaux de l'Institut Mathématique de l'Académie des Sciences de la RSS d'Ukraine, 1946, no. 8, 13-69(1947).

[17] T.Levi-Civita:Détermination rigoureuse des ondes permanentes d'ampleur finie, Math.Annalen 93(1925), 264-314.

[18] V.I.Nalimov:Cauchy-Poisson problem, Continuum Dynamics, Hydro-dynamics Institut, Siberian Section, URSS Akad. Nauk, 18(1974), 104-210.

[19]_____:A priori estimates of solutions of elliptic equations in the class of analytic functions and their applications to the Cauchy-Poisson problem, Soviet Math.Dokl., 10(1969), 1350-1354.

[20] L.Nirenberg:An abstract form of the nonlinear Cauchy-Kowalevski theorem, J.Diff.Geometry, 6(1972), 561-576.

[21] T.Nishida:Anote on a theorem of Nirenberg, J.Diff.Geometry, 12 (1977), 629-633.

[22] T.Nishida:Nonlinear hyperbolic equations and related topics in fluid dynamics, Publications Mathématiques d'Orsay, Université Paris-Sud, Département de Mathématiques, Orsay, 1978.

[23] T.Nishida:Equations of Fluid Dynamics - Free Surface Problems, Comm. Pure Appl.Math., 39(1986), 221-238.

[24] L.V.Ovsjannikov:Problème de Cauchy nonlinéaire dans l'échelle des espaces de Banach, Dokl.Akad.Nauk URSS, 200(1971), 789-792.

[25] _____:To the shallow water theory foundation, Arch. Mech.,(Varsovie), 26(1974), 407-422.

[26] _____:Cauchy problem in a scale of Banach spaces and its application to the shallow water theory justification, Lect.Note in Math., Springer-Verlag, 503(1976), 426-437.

[27] J.Scott Russell:Report on waves, in "Report of the fourteenth meeting of the British Association for the Advancement of Science held at York in September 1844", John Murray, London, 1845, 311-390.

[28] M.Shinbrot:The initial value problem for surface waves under gravity, I.The simplest case, Indiana Univ.Math.J., 25(1976), 281-300.

[29] J.J.Stoker:Water waves, the mathematical theory with applications, Interscience, New York, 1957.

[30] G.G.Stokes:On the theory of oscillatory waves, Trans.Camb. Phil. Soc., 8(1847), 441-455.

[31] D.J.Struik:Détermination rigoureuse des ondes irrotationnelles périodiques dans un canal à profondeur finie, Math.Annalen, 95(1925), 595-634.

[32] F.Ursell:The long wave paradox in the theory of gravity waves, Proc. Phil. Soc. Cambridge, 49(1953), 685-694.

[33] L.C.Woods:The theory of subsonic plane flow, Cambridge University Press, 1961.

[34] H.Yosihara:Gravity waves on the free surface of an incompressible perfect fluid of finite depth, Publ.RIMS, Kyoto Univ., 18(1982), 49-96.

[35] _____:Gravity waves on the free surface of an incompressible perfect fluid of finite depth (Revised unpublished manuscript), april 1982.

[36] _____:Capillary-gravity waves for an incompressible ideal fluid, J.Math.Kyoto Univ., 23(1983), 649-694.

Problème de Cauchy pour certains systèmes de Leray-Volevich du type de Schrödinger

JIRO TAKEUCHI

Faculty of Industrial Science and Technology, Science University of Tokyo, Campus Oshamanbe, Hokkaido, 049-3514, Japan

e-mail : takeuchi@rs.kagu.sut.ac.jp

Dédié au Professeur Jean Vaillant

Résumé. Nous donnons des conditions suffisantes et des conditions nécessaires afin que le problème de Cauchy pour certains opérateurs du type de Schrödinger soit bien posé dans un espace de Fréchet H^∞. Les opérateurs qui concernent sont certains systèmes de Leray-Volevich d'opérateurs différentiels linéaires. La méthode que nous utilise dans cet article est une L^2-symétrisation indépendante du temps d'opérateurs à la perte de régularité ; c'est une généralisation de la L^2-symétrisation indépendante du temps d'opérateurs sans perte de régularité obtenue par Takeuchi ([64]).

Abstract. We give sufficient conditions and necessary conditions for the Cauchy problem for certain operators of Schrödinger type to be well posed in a Fréchet space H^∞. Operators of which we treat are certain Leray-Volevich systems of linear partial differential operators. The method that we use in this article is a time-independent L^2-symmetrization of operators with loss of regularity which is a generalization of the time-independent L^2-symmetrization of operators without loss of regularity obtained by Takeuchi ([64]).

Mathematics Subject Classification 2000 : 35G10, 35Q40

Keywords and phrases : Cauchy problem, Schrödinger type equation, Leray-Volevich system of Schrödinger type, Time-independent L^2-symmetrization of operators

1 INTRODUCTION ET ÉNONCÉ DES RÉSULTATS

1.1 Conditions suffisantes — On considère un système d'équations aux dérivées partielles :

$$(1.1) \qquad P(x, D_x, D_t)u(x,t) = D_t u(x,t) - A(x, D_x)u(x,t) = 0,$$

où $x = (x_1, \ldots, x_n) \in \mathbb{R}^n$, $t \in \mathbb{R}^1$, $D_x = (D_1, \ldots, D_n)$, $D_j = i^{-1}\partial/\partial x_j$ $(1 \le j \le n)$, $D_t = i^{-1}\partial/\partial t$, $A(x, D_x) = (a_{jk}(x, D_x))$ est une matrice $m \times m$ dont la (j, k)-composante $a_{jk}(x, D_x)$ est un opérateur différentiel linéaire d'order r_{jk} à coefficients $\mathcal{B}^\infty(\mathbb{R}^n)$ et $u(x,t) = {}^t(u_1(x,t), \ldots, u_m(x,t))$ est un vecteur d'inconnues.

On impose une hypothèse *a priori* sur les ordres r_{jk} des opérateurs $a_{jk}(x, D_x)$:

Hypothèse (L-V) — Il existe une suite $\{s_1, \ldots, s_m\}$ de m entiers non-négatifs telle que l'on ait $r_{jk} \le s_k - s_j + 2$ $(j, k = 1, \ldots, m)$.

D'après Volevich [66], il est toujours possible de trouver de telles suites $\{s_1, \ldots, s_m\}$ si l'on suppose que

$$\max_\sigma \left(\frac{1}{m} \sum_{j=1}^m r_{j\sigma(j)} \right) \le 2,$$

où σ parcourt toutes les permutations de $\{1, \ldots, m\}$.

Cette formulation de $\{s_1, \ldots, s_m\}$ est due à Mizohata [39] ; un tel système $D_t I - A(x, D_x)$ est un des systèmes de Leray-Volevich (voir Leray [34], Volevich [66], [67], Gårding-Kotake-Leray [12], Hufford [18], Wagschal [68] et Miyake [37]).

Si une suite $\{s_1, \ldots, s_m\}$ de m entiers satisfait à la condition $r_{jk} \leq s_k - s_j + 2$ $(j, k = 1, \ldots, m)$, alors, pour tout entier p, $\{s_1 + p, \ldots, s_m + p\}$ satisfait à la même condition. Donc, on peut choisir comme s_j $(1 \leq j \leq m)$ un entier non négatif et le plus petit satisfaisant à la condition $r_{jk} \leq s_k - s_j + 2$ $(j, k = 1, \ldots, m)$.

Désormais, on fixe $\{s_1, \ldots, s_m\}$ dans l'hypothèse (L-V). Dans cette situation, l'opérateur $a_{jk}(x, D_x)$ peut s'écrire comme suit :

$$(1.2) \qquad a_{jk}(x, D_x) = \sum_{|\alpha| \leq s_k - s_j + 2} a_{jk\alpha}(x) D_x^\alpha \quad (1 \leq j, k \leq m),$$

avec les coefficients $a_{jk\alpha}(x) \in \mathcal{B}^\infty(\mathbb{R}^n)$, où $a_{jk}(x, D_x) \equiv 0$ si $s_k - s_j + 2 < 0$.

Nous allons donner des conditions afin que le problème de Cauchy pour le futur et pour le passé en même temps

$$(*) \qquad \begin{cases} D_t u(x, t) - A(x, D_x) u(x, t) = f(x, t) \quad \text{sur} \quad \mathbb{R}^n \times [-T, T] \quad (T > 0) \\ u(x, 0) = u_0(x) \quad \text{dans} \quad \mathbb{R}^n \end{cases}$$

soit bien posé dans $H^\infty(\mathbb{R}^n) = \bigcap \{H^s(\mathbb{R}^n); s \in \mathbb{R}\}$.

DÉFINITION 1.1 — (1) On dit que le problème de Cauchy $(*)$ pour le futur et pour le passé en même temps est bien posé dans $H^\infty(\mathbb{R}^n)$ si et seulement si pour tout $u_0(x) \in (H^\infty(\mathbb{R}^n))^m$ et tout $f(x, t) \in (C_t^\infty([-T, T]; H^\infty(\mathbb{R}^n)))^m$, il existe une solution unique $u(x, t)$ du problème de Cauchy $(*)$ telle que

$$u(x, t) \in (C_t^\infty([-T, T]; H^\infty(\mathbb{R}^n)))^m$$

et que de plus on ait l'inégalité d'énergie suivante : pour tout $l \in \mathbb{R}^1$, il existe $l' \in \mathbb{R}^1$ et une constante $C(l, l', T)$ qui ne dépendent pas de $u(t)$ tels que

$$|||u(t)|||_{(l)} \leq C(l, l', T) \left\{ |||u(0)|||_{(l')} + \left| \int_0^t |||f(s)|||_{(l')} ds \right| \right\}, \quad t \in [-T, T],$$

où

$|||u(t)|||_{(l)}^2 = \sum_{j=1}^m ||\langle D_x \rangle^{s_j} u_j(t)||_{(l)}^2$, $u_j(t) = u_j(\cdot, t)$ est la j-composante de $u(t) = u(\cdot, t)$ et $||u_j(t)||_{(l)}$ est la $H^l(\mathbb{R}^n)$-norme de $u_j(t)$.

Si on peut prendre $l' = l$, on dit que le problème de Cauchy $(*)$ est bien posé dans $H^l(\mathbb{R}^n)$ pour tout $l \in \mathbb{R}^1$.

(2) Soit κ_0 un nombre non négatif. On dit que le problème de Cauchy pour le futur et pour le passé en même temps $(*)$ est bien posé dans $H^\infty(\mathbb{R}^n)$, à la perte de régularité κ_0 près, si et seulement si pour tout $u_0(x) \in (H^\infty(\mathbb{R}^n))^m$ et pour $f(x,t) \equiv 0$, il existe une solution unique $u(x,t)$ du problème de Cauchy telle que

$$u(x,t) \in (C_t^\infty([-T,T]; H^\infty(\mathbb{R}^n)))^m$$

et que de plus, pour tout $l \in \mathbb{R}$, on ait l'inégalité d'énergie suivante :

$$|||u(t)|||_{(l-\kappa_0)} \leq C(l,T)|||u(0)|||_{(l)}, \quad t \in [-T,T].$$

où $C(l,T)$ est une constante positive qui ne dépend que des constantes l et T et de l'opérateur $P(x,D_x,D_t)$.

Dans le cas où $\kappa_0 = 0$, on dit que le problème de Cauchy $(*)$ est bien posé dans $H^l(\mathbb{R}^n)$ pour tout $l \in \mathbb{R}^1$.

Nos conditions s'expriment comme suit.

La première condition pour la partie principale est :

CONDITION (A.1) — On a $a_{jk\alpha}(x) = a_{jk\alpha}$ (Constante) pour $|\alpha| = s_k - s_j + 2$, $1 \leq j, k \leq m$.

On note $a_{jk}^0(\xi)$ le symbole principal de l'opérateur $a_{jk}(x, D_x)$:

$$(1.3) \qquad\qquad a_{jk}^0(\xi) = \sum_{|\alpha|=s_k-s_j+2} a_{jk\alpha}\,\xi^\alpha$$

et on pose $A_2(\xi) = (a_{jk}^0(\xi))$ qui s'appelle la *matrice caractéristique*.

La deuxième condition pour la partie principale est :

CONDITION (A.2) — Les valeurs propres de la matrice caractéristique $A_2(\xi)$ sont non nulles, réelles et distinctes pour $\xi \in \mathbb{R}^n \setminus 0$:

$$\det(\tau I - A_2(\xi)) = \prod_{j=1}^m (\tau - \lambda_j^0(\xi)) ;$$

$$\lambda_j^0(\xi) \neq \lambda_k^0(\xi) \;\; (j \neq k, \, \xi \neq 0) ; \;\; \lambda_j^0(\xi) \neq 0 \;\; (\xi \neq 0, 1 \leq j \leq m).$$

N.B — La valeur propre $\lambda_j^0(\xi)$ est positivement homogène de degré 2 en ξ pour $\xi \in \mathbb{R}^n \setminus 0$. La condition (A.2) et l'identité d'Euler impliquent que

$$\xi \cdot \nabla_\xi \lambda_j^0(\xi) = 2\lambda_j^0(\xi) \neq 0 \ (\xi \neq 0) \ \text{i.e.,} \ \nabla_\xi \lambda_j^0(\xi) \neq 0 \ (\xi \neq 0).$$

DÉFINITION 1.2 — On dit que un opérateur $D_t I - A(x, D_x)$ satisfaisant à l'hypothèse (L-V) est un *système de Leray-Volevich du type de Schrödinger* si et seulement si les valeurs propres de la matrice caractéristique $A_2(\xi)$ sont réelles pour tout $\xi \in \mathbb{R}^n$.

Afin que le problème de Cauchy pour l'opérateur $D_t I - A(x, D_x)$ pour le futur et pour le passé en même temps soit bien posé dans $H^\infty(\mathbb{R}^n)$ au sens de la définition 1.1, il est nécessaire que les valeurs propres de la matrice caractéristique $A_2(x, \xi)$ soient réelles pour tout $(x, \xi) \in \mathbb{R}^n \times \mathbb{R}^n$ même si les coefficients de $A_2(x, \xi)$ dépendent de x (Takeuchi [51], voir aussi Petrowsky [44], Mizohata [38]).

Pour caractériser l'opérateur du type de Schrödinger, il faut considérer non seulement le problème de Cauchy pour le futur ou pour le passé, mais aussi le problème de Cauchy pour le futur et pour le passé en même temps ; c'est pour exclure les systèmes « paraboliques ».

Pour $\omega \in S^{n-1}$, on note $l_j^0(\omega)$ le zéro-vecteur à gauche de $(\lambda_j^0(\omega)I - A_2(\omega))$ et $r_j^0(\omega)$ le zéro-vecteur à droite, c'est-à-dire que $l_j^0(\omega)$ est une matrice $1 \times m$ et $r_j^0(\omega)$ une matrice $m \times 1$ telles que

$$l_j^0(\omega)(\lambda_j^0(\omega)I - A_2(\omega)) = 0, \quad (\lambda_j^0(\omega)I - A_2(\omega))r_j^0(\omega) = 0.$$

Pour $\xi \in \mathbb{R}^n \setminus 0$, on définit $l_j^0(\xi)$ et $r_j^0(\xi)$ comme suit :

$$l_j^0(\xi) = l_j^0(\omega), \quad r_j^0(\xi) = r_j^0(\omega), \quad \xi \in \mathbb{R}^n \setminus 0, \quad \omega = \xi/|\xi|.$$

Et de plus, on impose la condition suivante :

(1.4) $$l_j^0(\xi)r_j^0(\xi) = 1, \quad \xi \in \mathbb{R}^n \setminus 0 \, ;$$

d'où

(1.5) $$l_j^0(\xi)r_k^0(\xi) = \delta_{jk}, \quad \xi \in \mathbb{R}^n \setminus 0$$

d'après la condition (A.2).

On note $a_{jk}^1(x, \xi)$ le symbole sous-principal de l'opérateur $a_{jk}(x, D_x)$:

(1.6)
$$a_{jk}^1(x,\xi) = \sum_{|\alpha|=s_k-s_j+1} a_{jk\alpha}(x)\xi^\alpha$$

et on pose $A_1(x,\xi) = (a_{jk}^1(x,\xi))$ qui s'appelle la *martice sous-principale*.

Remarquons que $l_j^0(\xi)A_1(x,\xi)r_j^0(\xi)$ est une scalaire.

DÉFINITION 1.3 — Soit $G(x)$ est une fonction continue dans $\mathbb{R}^n$ à valeurs réelles. On dit que $G(x)$ est un *poids modéré* si $G(x)$ satisfait aux conditions suivantes :

(1) $0 < G(x) \le C_0\langle x\rangle$ $(C_0 > 0)$ où $\langle x\rangle = (1+|x|^2)^{1/2}$;

(2) (sous-multiplicativité) $G(x+y) \le G(x)G(y)$.

Les conditions pour le symbole sous-principal s'expriment comme suit :

CONDITION (B.1) — Pour tout j $(1 \le j \le m)$, il existe $\kappa_j(x,\xi)$, fonction $\mathcal{B}^1$ définie dans $\mathbb{R}^n \times \mathbb{R}^n$ à valeurs réelles et $G_j(x)$, poids modéré au sens de la définition 1.3, tels que l'on ait $\kappa_j(x,\xi) = \kappa_j^\pm(\xi)$ si $\pm x \cdot \xi \ge \delta_0\langle x\rangle|\xi|$, où $0 < \delta_0 < 1$, et que

$$\sup_{(x,\omega)\in\mathbb{R}^n\times S^{n-1}} \left| \int_0^{-\mu_j(x,\omega)} \mathrm{Im}\left\{ l_j^0(\omega)A_1(x + s(\nabla_\xi\lambda_j^0)(\omega),\omega)r_j^0(\omega) \right\} ds \right.$$
$$\left. -\kappa_j(x,\omega) \log \frac{G_j\big(x - \mu_j(x,\omega)(\nabla_\xi\lambda_j^0)(\omega)\big)}{G_j(x)} \right| < +\infty,$$

où S^{n-1} est la sphère unité dans $\mathbb{R}^n$ et $\mu_j(x,\omega) = x\cdot(\nabla_\xi\lambda_j^0)(\omega)/|(\nabla_\xi\lambda_j^0)(\omega)|^2$.

CONDITION (B.2) — Pour tout j $(1 \le j \le m)$ et pour tout multi-indice ν $(|\nu| \ge 1)$, on a

$$\sup_{(x,\omega)\in\mathbb{R}^n\times S^{n-1}} \int_0^{+\infty} \left| D_x^\nu\{\mathrm{Im}\, l_j^0(\omega)A_1(x + s(\nabla_\xi\lambda_j^0)(\omega),\omega)r_j^0(\omega)\} \right| ds < +\infty.$$

CONDITION (B.3) — Pour tout j $(1 \le j \le m)$ et pour tout multi-indice ν, on a

$$\sup_{(x,\omega)\in\mathbb{R}^n\times S^{n-1}}\left\{\langle x\rangle\left|D_x^\nu\bigl(\operatorname{Im}l_j^0(\omega)A_1(x,\omega)r_j^0(\omega)\bigr)\right|\right\}<+\infty.$$

CONDITION (B.4) — Pour tout j $(1\le j\le m)$ et pour tout multi-indice ν $(|\nu|\ge 1)$, on a

$$\sup_{(x,\omega)\in\mathbb{R}^n\times S^{n-1}}\left\{\langle x\rangle\left|D_x^\nu\bigl(\operatorname{Re}l_j^0(\omega)A_1(x,\omega)r_j^0(\omega)\bigr)\right|\right\}<+\infty.$$

Les conditions (B.1) à (B.4) ne dépendent pas du choix du zéro-vecteur à gauche et du zéro-vecteur à droite satisfaisant (1.4).

Nos résultats s'énoncent ainsi :

THÉORÈME 1.3 — *Supposons les conditions (A.1), (A.2), (B.1) à (B.4) vérifiées. Alors, le problème de Cauchy* $(*)$ *pour le futur et pour le passé en même temps est bien posé dans* $H^\infty(\mathbb{R}^n)$, *à la perte de régularité* κ_0 *près, où*

$$\kappa_0=\max\{|\kappa_j^\pm(\omega)|\,;\,\omega\in S^{n-1},1\le j\le m\}.$$

COROLLAIRE 1.4 — *Supposons que les conditions (A.1), (A.2), (B.1) avec* $\kappa_j(\omega)\equiv 0$ $(1\le j\le m)$, *(B.2) à (B.4) soient vérifiées. Alors, le problème de Cauchy* $(*)$ *pour le futur et pour le passé en même temps est bien posé dans* $H^l(\mathbb{R}^n)$ *pour tout* $l\in\mathbb{R}^1$ *au sens de la définition 1.1.*

Pour démontrer le théorème 1.3, nous proposons une méthode de $\langle\!\langle L^2$-symétrisation indépendante du temps d'opérateurs à la perte de régularité$\rangle\!\rangle$. Cette méthode est comme suit : D'abord, on diagonalise ce système ; ensuite, on réduit ce système diagonal à un système diagonal et L^2-symétrique par un opérateur pseudo-différentiel indépendant du temps ; nous le verrons à la section 2. Notre résultat principal est le théorème 2.9. Ce résultat de L^2-symétrisation indépendante du temps est plus fort que celui du théorème 1.3.

1.2 Conditions nécessaires — Afin d'obtenir des conditions nécessaires, on impose la condition suivante plus faible que la condition (A.2) :

CONDITION (A.2)′ — Les valeurs propres $\lambda_j^0(\xi)$ $(1\le j\le m)$ de la matrice caractéristique $A_2(\xi)$ sont réelles et distinctes pour $\xi\in\mathbb{R}^n\setminus 0$.

N.B — La valeur propre $\lambda_j^0(\xi)$ peut s'annuler pour certains $\xi \in \mathbb{R}^n$. Mais, d'après l'identité d'Euler, on a

$$\nabla_\xi \lambda_j^0(\xi) \neq 0 \text{ sur } \{\xi \in \mathbb{R}^n; \lambda_j^0(\xi) \neq 0\}.$$

La condition (B.1) implique la condition suivante :

CONDITION (B.1)$'$ — Il existe une constante non négative κ_0 et une constante positive C telles que, pour tout $\rho \in \mathbb{R}^1$, on ait

$$\sup_{\substack{(x,\omega)\in\mathbb{R}^n\times S^{n-1} \\ 1\leq j\leq m}} \left| \int_0^\rho \mathrm{Im}\{ l_j^0(\omega)A_1(x + s(\nabla_\xi\lambda_j^0)(\omega),\omega)r_j^0(\omega)\} \, ds \right|$$

$$\leq \kappa_0 \log \langle\rho\rangle + C.$$

Nos résultats s'énoncent ainsi :

THÉORÈME 1.5 — *Supposons les conditions* (A.1) *et* (A.2)$'$ *vérifiées. Alors, afin que le problème de Cauchy* (∗) *pour le futur et pour le passé en même temps soit bien posé dans* $H^\infty(\mathbb{R}^n)$ *au sens de la définition 1.1, il est nécessaire que la condition* (B.1)$'$ *soit vérifiée.*

COROLLAIRE 1.6 — *Supposons les conditions* (A.1) *et* (A.2)$'$ *vérifiées. Afin que le problème de Cauchy pour le futur et pour le passé en même temps* (∗) *soit bien posé dans* $H^l(\mathbb{R}^n)$ *pour tout* $l \in \mathbb{R}^1$, *il est nécessaire que la condition* (B.1)$'$ *avec* $\kappa_0 = 0$ *soit vérifiée.*

Pour démontrer le théorème 1.5, on construit des solutions asymptotiques du problème de Cauchy (∗) selon Takeuchi [52], [63] (voir aussi Birkhoff [2], Leray [35], Maslov [36] et Mizohata [40] à [43]). Nous donnerons la démonstration du théorème 1.5 dans un autre article parce que la démonstration du théorème 1.5 est assez longue.

Le théorème suivant est une conséquence du théorème 1.5.

THÉORÈME 1.7 — *Supposons les conditions* (A.1) *et* (A.2)$'$ *vérifiées. De plus, supposons aussi qu'il existe* $j \in \{1, \ldots, m\}$, $\hat{\omega} \in S^{n-1}$ *et* $C_0 > 0$ *tels que, pour tout* $x \in \mathbb{R}^n$, *on ait*

$$\left| \mathrm{Im}\left\{ l_j^0(\hat{\omega})A_1(x,\hat{\omega})r_j^0(\hat{\omega})\right\}\right| \geq C_0 \langle x\rangle^{-\delta}.$$

Alors, une condition nécessaire afin que le problème de Cauchy pour le futur et pour le passé en même temps (∗) soit bien posé dans $H^\infty(\mathbb{R}^n)$ est que $\delta \geq 1$.

COROLLAIRE 1.8 — *Supposons les conditions* (A.1) *et* (A.2)′ *vérifiées. De plus, supposons aussi qu'il existe $j \in \{1, \ldots, m\}$, $\hat{\omega} \in S^{n-1}$ et $C_0 > 0$ tels que, pour tout $x \in \mathbb{R}^n$, on ait*

$$\left| \operatorname{Im} \left\{ l_j^0(\hat{\omega}) A_1(x, \hat{\omega}) r_j^0(\hat{\omega}) \right\} \right| \geq C_0 \langle x \rangle^{-\delta}.$$

Alors, une condition nécessaire afin que le problème de Cauchy pour le futur et pour le passé en même temps (∗) soit bien posé dans $H^l(\mathbb{R}^n)$ pour tout $l \in \mathbb{R}^1$ est que $\delta > 1$.

1.3. Notes historiques — Takeuchi [49], pour la première fois, a traité du problème de Cauchy pour l'opérateur de Schrödinger avec un potentiel vectoriel à valeurs complexes dans le cas où $n = 1$ (voir aussi Takeuchi [50], [51]). Mizohata ([40] à [42]) a donné des conditions suffisantes et des conditions nécessaires, qui sont le point de départ de nos études, afin que le problème de Cauchy (∗) soit bien posé dans L^2 (voir aussi Mizohata [43]). Après Mizohata ([40] à [42]), Ichinose ([19] et [20]) a considéré le problème de Cauchy pour le même opérateur dans le cadre H^∞. Takeuchi ([52], [55] à [57]) a aussi considéré le problème de Cauchy pour l'opérateur du type de Schrödinger dans le cadre H^∞. Baba ([1]) a traité du problème de Cauchy dans H^∞ par la méthode un peu différente. Hara [13], qui généralise le résultat d'Ichinose [19], a donné une condition nécessaire afin que le problème de Cauchy pour l'opérateur du type de Schrödinger soit bien posé dans H^∞. Tarama [65] a donné d'autres considérations sur le problème de Cauchy pour l'opérateur du type de Schrödinger. Doi ([8], [9]), dans lequel il a traité d'un opérateur plus général, a ajouté une nouvelle considération (dite des « effets régularisants ») de solutions) importante pour quelques équations non linéaires dispersives et du type de Schrödinger (voir Constantin-Saut [4], [5], Hayashi-Nakamitsu-Tsutsumi [14], Kato [30], [31], Kenig, Ponce et Vega [32], Yajima [69]). Kajitani et Baba [29] a traité du problème de Cauchy pour l'opérateur du type de Schrödinger dans la classe de Gevrey. Ichinose ([21] à [23]) a traité du problème de Cauchy pour l'opérateur du type de Schrödinger dans une variété riemannienne (Voir aussi Ichinose [24] et [25]). Kajitani [28] a traité du problème de Cauchy dans le cadre H^∞ pour un opérateur plus général.

2 L^2-SYMÉTRISATION INDÉPENDANTE DU TEMPS

2.1 Diagonalisation du système — Soit $A(x, D_x) = \left(a_{jk}(x, D_x)\right)$ est la matrice $m \times m$ d'opérateurs différentiels linéaires définie dans la section 1. On considère un système :

$$(2.1) \qquad D_t u(x,t) - A(x, D_x)u(x,t) = f(x,t).$$

On note $J(\xi)$ le symbole de l'opérateur pseudo-différentiel diagonal $J(D_x)$:

$$(2.2) \qquad J(\xi) = \operatorname{diag}\left(\langle\xi\rangle^{s_1}, \ldots, \langle\xi\rangle^{s_m}\right).$$

En notant

$$\tilde{u}(x,t) = J(D_x)u(x,t),$$
$$\tilde{A}(x, D_x) = J(D_x)A(x, D_x)J(D_x)^{-1},$$
$$\tilde{f}(x,t) = J(D_x)f(x,t),$$

on a un système suivant :

$$(2.3) \qquad D_t\tilde{u}(x,t) - \tilde{A}(x, D_x)\tilde{u}(x,t) = \tilde{f}(x,t).$$

On note

$$(2.4) \qquad \tilde{A}(x, D_x) = \tilde{A}_2(D_x) + \tilde{A}_1(x, D_x) + \tilde{A}_0(x, D_x),$$

où

$$\tilde{A}_2(D_x) = J(D_x)A_2(D_x)J(D_x)^{-1},$$
$$\tilde{A}_1(x, D_x) = J(D_x)A_1(x, D_x)J(D_x)^{-1}.$$

Le symbole $\tilde{A}_2(\xi)$ de l'opérateur $\tilde{A}_2(D_x)$ s'exprime comme suit :

$$(2.5) \qquad \tilde{A}_2(\xi) = J(\xi)A_2(\xi)J(\xi)^{-1} = A_2\left(\xi/\langle\xi\rangle\right)\langle\xi\rangle^2.$$

Le symbole $\tilde{A}_1(x, \xi)$ de l'opéreteur $\tilde{A}_1(x, D_x)$ s'exprime comme suit :

$$\tilde{A}_1(x, \xi) = J(\xi)A_1(x, \xi)J(\xi)^{-1}$$
$$+ \sum_{|\nu|\geq 1} \frac{1}{\nu!}J^{(\nu)}(\xi)A_{1(\nu)}(x, \xi)J(\xi)^{-1},$$

où

$$f^{(\alpha)}_{(\beta)}(x, \xi) = D_x^\beta(iD_\xi)^\alpha f(x, \xi).$$

Comme

$$\sum_{|\nu|\geq 1} \frac{1}{\nu!} J^{(\nu)}(\xi) A_{1(\nu)}(x,\xi) J(\xi)^{-1} \in \mathrm{Mat}\left(m; S_{1,0}^0(\mathbb{R}^n)\right),$$

on a

(2.6) $\tilde{A}_1(x,\xi)$

$$\equiv J(\xi) A_1(x,\xi) J(\xi)^{-1} \quad (\mathrm{mod} \quad \mathrm{Mat}\left(m; S_{1,0}^0(\mathbb{R}^n)\right))$$

$$= A_1(x, \xi/\langle\xi\rangle)\langle\xi\rangle.$$

LEMME 2.1 — *Soit* $a(x,\xi) \in S_{1,0}^d(\mathbb{R}^n)$. *Alors on a*

$$a(x, \xi/\langle\xi\rangle) - a(x, \xi/|\xi|)\chi_\infty(\xi) \in S_{1,0}^{d-2}(\mathbb{R}^n),$$

où $\chi_\infty(\xi) \in C^\infty(\mathbb{R}^n)$, $\chi_\infty(\xi) = 1$ $(|\xi| \geq 1)$, $\chi_\infty(\xi) = 0$ $(|\xi| \leq 1/2)$.

Démonstration : — Voir le lemme 3.12 de Takeuchi [63].   □

D'après (2.5), (2.6) et le lemme 2.1, on a

$$\tilde{A}(x,\xi) \equiv A_2(\xi/|\xi|)|\xi|^2\chi_\infty(\xi) + A_1(x, \xi/|\xi|)|\xi|\chi_\infty(\xi)$$

$$(\mathrm{mod}\ \mathrm{Mat}(m; S_{1,0}^0(\mathbb{R}^n))).$$

Grâce à la condition (A.2), on peut diagonaliser le système (2.3) :

PROPOSITION 2.2 (Diagonalisation parfaite) — *Supposons que les conditions* (A.1) *et* (A.2) *soient vérifiées. Alors, il existe un opérateur pseudo-différentiel diagonal* $\mathcal{D}(x, D_x) \in \mathrm{Mat}\left(m; \mathrm{OP}S_{1,0}^2(\mathbb{R}^n)\right)$ *et un opérateur pseudo-différentiel inversible* $N(x, D_x) \in \mathrm{Mat}\left(m; \mathrm{OP}S_{1,0}^0(\mathbb{R}^n)\right)$ *tels que*
(1) *l'on ait l'équation suivante :*

$$N(x, D_x)J(D_x)\big(D_t - A(x, D_x)\big)J(D_x)^{-1}N(x, D_x)^{-1} \equiv (D_t - \mathcal{D}(x, D_x))$$

$$\big(\mathrm{mod}\ \mathrm{Mat}\left(m; \mathrm{OP}S_{1,0}^{-d}(\mathbb{R}^n)\right)\big),$$

où d *est un entier plus grand que* $2\kappa_0$ *et* $\kappa_0 = \max\{|\kappa_j^\pm(\omega)|\,;\,\omega \in S^{n-1},$
$1 \leq j \leq m\}$,

(2) *en notant* $\sigma(\mathcal{D})(x,\xi) = (\delta_{jk}\lambda_k(x,\xi))$ *le symbole de l'opérateur* $\mathcal{D}$,
on ait

(2.7) $$\lambda_j(x,\xi) = \lambda_j^0(\xi) + \lambda_j^1(x,\xi) + \mu_j(x,\xi),$$

où

(2.8) $\lambda_j^1(x,\xi) = l_j^0(\xi)\tilde{\tilde{A}}_1(x,\xi)r_j^0(\xi) \in S_{1,0}^1(\mathbb{R}^n),$

(2.9) $\tilde{\tilde{A}}_1(x,\xi) = A_1(x,\xi')|\xi|\chi_\infty(\xi) \in \text{Mat}\left(m; S_{1,0}^1(\mathbb{R}^n)\right),$

$\mu_j(x,\xi) \in S_{1,0}^0(\mathbb{R}^n)$, $l_j^0(\xi)$ *est le zéro-vecteur à gauche de* $(\lambda_j^0(\xi)I - A_2(\xi')|\xi|^2)$
et $r_j^0(\xi)$ *le zéro-vecteur à droite de* $(\lambda_j^0(\xi)I - A_2(\xi')|\xi|^2)$, $\xi' = \xi/|\xi|$, $\chi_\infty(\xi) \in$
$C^\infty(\mathbb{R}^n)$, $\chi_\infty(\xi) = 1$ $(|\xi| \geq 1)$, $\chi_\infty(\xi) = 0$ $(|\xi| \leq 1/2)$.

Démonstration : — En répétant le même procédé de la preuve de la proposition 3.13 de Takeuchi [63], on a la proposition 2.2.   □

2.2 L^2-symétrisation indépendante du temps du système — Dans
cette sous-section, on toujours suppose les conditions (A.1), (A.2), (B.1) à (B.4) vérifiées.

On définit le symbole $K(x,\xi)$ de l'opérateur pseudo-différentiel $K(x,D_x)$ comme suit :

$$K(x,\xi) = \text{diag}\left(k_1(x,\xi),\ldots,k_m(x,\xi)\right),$$

$$k_j(x,\xi) = \exp\left(\varphi_j(x,\xi)\right),$$

$$\varphi_j(x,\xi) = \chi_R(\xi)\psi_R(x,\xi)\Phi_j(x,\xi,\mu_j(x,\xi)),$$

$$\Phi_j(x,\xi,t) = \int_0^t \text{Im}\,\lambda_j^1\left(x - s(\nabla_\xi\lambda_j^0)(\xi),\xi\right)ds$$

$$= \int_0^t \text{Im}\,l_j^0(\xi)A_1\left(x - s(\nabla_\xi\lambda_j^0)(\xi),\xi\right)r_j^0(\xi)\,ds,$$

$$\mu_j(x,\xi) = \left(x\cdot\nabla_\xi\lambda_j^0(\xi)\right)/\left|\nabla_\xi\lambda_j^0(\xi)\right|^2,$$

$$\chi_R(\xi) = \chi(\xi/R), \quad \psi_R(x,\xi) = \psi(R\langle x\rangle\langle\xi\rangle^{-1}),$$

où $\chi(\xi) \in C^\infty(\mathbb{R}^n)$, $\chi(\xi) = 1$ $(|\xi| \geq 2)$, $\chi(\xi) = 0$ $(|\xi| \leq 1)$, $\psi(s) \in C^\infty(\mathbb{R}^1)$,
$\psi(s) = 1$ $(s \leq 1)$, $\psi(s) = 0$ $(s \geq 2)$ et $R \geq 1$.

D'abord, on a les lemmes suivants.

LEMME 2.3 — *Supposons les conditions* (A.1), (A.2) *et* (B.1) *vérifiées. Alors, pour* $1 \leq j \leq m$, *on a les estimations :*

$$\left|\text{Re}\,\varphi_j(x,\xi)\right| \leq \kappa_0\log\langle\xi\rangle + C,$$

où $\kappa_0 = \max\left\{|\kappa_j^+(\omega)|, |\kappa_j^-(\omega)| \, ; \, \omega \in S^{n-1}, 1 \leq j \leq m\right\}$, la fonction $\kappa_j^\pm(\omega)$ est définie dans la condition (B.1) et C est une constante positive.

LEMME 2.4 — *Supposons les conditions* (A.1), (A.2), (B.1) *à* (B.3) *vérifiées. Alors, pour* $1 \leq j \leq m$, $|\alpha| + |\beta| \geq 1$ *et* $R \geq 1$, *on a les estimations*:

$$\left| D_x^\beta D_\xi^\alpha \varphi_j(x, \xi) \right| \leq C_{\alpha\beta} R^{-|\alpha|},$$

où $C_{\alpha\beta}$ *est une constante positive indépendante de* R.

En combinant le lemme 2.3 et le lemme 2.4, on a la proposition suivante.

PROPOSITION 2.5 — *Supposons les conditions* (A.1), (A.2), (B.1) *à* (B.3) *vérifiées. Alors, pour* $1 \leq j \leq m$ *et* $R \geq 1$, *on a les estimations*:

$$\left| D_x^\beta D_\xi^\alpha k_j(x, \xi) \right| \leq C_{\alpha\beta} R^{-|\alpha|} \langle \xi \rangle^{\kappa_0},$$

où $\kappa_0 = \max\left\{|\kappa_j^+(\omega)|, |\kappa_j^-(\omega)| \, ; \, \omega \in S^{n-1}, 1 \leq j \leq m\right\}$, *et* $C_{\alpha\beta}$ *est une constante positive indépendante de* R, *c'est-à-dire que* $k_j(x, \xi, t) \in S_{0,0}^{\kappa_0}(\mathbb{R}^n)$ *uniformément pour* $R \geq 1$.

PROPOSITION 2.6 — *Supposons les conditions* (A.1), (A.2), (B.1) *à* (B.3) *vérifiées. Alors, si* R *est suffisamment grand, l'opérateur* $K(x, D_x)$ *est inversible dans* $\mathrm{Mat}\,(m; \mathrm{OP}S_{0,0}^{\kappa_0}(\mathbb{R}^n))$, *où* $\kappa_0 = \max\left\{|\kappa_j^+(\omega)|, |\kappa_j^-(\omega)| \, ; \, \omega \in S^{n-1}, 1 \leq j \leq m\right\}$.

D'après le théorème de Calderón-Vaillancourt [3], les opérateurs $K(x, D_x)$ et $K(x, D_x)^{-1}$ sont continus de $(H^s(\mathbb{R}^n))^m$ dans $(H^{s-\kappa_0}(\mathbb{R}^n))^m$.

L'équation du lemme suivant était le point de départ du problème.

LEMME 2.7 — *Supposons les conditions* (A.1), (A.2), (B.1) *à* (B.3) *vérifiées. Alors, le symbole* $k_j(x, \xi)$ *de l'opérateur* $k_j(x, D_x)$ *satisfait à l'équation suivante*:

$$\left(\nabla_\xi \lambda_j^0(\xi) \cdot D_x - i \operatorname{Im} \lambda_j^1(x, \xi)\right) k_j(x, \xi) \equiv 0 \quad (\mathrm{mod}\ S^{-\infty}(\mathbb{R}^n)).$$

La proposition suivante est au cœur du problème.

PROPOSITION 2.8 — *Supposons les conditions* (A.1), (A.2), (B.1) *à* (B.4) *vérifiées. Alors, on a*

$$K(x, D_x)^{-1}[K(x, D_x), \mathcal{D}_1^R(x, D_x)] \in \mathrm{Mat}\,\left(m; \mathrm{OP}S_{0,0}^0(\mathbb{R}^n)\right),$$

où $[A, B] = AB - BA$ *est le commutateur d'opérateurs* A *et* B,

$$\mathcal{D}_1^R(x, \xi) = \mathrm{diag}\left(\mathrm{Re}\,\lambda_1^1(x, \xi), \cdots, \mathrm{Re}\,\lambda_m^1(x, \xi)\right),$$

le symbole $\lambda_j^1(x, \xi)$ *est défini par* (2.7) *à* (2.9).

En résumé, nous obtenons le théorème suivant qui est le but de nos études :

THÉORÈME 2.9 (L^2-symétrisation) — *Supposons les conditions* (A.1), (A.2), (B.1) *à* (B.4) *vérifiées. Alors, il existe un opérateur pseudo-différentiel matriciel* $K(x, D_x)$ *tels que,*

(1) $K(x, D_x)$ *appartienne à* $\mathrm{Mat}\left(m; \mathrm{OPS}_{0,0}^{\kappa_0}(\mathbb{R}^n)\right)$,

(2) $K(x, D_x)$ *soit inversible dans* $\mathrm{Mat}\left(m; \mathrm{OPS}_{0,0}^{\kappa_0}(\mathbb{R}^n)\right)$,

(3) $K(x, D_x)$ *satisfasse à l'équation suivante*:

$$K(x, D_x)^{-1} J(D_x)(D_t - A(x, D_x)) J(D_x)^{-1} K(x, D_x)$$
$$\equiv (D_t - \mathcal{D}^R(x, D_x))$$
$$(\mathrm{mod}\ \mathrm{Mat}\left(m; \mathrm{OPS}_{0,0}^0(\mathbb{R}^n)\right)),$$

où $\kappa_0 = \max\left\{|\kappa_j^+(\omega)|, |\kappa_j^-(\omega)|\,;\, \omega \in S^{n-1}, 1 \le j \le m\right\}$, *le symbole* $J(\xi)$ *de l'opérateur* $J(D_x)$ *est défini par* (2.2),

$$\mathcal{D}^R(x, D_x) = \mathrm{diag}\left(\lambda_1^R(x, D_x), \cdots, \lambda_m^R(x, D_x)\right),$$
$$\lambda_j^R(x, \xi) = \lambda_j^0(\xi) + \mathrm{Re}\,\lambda_j^1(x, \xi)$$

et le symbole $\lambda_j^1(x, \xi)$ *est défini par* (2.7) *à* (2.9).

Démonstration : — On prend R assez grand pour que les conclusions des propositions 2.6 et 2.8 sont satisfaites. En combinant les propositions 2.5, 2.6 et 2.8, on a la conclusion du théorème 2.9.   □

3 PREUVE DES THÉORÈMES 1.3 ET 1.7

3.1. Preuve du théorème 1.3 — Dans cette sous-section, on toujours suppose les conditions (A.1), (A.2), (B.1) à (B.4) vérifiées.

En appliquant $N(x, D_x) J(D_x)$ à (2.1) à gauche, on a, d'après la proposition 2.2,

$$\{D_t - \mathcal{D}(x, D_x) + B_{-d}(x, D_x)\} v(x, t) = N(x, D_x) J(D_x) f(x, t),$$

où $v(x,t) = N(x, D_x)J(D_x)u(x,t)$, $B_{-d}(x, D_x) \in \mathrm{Mat}\,(m; \mathrm{OPS}_{1,0}^{-d}(\mathbb{R}^n))$.

Le théorème 2.9 implique

$$K(x, D_x)^{-1}(D_t - \mathcal{D}(x, D_x))K(x, D_x) = D_t - \mathcal{D}^R(x, D_x) + B_0(x, D_x),$$

où $B_0(x, D_x) \in \mathrm{Mat}\,(m; \mathrm{OPS}_{0,0}^0(\mathbb{R}^n))$.

En posant $v(x,t) = K(x, D_x)w(x,t)$ et vu que $d > 2\kappa_0$, on a un système suivant :

$$\left(D_t - \mathcal{D}^R(x, D_x) + B(x, D_x)\right)w(x,t) = \tilde{K}(x, D_x)J(D_x)f(x,t),$$

où

$$B(x, D_x) = K(x, D_x)^{-1}B_{-d}(x, D_x)K(x, D_x) + B_0(x, D_x) \in \mathrm{Mat}\,(m; \mathrm{OPS}_{0,0}^0(\mathbb{R}^n)),$$

$$\tilde{K}(x, D_x) = K(x, D_x)^{-1}N(x, D_x) \in \mathrm{Mat}\,(m; \mathrm{OPS}_{0,0}^{\kappa_0}(\mathbb{R}^n)).$$

Le problème de Cauchy pour l'opérateur $Q(x, D_x, D_t) = D_t - \mathcal{D}^R(x, D_x) + B(x, D_x)$ dans $\mathbb{R}^n \times [-T, T]$ est bien posé dans $(H^l(\mathbb{R}^n))^m$ pour tout $l \in \mathbb{R}^1$ et on a l'inégalité suivante pour $t \in [-T, T]$:

$$\|w(t)\|_{(l)} \leq C(l, T_0)\left\{\|w(0)\|_{(l)} + \left|\int_0^t \|\tilde{K}(\tau)J(D_x)f(\tau)\|_{(l)}\, d\tau\right|\right\}.$$

Vu le théorème de Calderón-Vaillancourt [3] et vu que les normes $\|v(t)\|_{(l)}$, $\|J(D_x)u(t)\|_{(l)}$ et $\|\|u(t)\|\|_{(l)}$ sont équivalentes, on a l'inégalité suivante :

$$\|\|u(t)\|\|_{(l-\kappa_0)} \leq C(l, T_0)\left\{\|\|u(0)\|\|_{(l)} + \left|\int_0^t \|\|f(\tau)\|\|_{(l+\kappa_0)}\, d\tau\right|\right\},$$

pour $t \in [-T, T]$; d'où la conclusion du théorème 1.3.

La démonstration du théorème 1.3 est complète. $\qquad\square$

3.2. Preuve du théorème 1.7 — Supposons les conditions (A.1) et (A.2)′ vérifiées. De plus, supposons aussi qu'il existe $\hat{\omega} \in S^{n-1}$, $j \in \{1, \ldots, m\}$ et $C_0 > 0$ tels que, pour tout $x \in \mathbb{R}^n$, on ait

(3.1) $$\left|\mathrm{Im}\left\{l_j^0(\hat{\omega})A_1(x, \hat{\omega})r_j^0(\hat{\omega})\right\}\right| \geq C_0 \langle x\rangle^{-\delta}.$$

Supposons que le problème de Cauchy pour le futur et pour le passé en même temps (∗) est bien posé dans H^∞. Alors, d'après le théorème 1.5, la condition (B.1)$'$ est vérifiée. La condition (B.1)$'$ et (3.1) entraînent que, pour tout $\rho > 0$,

(3.2)
$$0 < C_0 \int_0^\rho \langle x + s(\nabla_\xi \lambda_j^0)(\hat{\omega}) \rangle^{-\delta} ds$$

$$\leq \int_0^\rho \left| \mathrm{Im} \left\{ l_j^0(\hat{\omega}) A_1(x + s(\nabla_\xi \lambda_j^0)(\hat{\omega}), \hat{\omega}) r_j^0(\hat{\omega}) \right\} \right| ds$$

$$= \left| \int_0^\rho \mathrm{Im} \left\{ l_j^0(\hat{\omega}) A_1(x + s(\nabla_\xi \lambda_j^0)(\hat{\omega}), \hat{\omega}) r_j^0(\hat{\omega}) \right\} ds \right|$$

$$\leq \kappa_0 \log \langle \rho \rangle + C.$$

Si $\nabla_\xi \lambda_j^0(\hat{\omega}) = 0$, la condition (B.1)$'$ implique que, pour tout $x \in \mathbb{R}^n$,

$$\mathrm{Im} \left\{ l_j^0(\hat{\omega}) A_1(x, \hat{\omega}) r_j^0(\hat{\omega}) \right\} = 0,$$

qui est en contradiction avec (3.1).

Donc, la condition (B.1)$'$ et (3.1) entraînent que $(\nabla_\xi \lambda_j^0)(\hat{\omega}) \neq 0$.

En posant $x = 0$ dans (3.2), on a, pour tout $\rho > 0$,

$$0 < C_0 \int_0^\rho \langle s(\nabla_\xi \lambda_j^0)(\hat{\omega}) \rangle^{-\delta} ds$$

$$\leq \int_0^\rho \left| \mathrm{Im} \left\{ l_j^0(\hat{\omega}) A_1(s(\nabla_\xi \lambda_j^0)(\hat{\omega}), \hat{\omega}) r_j^0(\hat{\omega}) \right\} \right| ds$$

$$= \left| \int_0^\rho \mathrm{Im} \left\{ l_j^0(\hat{\omega}) A_1(s(\nabla_\xi \lambda_j^0)(\hat{\omega}), \hat{\omega}) r_j^0(\hat{\omega}) \right\} ds \right|$$

$$\leq \kappa_0 \log \langle \rho \rangle + C ;$$

d'où $\delta \geq 1$. La démonstration du théorème 1.7 est complète. $\qquad \square$

3.3. Preuve du corollaire 1.8 — Supposons les conditions (A.1) et (A.2)$'$ vérifiées. De plus, supposons aussi qu'il existe $\hat{\omega} \in S^{n-1}$, $j \in \{1, \ldots, m\}$ et $C_0 > 0$ tels que, pour tout $x \in \mathbb{R}^n$, on ait

(3.3)
$$\left| \mathrm{Im} \left\{ l_j^0(\hat{\omega}) A_1(x, \hat{\omega}) r_j^0(\hat{\omega}) \right\} \right| \geq C_0 \langle x \rangle^{-\delta}.$$

Supposons que le problème de Cauchy pour le futur et pour le passé en même temps $(*)$ est bien posé dans H^l. Alors, d'après le corollaire 1.6, la condition (B.1)$'$ avec $\kappa_0 = 0$ est vérifiée. La condition (B.1)$'$ avec $\kappa_0 = 0$ et (3.3) entraînent que, pour tout $\rho > 0$,

$$(3.4) \qquad 0 < C_0 \int_0^\rho \left\langle x + s(\nabla_\xi \lambda_j^0)(\hat{\omega}) \right\rangle^{-\delta} ds$$

$$\leq \int_0^\rho \left| \mathrm{Im} \left\{ l_j^0(\hat{\omega}) A_1(x + s(\nabla_\xi \lambda_j^0)(\hat{\omega}), \hat{\omega}) r_j^0(\hat{\omega}) \right\} \right| ds$$

$$= \left| \int_0^\rho \mathrm{Im} \left\{ l_j^0(\hat{\omega}) A_1(x + s(\nabla_\xi \lambda_j^0)(\hat{\omega}), \hat{\omega}) r_j^0(\hat{\omega}) \right\} ds \right| \leq C.$$

Si $\nabla_\xi \lambda_j^0(\hat{\omega}) = 0$, (3.4) implique que, pour tout $x \in \mathbb{R}^n$,

$$\mathrm{Im} \left\{ l_j^0(\hat{\omega}) A_1(x, \hat{\omega}) r_j^0(\hat{\omega}) \right\} = 0,$$

qui est en contradiction avec (3.3). Donc, la condition (B.1)$'$ avec $\kappa_0 = 0$ et (3.3) entraînent que $(\nabla_\xi \lambda_j^0)(\hat{\omega}) \neq 0$.

En posant $x = 0$ dans (3.4), on a, pour tout $\rho > 0$,

$$0 < C_0 \int_0^\rho \left\langle s(\nabla_\xi \lambda_j^0)(\hat{\omega}) \right\rangle^{-\delta} ds$$

$$\leq \int_0^\rho \left| \mathrm{Im} \left\{ l_j^0(\hat{\omega}) A_1(s(\nabla_\xi \lambda_j^0)(\hat{\omega}), \hat{\omega}) r_j^0(\hat{\omega}) \right\} \right| ds$$

$$= \left| \int_0^\rho \mathrm{Im} \left\{ l_j^0(\hat{\omega}) A_1(s(\nabla_\xi \lambda_j^0)(\hat{\omega}), \hat{\omega}) r_j^0(\hat{\omega}) \right\} ds \right| \leq C;$$

d'où $\delta > 1$. La démonstration du corollaire 1.8 est complète. $\qquad\square$

Remerciements — L'auteur tient à exprimer ici toute sa reconnaissance à Professeur Jean Vaillant qui a permis à l'auteur de faire de la recherche dans son Laboratoire de l'Université de Paris-VI depuis 1989. Grâce à son accueil et ses conseils, l'auteur a pu avancer dans ses études et a pu réaliser ce travail.

L'auteur tient à remercier Professeur Jean Leray qui s'est intéressé aux Notes de l'auteur proposées aux Comptes Rendus de l'Académie des Sciences de Paris ; il a donné beaucoup de conseils précieux pour les Notes de l'auteur et il a fait l'effort de les améliorer.

L'auteur souhaite également remercier Professeur Sigeru Mizohata qui a donné à l'auteur l'impact fort et qui a fait faire des progrès sur le problème de Cauchy pour l'opérateur du type de Schrödinger en proposant ses conditions importantes.

RÉFÉRENCES

[1] A. Baba, *The H^∞-wellposed Cauchy problem for Schrödinger type equations*, Tsukuba J. Math., **18** (1994), 101–117.

[2] G. D. Birkhoff, *Quantum mechanics and asymptotic series*, Bull. Amer. Math. Soc., **39** (1933), 681–700.

[3] A. P. Calderón and R. Vaillancourt, *A class of bounded pseudo-differential operators*, Proc. Nat. Acad. Sci. U.S.A., **69** (1972), 1185–1187.

[4] P. Constantin and J. C. Saut, *Local smoothing properties of dispersive equations*, J. Amer. Math. Soc., **1** (1988), 413–439.

[5] P. Constantin and J. C. Saut, *Local smoothing properties of Schrödinger equations*, Indiana Univ. Math. J., **38** (1989), 791–810.

[6] W. Craig, T. Kappler and W. Strauss, *Microlocal dispersive smoothing for the Schrödinger equation*, Comm. Pure Appl. Math., **48** (1995), 769–860.

[7] J. Dereziński et C. Gérard, *Scattering Theory of Classical and Quantum N-Particle Systems*, Texts and Monographs in Physics, Springer-Verlag, Berlin Heidelberg, 1997.

[8] S. Doi, *On the Cauchy problem for Schrödinger type equations and the regularity of solutions*, J. Math. Kyoto Univ., **34** (1994), 319–328.

[9] S. Doi, *Remarks on the Cauchy problem for Schrödinger type equations*, Comm. Partial Differential Equations, **21** (1996), 163–178.

[10] S. Doi, *Smoothing effects of Schrödinger evolution groups on Riemannian manifolds*, Duke Math. J., **82** (1996), 679–706.

[11] V. Enss, *Asymptotic completeness for quantum mechanical potential scattering, I. Short range potentials*, Commun. Math. Phys., **61** (1978), 285–291.

[12] L. Gårding, T. Kotake et J. Leray, *Uniformisation et développement asymptotique de la solution du problème de Cauchy linéaire à données holomorphes*, Bull. Soc. Math. France, **92** (1964), 263–361.

[13] S. Hara, *A necessary condition for H^∞-wellposed Cauchy problem of Schrödinger type equations with variable coefficients*, J. Math. Kyoto Univ., **32** (1992), 287–305.

[14] N. Hayashi, K. Nakamitsu and M. Tsutsumi, *On solutions of the initial value problem for the nonlinear Schrödinger equations*, J. Funct. Anal., **71** (1987), 218–245.

[15] E. Hille and R. S. Phillips, *Functional Analysis and Semi-Groups*, Amer. Math. Soc., Providence, 1957.

[16] L. Hörmander, *The Analysis of Linear Partial Differential Operators*, II, Springer-Verlag, Berlin, 1983.

[17] L. Hörmander, *The Analysis of Linear Partial Differential Operators*, III, Springer-Verlag, Berlin, 1985.

[18] G. Hufford, *On the characteristic matrix of a matrix of differential operators*, J. Diff. Equations, **1** (1965), 27–38.

[19] W. Ichinose, *Some remarks on the Cauchy problem for Schrödinger type equations*, Osaka J. Math., **21** (1984), 565–581.

[20] W. Ichinose, *Sufficient condition on H_∞ wellposedness for Schrödinger type equations*, Comm. Partial Differential Equations, 9 (1984), 33–48.

[21] W. Ichinose, *The Cauchy problem for Schrödinger type equations with variables coefficients*, Osaka J. Math., 24 (1987), 853–886.

[22] W. Ichinose, *On L^2 well-posedness of the Cauchy problem for Schrödinger type equations on the Riemannian manifold and Maslov theory*, Duke Math. J., 56 (1988), 549–588.

[23] W. Ichinose, *A note on the Cauchy problem for Schrödinger type equations on the Riemannian manifold*, Math. Japonica, 35 (1990), 205–213.

[24] W. Ichinose, *On the Cauchy problem for Schrödinger type equations and Fourier integral operators*, J. Math. Kyoto Univ., 33 (1993), 583–620.

[25] W. Ichinose, *On a necessary condition for L^2 well-posedness of the Cauchy problem for some Schrödinger type equations with a potential term*, J. Math. Kyoto Univ., 33 (1993), 647–663.

[26] H. Isozaki and H. Kitada, *Microlocal resolvent estimates for two-body Schrödinger operators*, J. Funct. Anal., 57 (1984), 270–300.

[27] H. Isozaki and H. Kitada, *Modified wave operators with time-independent modifiers*, J. Fac. Sci. Univ. Tokyo, Sec1A., 32 (1985), 77–104.

[28] K. Kajitani, *The Cauchy problem for Schrödinger type equations with variable coefficients*, J. Math. Soc. Japan, 50 (1998), 179–202.

[29] K. Kajitani and A. Baba, *The Cauchy problem for Schrödinger type equations*, Bull. Sc. Math., 2e série, 119 (1995), 459–473.

[30] T. Kato, *Non-linear Schrödinger equations*, Schrödinger Operators, Springer Lecture Notes in Physics, 345 (1989), 218–263.

[31] T. Kato, *On the Cauchy problem for the (generalized) Korteweg-de Vries equation*, Studies in Appl. Math. Adv. in Math., Suppl. Studies., 8 (1983), 93–128.

[32] C. E. Kenig, G. Ponce and L. Vega, *Smoothing effects and local existence theory for the generalized nonlinear Schrödinger equations*, Inventiones Math., 134 (1998), 489–545.

[33] H. Kumano-go, *Pseudo-Differential Operators*, MIT Press, 1981.

[34] J. Leray, *Hyperbolic Differential Equations*, Inst. Adv. Study, Princeton, 1953.

[35] J. Leray, *Lagrangian Analysis and Quantum Mechanics. A mathematical structure related to asymptotic expansions and the Maslov index*, MIT Press, 1981.

[36] V. P. Maslov, *Theory of Perturbations and Asymptotic Methods*, Moskow, 1965 (en russe); French translation from Russian, Dunod, Paris, 1970.

[37] M. Miyake, *On Cauchy-Kowalewski's theorem for general systems*, Publ. Res. Inst. Math. Sci. Kyoto Univ., 15 (1979), 315–337.

[38] S. Mizohata, *Some remarks on the Cauchy problem*, J. Math. Kyoto Univ., 1 (1961), 109–127.

[39] S. Mizohata, *On Kowalewskian systems*, Russ. Math. Surveys, 29-7 (1974), 223–235.

[40] S. Mizohata, *On some Schrödinger type equations*, Proc. Japan Acad., 57 (1981), 81–84.

[41] S. Mizohata, *Sur quelques équations du type Schödinger*, Séminaire J. Vaillant 1980-1981, Univ. Paris-VI.

[42] S. Mizohata, *Sur quelques équations du type Schödinger*, Journées 《 Équations aux Dérivées Partielles 》, Saint-Jean-de-Monts, Soc. Math. France, 1981.

[43] S. Mizohata, *On the Cauchy Problem*, Notes and Reports in Math., **3**, Academic Press, 1985.

[44] I. G. Petrowsky, *Über das Cauchysche Problem für ein System linearer partieller Differentialgleichungen im Gebiete der nicht-analytischen Funktionen*, Bull. Univ. État Moscou, **1** (1938), 1–74.

[45] M. Reed and B. Simon, *Methods of Modern Mathematical Physics*, III. *Scattering Theory*, (XI.17, p. 331–340), Academic Press, New York, 1979.

[46] E. Schrödinger, *Quantisierung als Eigenwertproblem* (Vierte Mitteilung), Ann. der Physik, **81** (1926), 109–139.

[47] E. Schrödinger, *Gesammelte Abhandlungen*, **3** : *Beiträge zur Quantentheorie*, Herausgegeben von der Österreichischen Akademie der Wissenschaften, Verlag der Österreichischen Akademie der Wissenschaften, Wien, 1984.

[48] B. Simon, *Phase space analysis of simple scattering systems : Extensions of some work of Enss*, Duke Math. J., **46** (1979), 119–168.

[49] J. Takeuchi, *A necessary condition for the well-posedness of the Cauchy problem for certain class of evolution equations*, Proc. Japan Acad., **50** (1974), 133–137.

[50] J. Takeuchi, *Some remarks on my paper "On the Cauchy problem for some non-kowalewskian equations with distinct characteristic roots"*, J. Math. Kyoto Univ., **24** (1984), 741–754.

[51] J. Takeuchi, *On the Cauchy problem for systems of linear partial differential equations of Schrödinger type*, Bull. Iron and Steel Technical College, **18** (1984), 25–34.

[52] J. Takeuchi, *A necessary condition for H^∞-wellposedness of the Cauchy problem for linear partial differential operators of Schrödinger type*, J. Math. Kyoto Univ., **25** (1985), 459–472.

[53] J. Takeuchi, *Le problème de Cauchy pour quelques équations aux dérivées partielles du type de Schrödinger*, C. R. Acad. Sci. Paris, Série I, **310** (1990), 823–826.

[54] J. Takeuchi, *Le problème de Cauchy pour quelques équations aux dérivées partielles du type de Schrödinger*, II, C. R. Acad. Sci. Paris, Série I, **310** (1990), 855–858.

[55] J. Takeuchi, *Le problème de Cauchy pour certaines équations aux dérivées partielles du type de Schrödinger*, III, C. R. Acad. Sci. Paris, Série I, **312** (1991), 341–344.

[56] J. Takeuchi, *Le problème de Cauchy pour certaines équations aux dérivées partielles du type de Schrödinger*, IV, C. R. Acad. Sci. Paris, Série I, **312** (1991), 587–590.

[57] J. Takeuchi, *Le problème de Cauchy pour certains systèmes de Leray-Volevič du type de Schrödinger*, V, C. R. Acad. Sci. Paris, Série I, **312** (1991), 799–802.

[58] J. Takeuchi, *Le problème de Cauchy pour certaines équations aux dérivées partielles du type de Schrödinger*, VI, C. R. Acad. Sci. Paris, Série I, **313**

(1991), 761–764.

[59] J. Takeuchi, *Le problème de Cauchy pour certaines équations aux dérivées partielles du type de Schrödinger*, VII, C. R. Acad. Sci. Paris, Série I, **314** (1992), 527–530.

[60] J. Takeuchi, *Le problème de Cauchy pour certaines équations aux dérivées partielles du type de Schrödinger, VIII ; symétrisations indépendantes du temps*, C. R. Acad. Sci. Paris, Série I, **315** (1992), 1055–1058.

[61] J. Takeuchi, *Le problème de Cauchy pour certaines équations aux dérivées partielles du type de Schrödinger, IX ; symétrisations indépendantes du temps*, C. R. Acad. Sci. Paris, Série I, **316** (1993), 1025–1028.

[62] J. Takeuchi, *Le problème de Cauchy pour certaines équations aux dérivées partielles du type de Schrödinger, X; symétrisations indépendantes du temps*, Proc. Japan Acad., Ser. A, **69** (1993), 189–192.

[63] J. Takeuchi, *Le problème de Cauchy pour certaines équations aux dérivées partielles du type de Schrödinger*, Thèse de Doctorat de l'Université Paris-VI, Octobre 1995, pp. 115.

[64] J. Takeuchi, *Symétrisations indépendantes du temps pour certains opérateurs du type de Schrödinger* (I), Preprint, 1998, pp.50, Bollettino Unione Matematica Italiana (to appear).

[65] S. Tarama, *On the H^∞-wellposed Cauchy problem for some Schrödinger type equations*, Mem. Fac. Eng. Kyoto Univ., **55** (1993), 143–153.

[66] L. R. Volevich, *On general systems of differential equations*, Soviet Math. Dokl., **1** (1960), 458–461.

[67] L. R. Volevich, *A problem of linear programming arising in differential equations*, Uspehi Mat. Nauk., **18-3** (1963), 155–162. (en russe)

[68] C. Wagschal, *Diverses formulations du problème de Cauchy pour un système d'équations aux dérivées partielles*, J. Math. Pures et Appl., **53** (1974), 51–69.

[69] K. Yajima, *On smoothing property of Schrödinger propagators*, Functional Analytic Methods for partial differential equations, Proc. Intern. Conf. on Functional Analysis and its Applications, Springer Lecture Notes in Math., **1450** (1990), 20–35.

[70] K. Yosida, *Functional Analysis*, Springer-Verlag, 1965.

Systèmes du type de Schrödinger à raciness caractéristiques multiples

DANIEL GOURDIN

Université de Paris 6, UFR 920 (Mathématiques)
4, place Jussieu, 75252 Paris Cedex 05, France

e-mail : gourdin@math.jussieu.fr

Dédié au Professeur Jean Vaillant

On considère un opérateur scalaire de 2-évolution au sens de Petrowsky et on suppose que les racines (en τ) du polynôme principal à coefficients constants sont réelles de multiplicités constantes.

On donne alors des conditions suffisantes pour que le problème de Cauchy pour le futur et le passé soit bien posé en même temps dans les espaces de Sobolev. Nos conditions sont en partie analogues aux conditions de Levi pour les systèmes et aux conditions de bonne décomposition dans les cas hyperboliques introduites par J. Vaillant puis De Paris (respectivement en 1968 et en 1972).

On généralise ensuite aux systèmes 2×2, 3×3 et certains systèmes $N \times N$. On montre que ces conditions restent *scalaires* comme dans le cas hyperbolique. Le cas des systèmes $N \times N$ du type de Schrödinger *généraux* à multiplicité deux n'est pas terminé.

La première partie de ce travail a fait l'objet d'une Note aux CRAS en 1997 en collaboration avec S. Ngnosse et J. Takeuchi. La deuxième partie est écrite avec M. Ghedamsi, M. Mechab et J. Takeuchi.

— PREMIÈRE PARTIE —

I. INTRODUCTION – ÉNONCÉ DES RÉSULTATS

On note $(x,t) \in \mathbb{R}^n \times \mathbb{R}^1$, $D_t = -i\partial/\partial t$, $D_x = (D_1, \ldots, D_n)$, $D_j = -i\partial/\partial x_j$,

$$a_k(x, D_x) = \sum_{|\alpha| \leq 2k} a_{\alpha k}(x) D_x^\alpha, \quad a_{\alpha k}(x) \in \mathcal{B}^\infty(\mathbb{R}^n).$$

On considère un opérateur de 2-évolution au sens de Petrowsky :

$$(1) \qquad P(x, D_x, D_t) = D_t^m + a_1(x, D_x) D_t^{m-1} + \cdots + a_m(x, D_x)$$

On suppose le symbol principal P_{2m} à coefficients constants :

$$(2) \qquad P(\xi, \tau) = \tau^m + a_1^0(\xi)\tau^{m-1} + \cdots + a_m^0(\xi),$$

$$a_j^0(\xi) = \sum_{|\alpha|=2j} a_{\alpha j}\, \xi^\alpha \quad (1 \leq j \leq m).$$

On note aussi

$$(3) \qquad P_{2m-k}(x, \xi, \tau) = a_1^k(x, \xi)\tau^{m-1} + \cdots + a_m^k(x, \xi),$$

$$a_j^k(x, \xi) = \sum_{|\alpha|=2j-k} a_{\alpha j}(x)\xi^\alpha \quad (k \geq 1).$$

$P_{2m-1}(x, \xi, \tau)$ est le symbole sous-principal de $P(x, D_x, D_t)$; $P_{2m}(\xi, \tau)$ et $P_{2m-k}(x, \xi, \tau)$ sont quasi-homogènes respectivement de degrés $2m$ et $2m - k$ de poids $(1, 2)$:

$$P_{2m}(r\xi, r^2\tau) = r^{2m} P_{2m}(\xi, \tau) \quad (r \in \mathbb{R}^1),$$

$$P_{2m-k}(x, r\xi, r^2\tau) = r^{2m-k} P_{2m-k}(x, \xi, \tau) \quad (r \in \mathbb{R}^1, k \geq 1),$$

Nous allons donner des conditions suffisantes pour que le problème de Cauchy, à la fois pour le futur et pour le passé, noté

$$(*) \qquad \begin{cases} P(x, D_x, D_t)u(x,t) = f(x,t) \text{ sur } \mathbb{R}^n \times [-T, T] \ (T > 0) \\ D_t^{j-1}u(x, 0) = g_j(x) \text{ dans } \mathbb{R}^n \ (1 \leq j \leq m) \end{cases}$$

soit bien posé dans

$$C^m([-T, T]; H^\infty(\mathbb{R}^n))$$

dans le cas des opérateurs *dégénérés* du type de Schrödinger.

On impose la condition suivante :

CONDITION (A.1) — Les racines caractéristiques en τ de $P_{2m}(\xi, \tau)$ sont réelles pour $\xi \in \mathbb{R}^n$.

On dit alors que $P(x, D_x, D_t)$ est du type de Schrödinger.

Remarquons que cette condition est nécessaire pour que le problème de Cauchy $(*)$ soit bien posé dans $C^m([-T,T]; H^\infty(\mathbb{R}^n))$. Lorsque les racines caractéristiques sont simples, Takeuchi a donné des conditions nécessaires et des conditions suffisantes pour que le problème de Cauchy $(*)$ soit bien posé dans les espaces de Sobolev (cf. aussi les résultats de Mizohata, Ichinose, Tarama, Kajitani). Ici nous supposons les racines multiples et de multiplicités constantes et nous proposons des conditions suffisantes de résolubilité du problème de Cauchy $(*)$ analogues en partie aux conditions de Levi et aux conditions de bonne décomposition donnée dans le cas hyperbolique.

1°) Cas général.

On impose les conditions supplémentaires suivantes.

CONDITION (A.2) — Le symbole principal $P_{2m}(\xi,\tau)$ possède une décomposition en facteurs $H_s(\xi,\tau)$ premiers entre eux (deux à deux) dans $C[\xi,\tau]$ de multiplicités constantes ν_s.

$$P_{2m}(x,\xi) = \prod_{s=1}^{\sigma} (H_s(\xi,\tau))^{\nu_s},$$

$$H_s(\xi,\tau) = \tau^{q_s} + \sum_{k=1}^{q_s} a_k^{0,s}(\xi)\tau^{q_s-k},$$

$$a_k^{0,s}(\xi) = \sum_{|\alpha|=2k} a_\alpha^{0,s}\xi^\alpha, \quad m = \sum_{s=1}^{\sigma} \nu_s q_s.$$

CONDITION (A.3) — Toutes les racines caractéristiques $\lambda_i^0(\xi)$ du radical $R(\xi,\tau) = \prod_{s=1}^{\sigma} H_s(\xi,\tau)$ sont non nulles et distinctes deux à deux pour $\xi \in \mathbb{R}^n \setminus 0$:

$$H_s(\xi,\tau) = \prod_{i=n_{s-1}+1}^{n_s} (\tau - \lambda_i^0(\xi)) \ (1 \le s \le \sigma), \quad 0 = n_0 < n_1 < \cdots < n_\sigma.$$

On adopte la nouvelle notation suivante des racines $\lambda_i^0(\xi)$:

$$P_{2m}(x,\xi) = \prod_{s=1}^{\sigma}(H_s(\xi,\tau))^{\nu_s} = \prod_{j=1}^{p}\prod_{i=1}^{k_j}(\tau - \lambda_j^i(\xi)),$$

$$1 \le k_1 \le \cdots \le k_j \le \cdots \le k_p = \sup\{\nu_s\,;1 \le s \le \sigma\} = \mu,$$

$$\lambda_j^1(\xi) = \cdots = \lambda_j^{k_j}(\xi) \quad (1 \le j \le p),$$

$$\lambda_j^i(\xi) \ne \lambda_{j'}^{i'}(\xi) \quad \text{si} \quad j \ne j', 1 \le i \le k_j, 1 \le i' \le k_{j'}.$$

Soient

$$I_i = \{j; k_j \ge i\} \ (1 \le i \le \mu), \ m_i = \mathrm{Card}(I_i),$$

$$M_i = \sum_{r=1}^{i} m_r, \quad R_i(\xi,\tau) = \prod_{j \in I_i} (\tau - \lambda_j^i(\xi)).$$

Alors, on a $P_{2m}(x,\xi) = \prod_{i=1}^{\mu} R_i(\xi,\tau)$.

CONDITION (A.4) (Bonne décomposition) — Il existe des opérateurs différentiels en t et pseudo-différentiels en x notés c_j et r_j ($1 \leq j \leq \mu$)

$$c_j(x, D_x, D_t) = \sum_{k=0}^{M_{j-1}-(j-1)} d_k^j(x, D_x) D_t^{M_{j-1}-(j-1)-k},$$

$$r_j(x, D_x, D_t) = D_t^{m_j} + \sum_{k=1}^{m_j} r_k^j(x, D_x) D_t^{m_j-k}$$

avec $d_k^j \in BL^{2k}(\mathbb{R}^n)$, $r_k^j \in BL^{2k}(\mathbb{R}^n)$ tels que

$$P - \sum_{j=1}^{\mu} c_j r_j r_{j+1} \cdots r_\mu = \sum_{k=0}^{m-\mu} d_k^{\mu+1}(x, D_x) D_t^{m-\mu-k},$$

r_j admettant $R_j(\xi,\tau)$ pour symbole principal et $d_k^{\mu+1} \in BL^{2k}(\mathbb{R}^n)$.

CONDITION (A.5) — Soit $r_k^{j,1}(x,\xi)$ le symbole sous-principal de $r_k^j(x, D_x)$ ($1 \leq k \leq m_j$, $1 \leq j \leq \mu$).

(i) il existe $\varepsilon > 0$, tel que, pour tout β, on a

$$|D_x^\beta \operatorname{Im} r_k^{j,1}(x,\xi)| \leq C_\beta \langle x \rangle^{-1-\varepsilon-\min\{|\beta|,1\}} \langle \xi \rangle^{2k-1}.$$

(ii) on a, pour tout $|\gamma| \geq 1$,

$$|D_x^\gamma \operatorname{Re} r_k^{j,1}(x,\xi)| \leq C_\gamma \langle x \rangle^{-1} \langle \xi \rangle^{2k-1}.$$

Cette conditions sont invariantes par passage à l'adjoint.

Nous obtenons le théorème suivant.

THÉORÈME 1 — *Supposons les conditions (A.1) à (A.5) vérifiées. Alors, pour tout $g_1(x), \ldots, g_m(x)$ dans $H^\infty(\mathbb{R}^n)$ et tout $f(x,t) \in C^1([-T,T]; H^\infty(\mathbb{R}^n))$, il existe une solution unique du problème de Cauchy $(*)$ à la fois pour le futur et pour le passé telle que*

$$u(t) = u(\cdot, t) \in C^m([-T,T]; H^\infty(\mathbb{R}^n)).$$

De plus, pour tout $v(t) \in C^m([-T,T]; H^\infty(\mathbb{R}^n))$,

$$\|D^{l+m-\mu}v(t)\|_{(s)} \leq C(T)\Big\{ \sum_{j=0}^{m-1} \|D_t^j v(0)\|_{s+2(l+m-1-j)} + \|D^{l-1}Pv(0)\|_{(s)} +$$

$$+ \Big| \int_0^t \|D^l Pv(\tau)\|_{(s)} d\tau \Big| \Big\}, \quad t \in [-T,T],$$

où

$$\|D^l u(t)\|_{(s)} = \sup_{0 \leq j \leq l} \|D_t^j u(t)\|_{s+2(l-j)}$$

et $\|u(t)\|_s$ *est la* $H^s(\mathbb{R}^n)$*-norme de* $u(t) = u(\cdot, t)$ *avec la convention* $\|D^{l-1} u(0)\|_{(s)} = 0$ *si* $l = 0$.

2°) Cas particulier (multiplicité ≤ 2).

CONDITION (A.2)′ — Soient m_0 et m_1 deux nombres entiers positifs tels que $m = m_0 + m_1$ et $1 \leq m_0 \leq m_1$. Le symbole principal $P_{2m}(\xi, \tau)$ admet dans $\mathbb{C}[\xi, \tau]$ la décomposition en facteurs premiers $H_s(\xi, \tau)$ ($s = 0, 1$), unitaires en τ, quasi-homogènes de degrés respectifs $2m_0$ et $2(m_1 - m_0)$ de poids $(1, 2)$ notée

$$P_{2m}(\xi, \tau) = [H_0(\xi, \tau)]^2 H_1(\xi, \tau).$$

CONDITION (A.3)′ — Les racines $\lambda_j^0(\xi)$ ($1 \leq j \leq m_1 = m - m_0$) en τ du radical $R(\xi, \tau) = H_0(\xi, \tau) H_1(\xi, \tau)$ sont non nulles et distinctes deux à deux.

On note

$$Q_{2m_0}(\xi, \tau) = H_0(\xi, \tau) = \prod_{j=1}^{m_0} (\tau - \lambda_j^0(\xi)),$$

$$R_{2m_1}(\xi, \tau) = H_0(\xi, \tau) H_1(\xi, \tau) = \prod_{j=1}^{m-m_0} (\tau - \lambda_j^0(\xi)).$$

CONDITION (A.4)′ — Le polynôme $H_0(\xi, \tau)$ divise les polynômes $P'_{2m-k}(x, \xi, \tau)$ ($k = 1, 2, 3$) dans $\mathcal{B}[\xi, \tau]$ avec $\mathcal{B} = \mathcal{B}^\infty(\mathbb{R}^n)$ et

$$P'_{2m-1} = P_{2m-1},$$

$$P'_{2m-2} = P_{2m-2} - Q_{2m_0-1} R_{2m_1-1} - \sum_{|\alpha|=1} H_0^{(\alpha,0)} R_{2m_1-1(\alpha)},$$

$$P'_{2m-3} = P_{2m-3} - \{ Q_{2m_0-1} R_{2m_1-2} + Q_{2m_0-2} R_{2m_1-1} \}$$
$$- \sum_{|\alpha|=1} \{ H_0^{(\alpha,0)} R_{2m_1-2(\alpha)} + Q_{2m_0-1}^{(\alpha,0)} R_{2m_1-1(\alpha)} \}$$
$$- \sum_{|\alpha|=2} \frac{1}{\alpha!} H_0^{(\alpha,0)} R_{2m_1-1(\alpha)}.$$

où $\mathcal{B}^\infty(\mathbb{R}^n)$ est l'anneau des fonctions de classe $C^\infty(\mathbb{R}^n)$ à dérivées bornées sur $\mathbb{R}^n$, et en notant $P'_{2m-k} = H_0 P''_{2m_1-k}$ ($1 \leq k \leq 3$), Q_{2m_0-k} et R_{2m_1-k} sont respectivement le quotient et le reste de la division euclidienne de P''_{2m_1-k} par H_1.

CONDITION (A.5)′ — Pour $|\alpha| = 2j - 1$, $1 \leq j \leq m_1 = m - m_0$,

(i) il existe $\varepsilon > 0$, tel que, pour tout β

$$\left| D_x^\beta \operatorname{Im} a_{\alpha j}(x) \right| \leq C_\beta \left\langle x \right\rangle^{-1-\varepsilon-\min\{|\beta|,1\}} \quad (\varepsilon > 0),$$

(ii) on a, pour tout γ ($|\gamma| \geq 1$),

$$\left| D_x^\gamma \operatorname{Re} a_{\alpha j}(x) \right| \leq C_\gamma \left\langle x \right\rangle^{-1}.$$

PROPOSITION 1 — *Supposons les conditions* (A.1) *et* (A.2)$'$ *à* (A.5)$'$ *vérifiées. Alors, il existe des opérateurs différentiels* $C_k(x, D_x)$ *d'ordre* $2k$ *à coefficients dans* $\mathcal{B}^\infty(\mathbb{R}^n)$ ($0 \leq k \leq m-2$) *tels que*

$$P(x, D_x, D_t) - Q(x, D_x, D_t) R(x, D_x, D_t) = \sum_{k=0}^{m-2} C_k(x, D_x) D_t^{m-2-k}$$

où

$$Q(x, \xi, \tau) = Q_{2m_0}(\xi, \tau) + \sum_{k=1}^{2m_0} Q_{2m_0 - j}(x, \xi, \tau),$$

$$R(x, \xi, \tau) = R_{2m_1}(\xi, \tau) + \sum_{k=1}^{2m_1} R_{2m_1 - j}(x, \xi, \tau).$$

Grâce à la proposition 1, nous obtenons le théorème suivant.

THÉORÈME 1$'$ — *Supposons les conditions* (A.1) *et* (A.2)$'$ *à* (A.5)$'$ *vérifiées. Alors, on a la même conclusion que celle du théoreme 1 avec* $\mu = 2$.

II. REMARQUES

La démonstration repose sur des méthodes utilisées en hyperbolicité à caractéristiques multiples de Gourdin et Mechab (1984) et des travaux de J. Takeuchi (1995).

Dans le cas particulier 2°) (multiplicité ≤ 2), la condition (A.4) équivaut à l'existence de deux opérateurs de 2-évolution :

$$Q(x, D_x, D_t) = D_t^{m_0} + b_1(x, D_x) D_t^{m_0-1} + \cdots + b_{m_0}(x, D_x),$$

$$R(x, D_x, D_t) = D_t^{m-m_0} + c_1(x, D_x) D_t^{m-m_0-1} + \cdots + c_{m-m_0}(x, D_x)$$

d'ordres généralisés $2m_0$ et $2(m - m_0)$ tels que

$$P(x, D_x, D_t) - Q(x, D_x, D_t) R(x, D_x, D_t) = \sum_{k=0}^{m-2} d_k(x, D_x) D_t^{m-2-k}$$

d'ordres généralisés $2(m - 2)$.

En notant $Q_{2m_0} = H_0$, $Q_{2(m-m_0)} = H_0 H_1$,

$$Q(x, \xi, \tau) = Q_{2m_0}(\xi, \tau) + \sum_{k=1}^{3} Q_{2m_0-j}(x, \xi, \tau),$$

$$R(x, \xi, \tau) = R_{2m_1}(\xi, \tau) + \sum_{k=1}^{3} R_{2m_1-j}(x, \xi, \tau),$$

la décomposition de leur symboles respectifs en symboles quasi-homogènes Q_s et R_s de degrés s en (ξ, τ) de poids $(1, 2)$, la condition (A.4) équivaut à la divisibilité par H_0 de trois polynômes P'_{2m-1}, P'_{2m-2}, P'_{2m-3} (et non d'un seul comme dans le cas hyperbolique) car $P'_{2m-k} = Q_{2m_0-k}R_{2(m-m_0)} + R_{2(m-m_0)-k}Q_{2m_0}$ $(1 \leq k \leq 3)$ et par conséquent $P''_{2m_1-k} = Q_{2m_0-k}H_1 + R_{2m_1-k}$ $(1 \leq k \leq 3)$.

— DEUXIÈME PARTIE —

III. ETUDE D'UNE AUTRE CONDITION (A.4)″ IMPLIQUANT LA CONDITION (A.4) POUR L'OPÉRATEUR (1) SCALAIRE

Il suffit de remplacer R_{2m_1-1} par sa valeur $P''_{2m_1-1} - Q_{2m_0-1}H_1$ donnée par la première relation de divisibilité $P'_{2m_1-1} = H_0 P''_{2m_1-1}$, dans les deux autres relations $P'_{2m_1-2} = H_0 P''_{2m_1-2}$, $P'_{2m_1-3} = H_0 P''_{2m_1-3}$.

La deuxième relation $P'_{2m_1-2} = H_0 P''_{2m_1-2}$ donne l'équation d'inconnue Q_{2m_0-1} :

$$\left\{ H_1 (Q_{2m_0-1})^2 - P''_{2m_1-1} Q_{2m_0-1} + P_{2m-2} \right.$$
$$\left. - \sum_{|\alpha|=1} H_0^{(\alpha,0)} P''_{2m_1-1(\alpha)} + H_1 \sum_{|\alpha|=1} H_0^{(\alpha,0)} Q_{2m_0-1(\alpha)} \right\}(\xi, \lambda_j^0(\xi)) = 0$$

$$(1 \leq j \leq m_0)$$

qui est une équation différentielle non linéaire ordinaire le long des bicaractéristiques associées à H_0.

Le système différentiel équivalent est : en supposant

$$\left(\frac{\partial H_0}{\partial \xi_1} \right)(\xi, \lambda_j^0(\xi)) \neq 0,$$

$$\frac{dx_k}{dx_1} = \frac{\partial H_0}{\partial \xi_k}(\xi, \lambda_j^0(\xi)) \Big/ \frac{\partial H_0}{\partial \xi_1}(\xi, \lambda_j^0(\xi)) \quad (2 \leq k \leq n),$$

$$\frac{dQ_{2m_0-1}}{dx_1} = \left[\left\{ -(Q_{2m_0-1})^2 + \frac{P''_{2m_1-1}}{H_1} Q_{2m_0-1} \right. \right.$$
$$\left. \left. + \frac{-P_{2m-2} + \sum_{|\alpha|=1} H_0^{(\alpha,0)} P''_{2m_1-1(\alpha)}}{H_1} \right\} \Big/ \frac{\partial H_0}{\partial \xi_1} \right](\xi, \lambda_j^0(\xi))$$

avec les conditions initiales

$$x_2(0) = y_2, \ldots, x_n(0) = y_n,$$
$$Q_{2m_0-1}(x_1 = 0) = \varphi(y).$$

D'où

$$x_2 = x_1\left[\frac{\partial H_0}{\partial \xi_2}(\xi, \lambda_j^0(\xi))\Big/\frac{\partial H_0}{\partial \xi_1}(\xi, \lambda_j^0(\xi))\right] + y_2,$$

$$\vdots$$

$$x_n = x_n\left[\frac{\partial H_0}{\partial \xi_n}(\xi, \lambda_j^0(\xi))\Big/\frac{\partial H_0}{\partial \xi_1}(\xi, \lambda_j^0(\xi))\right] + y_n,$$

et l'équation différentielle ordinaire

$$\frac{dz}{dx_1}(x_1) = Az^2 + Bz + C \equiv F(x_1, y, \xi, z),$$

où

$$z(x_1) = Q_{2m_0-1}\left(x_1, x_1\frac{\frac{\partial H_0}{\partial \xi_2}}{\frac{\partial H_0}{\partial \xi_1}}(\xi, \lambda_j^0(\xi)) + y_2, \ldots, x_1\frac{\frac{\partial H_0}{\partial \xi_n}}{\frac{\partial H_0}{\partial \xi_1}}(\xi, \lambda_j^0(\xi)) + y_n, \xi, \lambda_j^0(\xi)\right),$$

$$A = \left(\frac{\partial H_0}{\partial \xi_1}(\xi, \lambda_j^0(\xi))\right)^{-1},$$

$$B = P''_{2m_1-1}(x, \xi, \lambda_j^0(\xi))\Big/H_1(\xi, \lambda_j^0(\xi))\frac{\partial H_0}{\partial \xi_1}(\xi, \lambda_j^0(\xi)),$$

$$C = \left[-P_{2m-2} + \sum_{|\alpha|=1} H_0^{(\alpha,0)}P''_{2m_1-1(\alpha)}\right](x, \xi, \lambda_j^0(\xi))\Big/\left[H_1\frac{\partial H_0}{\partial \xi_1}\right](\xi, \lambda_j^0(\xi)).$$

On a $\|F\| \leq \|A\||z|^2 + \|B\||z| + \|C\|$. Comme

$$\int^\infty \frac{d|z|}{\|A\||z|^2 + \|B\||z| + \|C\|} < \infty,$$

cette équation différentielle non linéaire n'a pas de solution globale en général ([2] p.379–380). Pour qu'il y ait une solution globale, il faut supposer Q_{2m_0-1} constant par rapport à x et donc P_{2m-1} et P_{2m-2} aussi constants par rapport à x. Alors, la condition (A.5)' implique que Q_{2m_0-1} et R_{2m_1-1} doivent être réels donc que le discriminant de l'équation algébrique (en $X = Q_{2m_0-1}$)

$$\left[-X^2 + \frac{P''_{2m_1-1}}{H_1}X - \frac{P_{2m-2}}{H_1}\right](\xi, \lambda_j^0(\xi)) = 0 \quad (1 \leq j \leq m_0)$$

soit positif ou nul.

Donc, on peut remplacer de la condition (A.4)$'$ par la condition suivante.

CONDITION (A.4)$''$ —

(i) (divisibilité de $P_{2m-1} = P'_{2m-1}$ par H_0) $P'_{2m-1} = H_0 P''_{2m_1-1}$,

(ii) on a l'inégalité suivante :

$$\left[(P''_{2m_1-1})^2 - 4H_1 P_{2m-2} \right](\xi, \lambda_j^0(\xi)) > 0 \quad (1 \leq j \leq m_0).$$

Dans ces conditions, on obtient Q_{2m_0-1} et par suite R_{2m_1-1} que l'on reporte dans la troisième relation de divisibilité $P'_{2m-3} = H_0 P''_{2m_1-3}$; en tenant compte du fait que

$$R_{2m_1-1} = P''_{2m_1-2} - H_1 Q_{2m_0-2},$$

cette troisième relation de divisibilité s'écrit sous forme d'une équation dérivée partielle du premier ordre (équation différentielle linéaire ordinaire le long des bicaractéristiques associées à H_0) d'inconnue Q_{2m_0-2}. D'où Q_{2m_0-2} et R_{2m_0-2}. On remarque que la résolution est pseudo-différentiel en x pour Q et R.

On a le théorème suivant.

THÉORÈME 2 — *Supposons les conditions* (A.1), (A.2)$'$, (A.3)$'$ *et* (A.4)$''$ *vérifiées. Alors, le problème de Cauchy* (∗) *est bien posé dans*

$$C^m([-T,T]; H^\infty(\mathbb{R}^n))$$

et l'opérateur de Schrödinger vérifie la même inégalité d'énergie que celle du théoreme 1 avec $\mu = 2$.

IV.   ETUDE D'UN SYSTÈME DU PREMIER ORDRE DÉGÉNÉRÉ DU TYPE DE SCHRÖDINGER

Soit $h(D_x, D_t)$ l'opérateur matriciel 2×2 à coefficients constantes de 2-evolution

$$h(D_x, D_t) = D_t I_2 - A(D_x)$$

où $A(D_x)$ est d'ordre 2 en x.

On fait les trois hypothèses suivantes :

(i) La partie principale $H(\xi, \tau)$ de l'opérateur $h(D_x, D_t)$ est triangularisée sous la forme :

$$H(\xi, \tau) = \begin{pmatrix} \tau - \lambda_0(\xi) & \mu_0(\xi) \\ 0 & \tau - \lambda_0(\xi) \end{pmatrix}$$

avec $\lambda_0(\xi)$ et $\mu_0(\xi)$ deux formes quadratiques en ξ définies positives ou négatives à coefficients réels.

(ii) La partie sous-principale $H^*(\xi, \tau)$ de l'opérateur $h(D_x, D_t)$ est une matrice 2×2 de formes linéaires en ξ à coefficients réels :

$$H^*(\xi, \tau) = \begin{pmatrix} \alpha(\xi) & \beta(\xi) \\ \gamma(\xi) & \delta(\xi) \end{pmatrix}.$$

(iii) La partie quasi homogène $H^{**}(\xi, \tau)$ de degré 0 de l'opérateur $h(D_x, D_t)$ est une matrice 2×2 des constantes réels :

$$H^{**}(\xi, \tau) = H^{**} = \begin{pmatrix} \alpha' & \beta' \\ \gamma' & \delta' \end{pmatrix}.$$

Nous obtenons le théorème suivant :

THÉORÈME 3 — *Si $\gamma(\xi) = 0$ et $(\delta(\xi) - \alpha(\xi))^2 - 4\mu(\xi)\gamma'$ est une forme quadratique définie positive, alors le problème de Cauchy*

$$\begin{cases} h(D_x, D_t)u(x,t) = f(x,t) & sur \ \mathbb{R}^n \times [-T, T] \ \ (T > 0) \\ u(x, 0) = u_0(x) & dans \ \mathbb{R}^n \end{cases}$$

est bien posé pour le futur et le passé au sens où : pour tout $u_0(x) \in [H^\infty(\mathbb{R}^n)]^2$ et tout $f(x,t) \in C^1([-T, T]; [H^\infty(\mathbb{R}^n)]^2)$, il existe une solution unique

$$u(x,t) \in C^1([-T, T]; [H^\infty(\mathbb{R}^n)]^2)$$

et de plus on a

$$\|v(t)\|_{L^2(\mathbb{R}^n)} \le C(T) \Bigg\{ \|v(0)\|_{H^2(\mathbb{R}^n)} + \|h(D_x, D_t)v(0)\|_{L^2(\mathbb{R}^n)} +$$

$$+ \left| \int_0^t \Big[\|D_t h(D_x, D_s)v(s)\|_{L^2(\mathbb{R}^n)} + \|h(D_x, D_s)v(s)\|_{H^2(\mathbb{R}^n)} \Big] ds \right| \Bigg\},$$

pour tout $t \in [-T, T]$ et pour tout $v(t) \in C^1([-T, T]; [H^\infty(\mathbb{R}^n)]^2)$.

La preuve de ce théorème repose sur la proposition suivante :

PROPOSITION 2 — *Il existe $B^*(\xi)$ matrice 2×2 de formes linéaires en ξ et $B^*(\xi)$ matrice 2×2 de constante, $Q(x, D_x, D_t)$ et $R(x, D_x, D_t)$ deux opụrateurs de 2-évolution différentiel en t et pseudo-différentiels en x de symbole principal comme $H_0 I_2$, $d_0(x, D_x) \in BL^0(\mathbb{R}^n)$ tels que*

$$P = h(D_x, D_t)[B(D_x, D_t) + B^*(D_x) + B^{**}(D_x)]$$
$$= Q(x, D_x, D_t)R(x, D_x, D_t) + d_0(x, D_x)$$

où

$$B(\xi, \tau) = \begin{pmatrix} \tau - \lambda_0(\xi) & -\mu_0(\xi) \\ 0 & \tau - \lambda_0(\xi) \end{pmatrix} = H^{co}(\xi, \tau),$$

Q et R admettent des symboles sous-principaux réels indépendante de x.

Par un raisonnement analogue à celui de [4] on déduit le théorème 3.

REMARQUE 1 — Le polynôme sous-caractéristique

$$\mathcal{K}(\xi, \tau) = \sum_{1 \leq i,j \leq 2} B_{1,i} H^{*}_{i,j} B_{j,2}$$

$(B_{1,2}(\xi, \lambda_0(\xi)) \neq 0)$ du système (cf. [3]) est

$$\mathcal{K}(\xi, \tau) = \beta(\tau - \lambda_0)^2 - \mu_0(\alpha + \delta)(\tau - \lambda_0) + \gamma \mu_0^2.$$

Par consequent,

$$\delta = 0 \;\Leftrightarrow\; \mathcal{K}(\xi, \tau) \text{ est divisible par } (\tau - \lambda_0)$$

et

$$(\delta - \alpha)^2 + 4\mu_0 \gamma' > 0 \;\Leftrightarrow\; \left(\frac{1}{\mu_0}\right)^2 \left(\frac{\mathcal{K}}{H_0}\right)^2 (\xi, \lambda_0(\xi)) - 4\alpha\delta + 4\mu_0 \gamma' > 0.$$

V. SYSTÈME 3×3 DE SCHRÖDINGER À RACINES CARACTÉRISTIQUES DE MULTIPLICITÉ 1 ET 2

Soit $h(D_x, D_t)$ l'opérateur aux dérivées partielles (matriciel 3×3) à coefficients constantes de 2-evolution

$$h(D_x, D_t) = D_t I_3 - A(D_x)$$

où $A(D_x)$ est d'ordre 2 en x.

On fait les trois hypothèses suivantes :

(i) La partie principale $H(\xi, \tau)$ de l'opérateur $h(D_x, D_t)$ est triangularisée sous la forme :

$$H(\xi, \tau) = \begin{pmatrix} \tau - \lambda_0(\xi) & \mu_0(\xi) & \mu_1(\xi) \\ 0 & \tau - \lambda_0(\xi) & \mu_2(\xi) \\ 0 & 0 & \tau - \lambda_1(\xi) \end{pmatrix}$$

avec $\lambda_0(\xi), \lambda_1(\xi), \mu_0(\xi), \mu_1(\xi), \mu_2(\xi)$ cinq formes quadratiques en ξ à coefficients réels constants telles que $\lambda_0(\xi), \lambda_1(\xi), \mu_0(\xi)$ et $\lambda_0(\xi) - \lambda_1(\xi)$ soient définies, positives ou négatives.

(ii) La partie sous-principale de l'opérateur $h(D_x, D_t)$

$$H^{*}(\xi) = \begin{pmatrix} \alpha(\xi) & \beta(\xi) & \gamma(\xi) \\ \delta(\xi) & \varepsilon(\xi) & \varphi(\xi) \\ \psi(\xi) & \rho(\xi) & \sigma(\xi) \end{pmatrix}$$

est formée de neuf formes linéaires en ξ à coefficients réels.

(iii) La partie quasi homogène H^{**} de degré 0 de l'opérateur $h(D_x, D_t)$ est formée de neuf constantes réelles :

$$H^{**} = \begin{pmatrix} \alpha' & \beta' & \gamma' \\ \delta' & \varepsilon' & \varphi' \\ \psi' & \rho' & \sigma' \end{pmatrix}.$$

Remarquons que les hypothèses sont invariantes par passage à l'adjoint.

La matrice des cofacteurs de $H(\xi, \tau)$ est

$$B(\xi, \tau) = \begin{pmatrix} (\tau - \lambda_0)(\tau - \lambda_1) & -\mu_0(\tau - \lambda_1) & \mu_0\mu_2 - \mu_1(\tau - \lambda_0) \\ 0 & (\tau - \lambda_0)(\tau - \lambda_1) & -\mu_2(\tau - \lambda_0) \\ 0 & 0 & (\tau - \lambda_0)^2 \end{pmatrix}.$$

Le polynôme sous-caractéristique de l'opérateur $h(D_x, D_t)$ est (cf. [13])

$$\mathcal{K}(\xi, \tau) = \sum_{1 \leq i,j \leq 3} B_{1,i} H^*_{i,j} B_{j,2},$$

$(B_{12}(\xi, \lambda_0(\xi)) \neq 0)$.

En notant $H_k = \tau - \lambda_k$ $(k = 0, 1)$, nous obtenons le théorème suivant :

THÉORÈME 4 — *Si $\mathcal{K}$ est divisible par H_0 $(\Leftrightarrow \mu_2\psi - (\lambda_0 - \lambda_1)\delta = 0)$ et si*

$$\frac{1}{4\mu_0^2(\lambda_0 - \lambda_1)^3} \left(\frac{\mathcal{K}}{H_0}\right)^2 (\xi, \lambda_0(\xi)) > \mu_0[\mu_2\psi' - (\lambda_0 - \lambda_1)\delta'] + \mu_0(\varphi\psi - \sigma\delta)$$

$$+ \rho(\delta\mu_1 - \alpha\mu_2) + \alpha\varepsilon(\lambda_0 - \lambda_1),$$

alors le problème de Cauchy

$$\begin{cases} h(D_x, D_t)u(x, t) = f(x, t) & sur \ \mathbb{R}^n \times [-T, T] \ \ (T > 0) \\ u(x, 0) = u_0(x) & dans \ \mathbb{R}^n \end{cases}$$

admet une solution unique

$$u(x, t) \in C^1([-T, T]; [H^\infty(\mathbb{R}^n)]^3)$$

lorsque $u_0(x) \in [H^\infty(\mathbb{R}^n)]^3$ et $f(x, t) \in C^1([-T, T]; [H^\infty(\mathbb{R}^n)]^3)$ et on a la même inégalité d'énergie que celle dans le théoreme 3.

La preuve de ce théorème repose sur le même type de proposition que dans le paragraphe précédent. Cette fois, on expose $h(D_x, D_t)$ avec

$$b(D_x, D_t) = B(D_x, D_t) + B^*(D_x, D_t) + B^{**} + B^{***} + B^{****}$$

où les $B^{*\cdots*}$ sont pseudo-différentiels indépendants de x en D_x et différentiels en D_t ; Q et R matriciels triangulaires 3×3 ont respectivement pour symboles principaux

$H_0 I_3$ et $H_0 H_1 I_3$ et leurs symboles sous-principaux réels sont des matrices triangulaires dont les coefficients sont solutions d'équations algégriques du second degré pour les temps diagonaux et d'équations différentielles du premier ordre linéaire pour les temps non diagonaux $(X = (x, t))$.

VI. REMARQUES

Par le même type de démonstration qu'au §.V, on a un théorème analogue au théorème 2 du paragraphe 1 (en remplaçant P par h) avec les opérateurs différentiels matriciels $(N + 2) \times (N + 2)$ d'ordre m

$$h(x, D_x, D_t) = D_t^m I_{N+2} + \sum_{k=1}^{m} A_k(x, D_x) D_t^{m-k}$$

telle que

1) $A_j(x, D_x)$ soit un opérateur différentiels d'ordre $2j$ $(1 \leq j \leq m)$:

$$A_j(x, D_x) = \sum_{|\alpha| \leq 2j} A_{\alpha j}(x) D_x^{\alpha}, \quad A_{\alpha j}(x) \in \mathcal{B}^{\infty}(\mathbb{R}^n);$$

2) La partie principale (quasi-homogène de degré $2m$)

$$H(\xi, \tau) = \tau^m I_{N+2} + \sum_{k=1}^{m} A_k^0(\xi) \tau^{m-j}, \quad A_j^0(\xi) = \sum_{|\alpha| = 2j} A_{\alpha j} \xi^{\alpha}$$

soit à coefficients réels constants et s'écrive :

$$H = \begin{pmatrix} H_0 & K & 0 & \cdots & 0 \\ 0 & H_0 & 0 & \cdots & 0 \\ \vdots & \ddots & H_1 & \ddots & \vdots \\ \vdots & & \ddots & \ddots & 0 \\ 0 & \cdots & \cdots & 0 & H_N \end{pmatrix}$$

où

$$H_k = \prod_{j=km+1}^{(k+1)m} (\tau - \lambda_j(\xi)) \ (0 \leq k \leq N), \quad K = \mu_0(\xi) \prod_{j=1}^{m-1} (\tau - \mu_j(\xi))$$

λ_j $(1 \leq j \leq (N+1)m)$ et μ_j $(1 \leq j \leq m-1)$ sont des formes quadratiques telles que λ_j $(1 \leq j \leq (N+1)m)$, $\lambda_j - \lambda_k$ $(j \neq k)$, μ_0, $\lambda_j - \mu_k$ $(1 \leq j \leq (N+1)m, 1 \leq k \leq m-1)$ soient définies positives ou négatives ;

3) Les parties qusi-homogènes H^* et H^{**} de degré $(2m - 1)$ et $(2m - 2)$ soient à coefficients constants réeles et le polynôme sous-caractéristique

$$\mathcal{K}(\xi, \tau) = \sum_{1 \leq p, q \leq N+2} B_{1p} H_{pq}^* B_{q2}$$

soit divisible par H_0 dans $\mathbb{R}[\xi, \tau]$.

4) $\mathcal{K} = H_0 \mathcal{K}'$ et pour $l = 1, \ldots, m$,

$$\left(\frac{\mathcal{K}'}{K \prod_{i=1}^{N} H_i} \right)^2 (\xi, \lambda_l(\xi)) > \left\{ 4 \prod_{i=1}^{N} H_i \left[-H_{2,1} K \prod_{i=1}^{N} H_i + H_{2,2}^* H_{2,2}^{**} \prod_{i=1}^{N} H_i + \right.\right.$$
$$\left.\left. + \sum_{j=3}^{N+2} H_{2,j}^* H_{j,1}^* K \prod_{k=1, k \neq j-2}^{N} H_k \right] \right\} (\xi, \lambda_l(\xi)).$$

(Les hypothèses sont invariantes par passage à l'ajoint.)

BIBLIOGRAPHIE

[1] J. Chazarain et A. Piriou, *Introduction à la théorie des équations aux dérivées partielles linéaires,* Gauthier-Villars, Paris, 1981.

[2] J. Dieudonné, *Calcul infinitésimal,* Collection Méthodes, Hermann, Paris, 1968.

[3] D. Gourdin, *Systèmes faiblement hyperboliques à caractéristiques multiples,* C. R. Acad. Sci. Paris, **278** (1974), pp. 269–272.

[4] D. Gourdin et M. Mechab, Propagation des singularités et opérateurs différentiels (ed. J. Vaillant), Travaux en Cours, Hermann, 1984, pp.121–147.

[5] D. Gourdin, S. Ngnosse et J. Takeuchi, *Problème de Cauchy pour certaines équations du type de Schrödinger,* C. R. Acad. Sci. Paris, **324** (1997), pp. 1111–1116.

[6] W. Ichinose, *On the Cauchy problem for Schrödinger type equations and Fourier integral operators,* J. Math. Kyoto Univ., **33** (1993), pp.583–620.

[7] K. Kajitani, *The Cauchy problem for Schrödinger type equations with variable coefficients,* J. Math. Soc. Japan, **50** (1998), pp.179–202.

[8] K. Kajitani et A. Baba, *The Cauchy problem for Schrödinger type equations,* Bull. Sc. Math., 2ème série, **119** (1995), pp.459–473.

[9] S. Mizohata, *Sur quelques équations du type de Schrödinger,* Journées EDP, Saint-Jean-de-Monts, Soc. Math. France, 1981.

[10] I. G. Petrowsky, *Über das Cauchysche Problem für ein System linearer partieller Differentialgleichungen im Gebiete der nicht-analytischen Funktionen,* Bull. Univ. État Moscou, **1** (1938), pp.1–74.

[11] J. Takeuchi, *Le problème de Cauchy pour certaines équations aux dérivées partielles du type de Schrödinger,* Thèse de Doctorat de l'Université Paris 6, Octobre 1995, 115 pp.

[12] S. Tarama, *On the H^∞-wellposed Cauchy problem for some Schrödinger type equations,* Mem. Fac. Eng. Kyoto Univ., **55** (1993), pp.143–153.

[13] L. Vaillant, *Données de Cauchy portées par une caractéristique double, dans le cas d'un système linéaires d'équations aux dérivées partielles, rôle des bicaractéristiques,* J. Math. Pures et Appl. **47** (1968), pp.1–40.

Smoothing effect in Gevrey classes
for Schrödinger equations

Kunihiko Kajitani
Institute of Mathematics
University of Tsukuba
305 Tsukuba Ibaraki Japan

- to the memory of Nobuhisa Iwasaki -

Introduction

We shall investigate Gevrey smoothing effects of the solutions to the Cauchy problem for Schrödinger type equations. Roughly speaking, we shall prove that if the initial data decay as $e^{-c<x>^\kappa}(0 < \kappa \leq 1, c > 0)$, then the solutions belong to Gevrey class $\gamma^{1/\kappa}$ with respect to the space variables. Let $T > 0$. We consider the following Cauchy problem,

$$(1) \qquad \frac{\partial}{\partial t}u(t,x) - i\Delta u(t,x) - b(t,x,D)u(t,x) = 0, t \in [-T,T], x \in R^n,$$

$$(2) \qquad u(0,x) = u_0(x), x \in R^n,$$

where

$$(3) \qquad b(t,x,D)u = \sum_{j=1}^{n} b_j(t,x)D_ju + b_0(t,x)u,,$$

and $D_j = -i\frac{\partial}{\partial x_j}$. We assume that the coefficients $b_j(t,x)$ satisfy

$$(4) \qquad |D_x^\alpha b_j(t,x)| \leq C_b(\rho_b\langle x\rangle)^{-|\alpha|}|\alpha|!^d,$$

for $(t,x) \in [-T,T] \times R^n, \alpha \in N^n$, where $\langle x\rangle = (1 + |x|^2)^{1/2}$. Moreover we assume that there is $\kappa \in (0,1]$ such that for $j = 1,2,\cdots,n,$

$$(5) \qquad \lim_{|x|\to\infty} Reb_j(t,x)\langle x\rangle^{1-\kappa} = 0, \text{uniformly in } t \in [-T,T].$$

269

For $\varepsilon \in R$ denote $\phi_\varepsilon = x\xi - i\varepsilon x\xi \langle x \rangle^{\sigma-1} \langle \xi \rangle^{\delta-1}$, where $\sigma + \delta = \kappa$ and define

$$I_{\phi_\varepsilon}(x, D)u(x) = \int_{R^n} e^{i\phi_\varepsilon(x,\xi)} \hat{u}(\xi) d\xi,$$

where $\hat{u}(\xi)$ stands for a Fourier transform of u and $\bar{d}\xi = (2\pi)^{-n} d\xi$. Then our main theorem is the following.

Theorem. *Assume (4)-(5) are valid and there is $\varepsilon_0 > 0$ such that $I_{\phi_\varepsilon} u_0 \in L^2(R^n)$ for $|\varepsilon| \leq \varepsilon_0$. Then if $d\kappa \leq 1$ and $d \geq 1$, there exists a solution of (1)-(2) satisfying that there are $C > 0, \rho > 0$ and $\delta > 0$ such that*

(6)
$$|\partial_x^\alpha u(t,x)| \leq C(\rho|t|)^{-|\alpha|} |\alpha|!^{\frac{1}{\kappa}} e^{\delta \langle x \rangle^\kappa},$$

for $(t,x) \in [-T,T]\backslash 0 \times R^n, \alpha \in N^n$.

Remark. (i) Kato T. and Yajima in [13] began to investigate the smoothing effect phenomena . A. Jensen in [6] and Hayashi,Nakamitsu & Tsutsumi in [5] showed that if $< x >^k u_0(x) \in L^2(R^n)$, the solution u of (1)-(2) belongs to H^k_{loc} for $t \neq 0$, Hayashi & Saitoh in [4] proved that if $e^{\delta < x >^2} u_0$ ($\delta > 0$) is in $L^2(R^n)$, the solution u is analytic in x for $t \neq 0$ and De Bouard, Hayashi & Kato in [1], Kato & Taniguti in [12] show that if u_0 satisfies $\|(x \cdot \nabla)^j u_0\| \leq C^{j+1} j!^d$ for $j = 0, 1.2...$, then the solution belongs to Gevrey $\gamma^{d/2}$ with respect to x for $t \neq 0$. Theorem 1 is proved by Kajitani in [9] and [11], when $\sigma = \kappa = 1$. In the forthcoming Part II, we shall discuss the Gevrey smoothing effect for Shrödinger equations with variable coefficients.

1   Weighted Sobolev spaces

We introduce some Sobolev spaces with weights. Let ρ, δ be real numbers and $\kappa \in (0,1]$. Define
$$\hat{H}^\kappa_\delta = \{u \in L^2_{loc}(R^n); e^{\delta < x >^\kappa} u(x) \in L^2(R^n)\}.$$

For $\rho \geq 0$ let define

$$H^\kappa_\rho = \{u \in L^2(R^n); Fu(\xi) \in \hat{H}_\rho(R^n_\xi)\},$$

where Fu stands for the Fourier transform of u. For $\rho < 0$ we define H^κ_ρ as the dual space of $H^\kappa_{-\rho}$. Then the Fourier transform F becomes bijective from H^κ_ρ to $\hat{H}^\kappa_\rho$. We define the operator $e^{\rho < D >^\kappa}$ mapping continuously from $H^\kappa_{\rho_1}$ to $H^\kappa_{\rho_1-\rho}$ as follows;

$$e^{\rho < D >^\kappa} u(x) = F^{-1}(e^{\rho < \xi >^\kappa} Fu(\xi))(x),$$

for $u \in H^\kappa_{\rho_1}$ and $e^{\delta < x >^\kappa}$ maps continuously from $\hat{H}^\kappa_{\delta_1}$ to $\hat{H}^\kappa_{\delta_1-\delta}$. We define for $\delta \geq 0$ and $\rho \in R$

(1.1)
$$H^\kappa_{\rho,\delta} = \{u \in H_\rho; e^{\rho < D >^\kappa} u \in \hat{H}^\kappa_\delta\}.$$

For $\delta < 0$ we define $H^\kappa_{\rho,\delta}$ as the dual space of $H^\kappa_{-\rho,-\delta}$. We note that $H^\kappa_{\rho,0} = H^\kappa_\rho, H^\kappa_{0,\delta} = \hat{H}^\kappa_\delta$ and $H^\kappa_{0,0} = L^2(R^n)$. Furthermore we define for $\rho \geq 0$ and $\delta \in R$

$$(1.2) \qquad \tilde{H}^\kappa_{\rho,\delta} = \{u \in \hat{H}^\kappa_\delta ; e^{\delta<x>^\kappa} u \in H^\kappa_\rho\}$$

and for $\rho < 0$ define $\tilde{H}^\kappa_{\rho,\delta}$ as the dual spase of $\tilde{H}^\kappa_{-\rho,-\delta}$. Denote by H' the dual space of a topological space H. Then $H^{\kappa'}_{\rho,\delta} = H^\kappa_{-\rho,-\delta}$ and $\tilde{H}^{\kappa'}_{\rho,\delta} = \tilde{H}^\kappa_{-\rho,-\delta}$ hold for any ρ and $\delta \in R$. We shall prove $H^\kappa_{\rho,\delta} = \tilde{H}^\kappa_{\rho,\delta}$ later on (see Proposition 3.8).

Lemma 1.1. *Let $\rho, \delta \in R$. Then*

(i)
$$H^\kappa_{\rho,\delta} = e^{-\rho<D>^\kappa} e^{-\delta<x>^\kappa} L^2 = e^{-\rho<D>^\kappa} \hat{H}^\kappa_\delta.$$

(ii)
$$\tilde{H}^\kappa_{\rho,\delta} = e^{-\delta<x>^\kappa} e^{-\rho<D>^\kappa} L^2 = e^{-\delta<x>^\kappa} H^\kappa_\rho.$$

Lemma 1.2 *Let $1 > \rho > 0, \delta \in R$ and $u \in \tilde{H}^\kappa_{\rho,\delta}$. Then*

$$(1.6) \qquad |D^\alpha_x u(x)| \leq C_n (1-\epsilon)^{-n/2} \|u\|_{\tilde{H}^\kappa_{\rho,\delta}} (\epsilon\rho)^{-|\alpha|} |\alpha|! e^{\delta<x>^\kappa}$$

for $x \in R^n, \alpha \in N^n$ and $0 < \epsilon < 1$.

We can prove these lemmas analogously to the case of $\kappa = 1$ which is proved in [11].

2 Almost analytic extension of symbols

Following Hörmander's notation we define the symbol classes of pseudo-differential operators. Let $m(x,\xi), \varphi(x,\xi), \psi(x,\xi)$ be weights and $g = \varphi^{-2}dx^2 + \psi^{-2}d\xi^2$ a Riemann metric. We denote by $S(m,g)$ the set of symbols $a(x,\xi)$ satisfying

$$|a^{(\alpha)}_{(\beta)}(x,\xi)| \leq C_{\alpha\beta} m(x,\xi)\psi^{-\alpha}\psi^{-|\beta|},$$

for $(x,\xi) \in R^{2n}, \alpha, \beta \in N^n$, where $a^{(\alpha)}_{(\beta)} = \partial^\alpha_\xi D^\beta_x a$. Let $d \geq 1$. Moreover we say that a function $a(x,\xi) \in S(m,g)$ belongs to $\gamma^d S(m,g)$, if $a(x,\xi)$ satisfies that there are $C_a > 0, \rho_a > 0$ such that

$$(2.1) \qquad |a^{(\alpha)}_{(\beta)}(x,\xi)| \leq C_a m(x,\xi)\rho_a^{-|\alpha+\beta|}|\alpha + \beta|!^d \psi^{-|\beta|}\varphi^{-|\alpha|}$$

for $(x,\xi) \in R^{2n}, \alpha, \beta \in N^n$. We denote $g_0 = dx^2 + d\xi^2$ and $g_1 = <x>^{-2}dx^2 + <\xi>^{-2}d\xi^2$. We remark that the symbol class $\gamma^1 S(m,g_i)(i = 0,1)$ is introduced in [11] when $d = 1$. Here we consider the case of $d > 1$. Let $\chi(t) \in C_0^\infty((0, \infty))$ be a monotone function satsfying that $\chi(t) = 0, t \leq 1/2, \chi(t) = 1, t \geq 1$, and

$$(2.2) \qquad |D_t^k \chi(t)| \leq C_0 \rho_0^{-k} k!^d,$$

for $t \in R, k \in N$. Then for a weight $w(x, \xi) \in \gamma^d S(m, g_1)$ and a parameter $b > 0$ we can see easily that $\chi(bw(x, \xi)) \in \gamma^d S(1, g_1)$ satisfying

$$(2.3) \qquad |D_x^\beta D_\xi^\alpha \chi(bw(x, \xi)))| \leq C_1 \rho_1^{-|\alpha+\beta|} |\alpha + \beta|!^d \langle x \rangle^{-|\beta|} \langle \xi \rangle^{-|\alpha|},$$

for $(x, \xi) \in R^{2n}, \alpha, \beta \in N^n, b \geq 1$.

Lemma 2.1. Let $d \geq 1, d_1 \geq 1$ and $\{p_k(x, \xi)\}_{k=1}^\infty$ be a sequence of symbols satisfying

$$(2.4) \qquad |p_{k(\beta)}^{(\alpha)}(x, \xi)| \leq m(x, \xi)((\langle x \rangle \langle \xi \rangle))^{-k} \rho_p^{-|\alpha+\beta|-k} |\alpha + \beta|!^d k!^{d_1} \langle x \rangle^{-|\beta|} \langle \xi \rangle^{-|\alpha|},$$

for $(x, \xi) \in R^{2n}, \alpha, \beta \in N^n$ and $k \geq 0$. Then there is $p(x, \xi) \in \gamma^d S(m, g_1)$ such that

$$(2.5) \qquad p(x, \xi) - \sum_{k=0}^{N-1} p_k(x, \xi) \in \gamma^d S(m(\langle x \rangle \langle \xi \rangle \rho_p)^{-N} N!^{d_1}, g_1),$$

for any integer $N \geq 0$.

Proof. This lemma is essentially a result of [2]. The case of $d = 1$ is explained in [11]. Here we prove the lemma in the case of $d > 1$. Let $b_k = \rho_p^{-1} k!^{\frac{d_1}{k}} M$ and $M \geq 2$. Define

$$(2.6) \qquad p(x, \xi) = \sum_{k=0}^\infty p_k(x, \xi) \chi(b_k(\langle x \rangle \langle \xi \rangle)^{-1}),$$

Then we have from (2.3)

$$|p_{(\beta)}^{(\alpha)}(x, \xi)| = |\sum_k \sum_{\alpha', \beta'} \binom{\alpha}{\alpha'} \binom{\beta}{\beta'} p_{k(\beta')}^{(\alpha')} (\chi(b_k(\langle x \rangle \langle \xi \rangle)^{-1}))_{(\beta-\beta')}^{(\alpha-\alpha')}|$$

$$\leq \sum_k \sum_{\alpha', \beta'} \binom{\alpha}{\alpha'} \binom{\beta}{\beta - \beta'} C_p m(x, \xi) \rho_k^{-|\alpha'+\beta'|} |\alpha' + \beta'|!^d k!^{d_1} \langle x \rangle^{-|\beta|} \langle \xi \rangle^{-|\alpha|}$$

$$\times M^{-k} C_0 \rho_0^{-|\alpha-\alpha'+\beta-\beta'|} |\alpha - \alpha' + \beta - \beta'|!^d$$

$$\leq 2 \frac{C_0 C_p \rho_0}{\rho_0 - \rho_p} m(x, \xi) \rho_p^{-|\alpha+\beta|} |\alpha + \beta|!^d \langle x \rangle^{-|\beta|} \langle \xi \rangle^{-|\alpha|},$$

for $(x, \xi) \in R^{2n}, \alpha, \beta \in N^n$. Here we used the following inequality

$$(2.7) \qquad \sum_{\alpha' \leq \alpha} \binom{\alpha}{\alpha'} \rho_p^{-|\alpha'|} |\alpha'|!^d \rho_0^{-|\alpha-\alpha'|} |\alpha - \alpha'|!^d \leq \frac{\rho_0 \rho_p^{-|\alpha|}}{\rho_0 - \rho_p} |\alpha|!^d,$$

for $\rho_0 > \rho_p$. Moreover we can write

$$p(x,\xi) - \sum_{k=0}^{N-1} p_k(x,\xi)$$

$$= \sum_{k=N}^{\infty} p_k(x,\xi)\chi(b_k(\langle x\rangle\langle\xi\rangle)^{-1}) + \sum_{k=0}^{N-1} p_k(x,\xi)(1 - \chi(b_k(\langle x\rangle\langle\xi\rangle)^{-1}))$$

$$=: I + II.$$

Noting that $\rho_p^{-k}k!^{d_1}(M\langle x\rangle\langle\xi\rangle)^{-N} \leq 1$ on $supp\chi(b_k(\langle x\rangle\langle\xi\rangle)^{-1})$ for $k \geq N$ and $\rho_p^{-k}k!^{d_1}(M\langle x\rangle\langle\xi\rangle)^{-N} \geq 1/2$ on $supp(1 - \chi(b_k(\langle x\rangle\langle\xi\rangle)^{-1}))$ for $k \leq N-1$ respectively, we can see that I and II belong to $\gamma^d S(m(\langle x\rangle\langle\xi\rangle\rho_p)^{-N}N!^d, g)$. Q.E.D.

Let $a(x,\xi) \in \gamma^d S(m, g_1)$, that is, $a(x,\xi)$ satisfies (2.1) with $\psi = \langle\xi\rangle$ and $\varphi = \langle x\rangle$. Denote $b_\alpha(x) = B\rho_a^{-1}4^n\langle x\rangle^{-1}|\alpha|!^{\frac{d-1}{|\alpha|}}$ for $x \in R^n$. We define an almost analytic extension of $a(x,\xi)$ as follows,

$$(2.8) \qquad a(x+iy,\xi+i\eta) = \sum_{\alpha,\beta} a_{(\beta)}^{(\alpha)}(x,\xi)(-y)^\beta(i\eta)^\alpha\chi(b_\beta(x)|y|)\chi(b_\alpha(\xi)|\eta|)(\alpha!\beta!)^{-1},$$

for $x,y,\xi,\eta \in R^n$, where $a_{(\beta)}^{(\alpha)}(x,\xi) = \partial_\xi^\alpha(-i\partial_x)^\beta a(x,\xi)$. Then we can prove analogously to Lemma 2.1(see Proposition 5.2 in [7]).

Proposition 2.2. *Let $a(x,\xi) \in \gamma^d S(m, g_1)$. Then the function $a(x+iy,\xi+i\eta)$ defined by (2.8) satisfies the following properties.*

(i) $$|D_x^\beta\partial_\xi^\alpha D_y^\gamma\partial_\eta^\delta a(x+iy,\xi+i\eta)|$$

$$\leq Cm(x,\xi)(C\rho_a)^{-|\alpha+\beta+\gamma+\delta|}\langle x\rangle^{-|\beta|}\langle\xi\rangle^{-\alpha}\langle y\rangle^{-|\gamma|}\langle\eta\rangle^{-|\delta|}|\alpha+\beta+\gamma+\delta|!^d,$$

(ii) $$|(\partial_{x_j} + i\partial_{y_j})D_x^\beta\partial_\xi^\alpha D_y^\gamma\partial_\eta^\delta a(x+iy,\xi+i\eta)|$$

$$\leq Cm(x,\xi)(C\rho_a)^{-|\alpha+\beta+\gamma+\delta|}e^{-c_0(\frac{\langle x\rangle}{|y|})^{\frac{1}{d-1}}}\langle x\rangle^{-|\beta|}\langle\xi\rangle^{-\alpha}\langle y\rangle^{-|\gamma|}\langle\eta\rangle^{-|\delta|}|\alpha+\beta+\gamma+\delta|!^d(c_0 > 0),$$

(iii) $$|(\partial_{\xi_j} + i\partial_{\eta_j})D_x^\beta\partial_\xi^\alpha D_y^\gamma\partial_\eta^\delta a(x+iy,\xi+i\eta)|$$

$$\leq Cm(x,\xi)(C\rho_a)^{-|\alpha+\beta+\gamma+\delta|}e^{-c_0(\frac{\langle\xi\rangle}{|\eta|})^{\frac{1}{d-1}}}\langle x\rangle^{-|\beta|}\langle\xi\rangle^{-\alpha}\langle y\rangle^{-|\gamma|}\langle\eta\rangle^{-|\delta|}|\alpha+\beta+\gamma+\delta|!^d(c_0 > 0),$$

for $x,y,\xi,\eta \in R^n$, and $\alpha,\beta,\gamma,\delta \in N^n$.

Let $\vartheta(x,\xi)$ be in $\gamma^d S(\delta_\vartheta\langle x\rangle^\kappa + \rho_\vartheta\langle\xi\rangle^\kappa, g_1)$, where $d\kappa \leq 1, d > 1$. Denote by $\vartheta(z,\zeta)$ defined by (2.8) the almost analytic extension of ϑ. Denote for $x,\xi,z,\zeta \in C^{2n}$,

$$(2.9) \qquad \bar{\nabla}_x\vartheta(x,z,\zeta) = \int_0^1 \nabla_x\vartheta(z+t(x-z),\zeta)dt,$$

(2.10) $$\tilde{\nabla}_\xi \vartheta(x,\xi,\zeta) = \int_0^1 \nabla_\xi \vartheta(z, \zeta + t(\xi - \zeta))dt.$$

Then we can prove the following lemma by repeating the argument of Lemma 5.4 in [7].

Lemma 2.3. *Let* $\vartheta(x,\xi)$ *be in* $\gamma^d S(\delta_\vartheta \langle x\rangle^\kappa + \rho_\vartheta \langle \xi\rangle^\kappa, g_1)$, *where* $d\kappa \le 1$ *and* $d > 1$. *Then there exist* $\Phi(x,z,\zeta)$ *and* $\tilde{\Phi}'(z,\xi,\zeta)$ *in* $C^\infty(C^{3n})$ *satisfying respectively,*

(2.11) $$\Phi - i\tilde{\nabla}_x \vartheta(x, z, \Phi) = \zeta,$$

(2.12) $$\Phi' - i\tilde{\nabla}_\xi \vartheta(\Phi', \xi, \zeta) = z.$$

for $x, z, \zeta \in C^n$ *and*

(2.13) $$|D_x^\beta D_z^\gamma D_\zeta^\alpha(\Phi(x,z,\zeta) - \zeta)| \le C_\Phi(\rho_\vartheta + \delta_\vartheta)\rho_\Phi^{-|\alpha+\beta+\gamma|}|\alpha+\beta+\gamma|!^d \langle\zeta\rangle^{\kappa-|\alpha|},$$

(2.14) $$|D_z^\beta D_\zeta^\gamma D_\xi^\alpha(\Phi'(z,\zeta,\xi) - z)| \le C_{\Phi'}(\rho_\vartheta + \delta_\vartheta)\rho_{\Phi'}^{-|\alpha+\beta+\gamma|}|\alpha+\beta+\gamma|!^d \langle z\rangle^{\kappa-|\beta|},$$

and

(2.15) $$|\bar{\partial}_{\zeta_j} D_x^\beta D_z^\gamma D_\zeta^\alpha \Phi(x,z,\zeta)| \le C_\Phi(\rho_\vartheta + \delta_\vartheta)e^{-(\frac{|Re\zeta|}{|Im\zeta|})^{\frac{1}{d-1}}}\rho_\Phi^{-|\alpha+\beta+\gamma|}|\alpha+\beta+\gamma|!^d \langle\zeta\rangle^{\kappa-|\alpha|},$$

(2.16) $$|\bar{\partial}_{z_j} D_z^\beta D_\zeta^\gamma D_\xi^\alpha \Phi'(z,\zeta,\xi)| \le C_{\Phi'}(\rho_\vartheta + \delta_\vartheta)e^{-(\frac{|Rez|}{|Imz|})^{\frac{1}{d-1}}}\rho_{\Phi'}^{-|\alpha+\beta+\gamma|}|\alpha+\beta+\gamma|!^d \langle z\rangle^{\kappa-|\alpha|},$$

for $z, \zeta, \xi \in C^n$ *and* $\alpha, \beta, \delta \in N^n$. *Moreover* $\Phi(x,\xi) = \Phi(x,x,\xi)$ *and* $\Phi'(x,\xi) = \Phi'(x,\xi,\xi)$ *satisfy*

(2.17) $$|D_\xi^\alpha D_x^\beta(\Phi(x,\xi) - \xi)| \le C_\Phi(\rho_\vartheta \langle\xi\rangle^\kappa \langle x\rangle^{-1} + \delta_\vartheta \langle x\rangle^{\kappa-1})\langle x\rangle^{-|\beta|}\langle\xi\rangle^{-|\alpha|}\rho_\Phi^{-|\alpha+\beta|}|\alpha+\beta|!^d,$$

(2.18) $$|D_\xi^\alpha D_x^\beta(\Phi'(x,\xi) - x)| \le C_{\Phi'}(\rho_\vartheta \langle\xi\rangle^{\kappa-1} + \delta_\vartheta \langle x\rangle^\kappa \langle\xi\rangle^{-1})\langle x\rangle^{-|\beta|}\langle\xi\rangle^{-|\alpha|}\rho_{\Phi'}^{-|\alpha+\beta|}|\alpha+\beta|!^d,$$

for $x, \xi \in R^n$ *and* $\alpha, \beta, \delta \in N^n$.

For simplicity denote $\gamma^{1/\kappa} S(e^{\delta\langle x\rangle^\kappa + \rho\langle\xi\rangle^\kappa}, g_0)$ by $A_{\rho,\delta}^\kappa$, where $g_0 = dx^2 + d\xi^2$. For $a_i \in A_{\rho_i,\delta_i}^\kappa (i = 1,2)$ we define a product of a_1 and a_2 as follows,

(2.19) $$(a_1 \circ a_2)(x,\xi) = os - \int\int_{R^{2n}} e^{-iy\eta} a_1(x, \xi+\eta)a_2(x+y,\xi)dy d\bar\eta,$$

$$= \lim_{\epsilon \to 0} \int\int_{R^{2n}} e^{-iy\eta - \epsilon(|y|^2 + |\eta|^2)} a_1(x, \xi+\eta)a_2(x+y,\xi)dy d\bar\eta,$$

where $d\bar\eta = (2\pi)^{-n}d\eta$. Then we can show the proposition below.

Proposition 2.4. *(i) Let* $\kappa \le 1$ *and* $a_i \in A_{\rho_i,\delta_i}^\kappa, i = 1,2$. *Then there is* $\epsilon_0 > 0$ *such that if* $|\rho_1|, |\delta_2| \le \epsilon_0$, *the product* $a_1 \circ a_2$ *belongs to* $A_{\rho_1+\rho_2,\delta_1+\delta_2}^\kappa$.

(ii) Let $a_i \in A^\kappa_{\rho_i, \delta_i}, i = 1, 2, 3$. Then if $|\rho_i| (i = 1, 2), |\delta_i| (i = 2, 3) \leq \epsilon_0/2$, we have $(a_1 \circ a_2) \circ a_3 = a_1 \circ (a_2 \circ a_3)$.

Proof. (i) Put $\tilde{a}_i = e^{-\rho_i \langle \zeta \rangle^\kappa - \delta_i \langle x \rangle^\kappa} a_i(x, \zeta)$ and $\tilde{a} = e^{-(\rho_1 + \rho_2) \langle \zeta \rangle^\kappa - (\delta_1 + \delta_2) \langle x \rangle^\kappa} a_1 \circ a_2$. Then (2.19) implies

$$(2.20) \qquad \tilde{a}(x, \zeta) = \lim_{\epsilon \to +0} \int \int_{R^{2n}} e^{-iy\eta - \epsilon(|y|^2 + |\eta|^2)} e^{\rho_1(\langle \zeta + \eta \rangle^\kappa - \langle \zeta \rangle^\kappa) + \delta_2(\langle x + y \rangle^\kappa - \langle x \rangle^\kappa)}$$

$$\times \tilde{a}_1(x, \zeta + \eta) \tilde{a}_2(x + y, \zeta) dy d\eta = \lim_{\epsilon \to +0} \tilde{a}_\epsilon(x, \zeta).$$

We shall show that $a_\epsilon(x, \zeta)$ converges in $A^\kappa_{0,0}$ tending $\epsilon \to 0$. Put

$$\varphi = y\eta + i(\rho_1(\langle \zeta + \eta \rangle^\kappa - \langle \zeta \rangle^\kappa) + \delta_2(\langle x + y \rangle^\kappa - \langle x \rangle^\kappa)),$$

$$L = \sum_{j=1}^{n} (a_j D_{y_j} + b_j D_{\eta_j}) + c,$$

$$a_j = \frac{\bar{\varphi}_{y_j}}{|\nabla_y \varphi|^2 + |\nabla_\eta \varphi|^2 + 1},$$

$$b_j = \frac{\bar{\varphi}_{\eta_j}}{|\nabla_y \varphi|^2 + |\nabla_\eta \varphi|^2 + 1},$$

and

$$c = \frac{1}{|\nabla_y \varphi|^2 + |\nabla_\eta \varphi|^2 + 1}.$$

Then $Le^{i\varphi} = e^{i\varphi}$. Hence we get from (2.20) for any positive integer N,

$$(2.21) \qquad a_\epsilon(x, \xi) = \int \int e^{-i\varphi} (L^t)^N e^{-\epsilon(|y|^2 + |\eta|^2)} \tilde{a}_1 \tilde{a}_2 dy d\eta.$$

Noting that $|\nabla_y \varphi|^2 + |\nabla_\eta \varphi|^2 + 1 \geq |y|^2 + |\eta|^2 + 1$, a_j satisfies

$$(2.22) \qquad |\partial_\xi^\alpha D_x^\beta \partial_\eta^\lambda a_j(x, y, \xi, \eta)| \leq \frac{C_0^{1 + |\alpha + \beta + \gamma + \lambda|}}{\langle y \rangle + \langle \eta \rangle} |\alpha + \beta + \gamma + \lambda|!^{1/\kappa}$$

and b_j and c also satisfy (2.12), we can see

$$\inf_N |(L^t)^N \partial_\xi^\alpha D_x^\beta e^{-\epsilon(|y|^2 + |\eta|^2)} \tilde{a}_1 \tilde{a}_2|$$

$$\leq \inf_N \frac{C_1^{1 + |\alpha + \beta| + N} (|\alpha + \beta| + N)!^{1/\kappa}}{(\langle y \rangle + \langle \eta \rangle)^N} \leq C_2^{1 + |\alpha + \beta|} |\alpha + \beta|!^{1/\kappa} e^{-\epsilon_0(\langle y \rangle^\kappa + \langle \eta \rangle^\kappa)},$$

where $C_2 > 0, \varepsilon_0 > 0$ are independent of ϵ. If $|\rho_1| + |\delta_1| < \varepsilon_0$, $Re(i\varphi) - \varepsilon_0(\langle y \rangle^\kappa + \langle \eta \rangle^\kappa) < 0$ and consequently we can see easily that $a_\epsilon(x, \zeta)$ converges in $A^\kappa_{0,0}$ tending $\epsilon \to 0$. Q.E.D.

Proposition 2.5 *Let $d \geq 1$ and $a_i \in \gamma^d S(\langle x \rangle^{m_i} \langle \xi \rangle^{\ell_i}, g_1), i = 1, 2$. Then $a_1 \circ a_2$ belongs to $S(\langle x \rangle^{m_1 + m_2} \langle \xi \rangle^{\ell_1 + \ell_2}, g_1)$ and moreover we can decompose*

$$(2.23) \qquad a_1 \circ a_2(x, \xi) = p(x, \xi) + r(x, \xi),$$

where $p(x, \xi) \in \gamma^d S(\langle x \rangle^{m_1+m_2} \langle \xi \rangle^{\ell_1+\ell_2}, g_1)$ satisfies that there are $C > 0$ and $\varepsilon_0 >$ such that

$$(2.24) \qquad p(x, \xi) - \sum_{|\gamma| < N} \gamma!^{-1} a_1^{(\gamma)}(x, \xi) a_{2(\gamma)}(x, \xi)$$

$$\in \gamma^d S(C^{1+N} N!^{2d-1} \langle x \rangle^{m_1+m_2-N} \langle \xi \rangle^{\ell_1+\ell_2-N}, g_1),$$

for any non negative integer N, and $r(x, \xi)$ belongs to $\gamma^d S(e^{-\varepsilon_0(\langle \xi \rangle \langle x \rangle)^{\frac{1}{2d-1}}}, g_0)$.

The proof of this proposition is given in [11] in the case of $d = 1$. We can prove analogously in the case of $d > 1$.

3   Pseudo-differential operators

Let $0 < \kappa \leq 1$. Now we want to define a pseudo differential operator $a(x, D)$ for a symbol $a(x, \xi) \in A_{\rho, \delta}^\kappa$, which operates from $H_{\rho', \delta'}^\kappa$ to $H_{\rho'-\rho, \delta'-\delta}^\kappa$. When ρ and δ are non positive, since $A_{\rho, \delta}^\kappa$ is contained in the usual symbol class $S_{0,0}^0$ (denote by $S_{\rho, \delta}^m$ the Hörmander's class), we can define

$$(3.1) \qquad a(x, D)u(x) = \int e^{ix\xi} a(x, \xi) \hat{u}(\xi) d\!\!\!\!\!\!\;\bar{}\,\xi,$$

for $u \in L^2(R^n)$ and for $a \in A_{\rho, \delta}^\kappa$. Moreover for $a_i \in A_{\rho_i, \delta_i}^\kappa, i = 1, 2$ (ρ_i and δ_i non positive) the symbol $\sigma(a_1(x, D) a_2(x, D))(x, \xi)$ of the product of $a_1(x, D)$ and $a_2(x, D)$ can be written as follows,

$$(3.2) \qquad \sigma(a_1(x, D) a_2(x, D))(x, \xi) = (a_1 \circ a_2)(x, \xi)$$

and we have

$$(3.3) \qquad a_1(x, D)(a_2(x, D)u)(x) = (a_1 \circ a_2)(x, D)u(x)$$

for $u \in L^2(R^n)$, where $a_1 \circ a_2$ is defined by (2.9). Next we shall show that (3.2) and (3.3) are valid for any ρ_i, δ_i. To do so, we need some preparations. Let $a \in A_{\rho, \delta}^\kappa$ and $u \in H_\rho^\kappa$. Then we can define $a(x, D)u(x)$ which belongs to $\hat{H}_\delta^\kappa$. In fact, put $\tilde{a}(z, \eta) = e^{-\delta \langle x \rangle^\kappa - \rho \langle \xi \rangle^\kappa} a(x, \xi)$. Then $\tilde{a}(z, \xi) \in A_{0,0}^\kappa$. Noting that $e^{\rho \langle \xi \rangle^\kappa} \hat{u}(\xi) \in L^2$, we can define

$$(3.4) \qquad e^{-\delta \langle x \rangle^\kappa} a(x, D)u(x) = \int e^{ix\xi} \tilde{a}(x, \xi) e^{\rho \langle \xi \rangle^\kappa} \hat{u}(\xi) d\!\!\!\!\!\!\;\bar{}\,\xi,$$

which is in L^2, that is, $a(x, D)u \in \hat{H}_\delta^\kappa$. For $\epsilon > 0$ we denote $\chi_\epsilon(x) = e^{-\epsilon \langle x \rangle^2}$ and $\chi_\epsilon(D) = e^{-\epsilon \langle D \rangle^2}$.

Lemma 3.1. *(i) Let $a \in A_{\rho, \delta}^\kappa(\rho, \delta \in R)$, $u \in L^2$ and $\epsilon_0 > 0$ chosen in Proposition 2.3. Then for any $\epsilon > 0$*

$$(3.5) \qquad a(x, D)(\chi_\epsilon(D)\chi_\epsilon(x)u)(x) = (a(x, \xi)\chi_\epsilon(\xi)) \circ \chi_\epsilon(x))(x, D)u(x)$$

and

(3.6)
$$(a\chi_\epsilon(\xi)) \circ \chi_\epsilon(x) \in A^\kappa_{\rho-\epsilon_0,\delta-\epsilon_0}.$$

(ii) Let $u \in L^2$ and $\epsilon_0 > 0$ chosen in Proposition 2.3. Then there is $\epsilon_1 > 0$ such that for any $\epsilon > 0$

(3.7)
$$e^{-\rho<D>^\kappa}(e^{-\delta<x>^\kappa}\chi_\epsilon(x)\chi_\epsilon(D)u)(x) = a_\epsilon(x,D)u(x),$$

where

(3.8)
$$a_\epsilon(x,\xi) = e^{-\rho<\xi>^\kappa} \circ (e^{-\delta<x>^\kappa}\chi_\epsilon(x)\chi_\epsilon(\xi)) \in A^\kappa_{-\rho-\epsilon_0,-\delta-\epsilon_0},$$

for $|\rho| \leq \epsilon_0$ and $|\delta| < \epsilon_1$.

Lemma 3.2. *Let $u \in H^\kappa_{\rho,\delta}$ and $|\rho|,|\delta| \leq \epsilon_0/2$ (ϵ_0 is given in Proposition 2.3). Then for any $\epsilon > 0$ there is $u_\epsilon \in H^\kappa_{\epsilon_0/2,\epsilon_0/2}$ such that*

(3.9)
$$\|u - u_\epsilon\|_{H^\kappa_{\rho,\delta}} < \epsilon.$$

Lemma 3.3. *Let $a \in A^\kappa_{\rho,\delta}, 0 < \epsilon'_0, \bar{\epsilon}_0 \leq \epsilon_0$ (ϵ_0 is given in Proposition 2.3) and $u \in H^\kappa_{\epsilon'_0,\bar{\epsilon}_0}$. Then there is $\epsilon_2 > 0$ independent of a, ρ and δ such that $a(x,D)u(x)$ belongs to $H^\kappa_{\epsilon'_0-\rho,\bar{\epsilon}_0-\delta}$ if $0 < \epsilon'_0 - \rho \leq min\{\epsilon_0, \epsilon_2\rho_a\}$ and $0 < \bar{\epsilon}_0 - \delta \leq \epsilon_0$.*

Lemma 3.4. *Let $a_i \in A^\kappa_{\rho_i,\delta_i}(i = 1, 2)$ and $u \in H^\kappa_{\epsilon'_0,\bar{\epsilon}_0}(\epsilon'_0, \bar{\epsilon}_0 > 0)$. Then if $|\rho_1| \leq \epsilon_0, |\delta_2| \leq \epsilon_0, 0 < \epsilon'_0 - \rho_2 \leq \epsilon_0 min\{1, \rho_{a_2}\}, 0 < \bar{\epsilon}_0 - \delta_2 \leq \epsilon_0, 0 < \epsilon'_0 - \rho_2 - \rho_1 \leq \epsilon_0 min\{1, \rho_{a_1}\}$ and $0 < \bar{\epsilon}_0 - \delta_2 - \delta_1 \leq \epsilon_0$ are valid (ϵ_0 is given in Proposition 2.3), we have*

(3.10)
$$a_1(x,D)(a_2(x,D)u)(x) = (a_1 \circ a_2)(x,D)u(x),$$

which is in $H^\kappa_{\epsilon'_0-\rho_1-\rho_2,\bar{\epsilon}_0-\delta_1-\delta_2}$.

The above lemmas 3.1-3.4 are proved in [10] for $d = 1$. For $d > 1$ we can prove analogously.

Let $a \in A^\kappa_{\rho,\delta}(|\rho|,|\delta| \leq \epsilon_0/4), u \in H^\kappa_{\epsilon_0/2,\epsilon_0/2}$ and $|\rho_1|,|\delta_1| < \epsilon_0/4$. Put $w = e^{\delta_1<x>^\kappa}e^{\rho_1<D>^\kappa}u$, which is in $H^\kappa_{\epsilon_0/2-\rho_1,\epsilon_0/2-\delta_1}$. Since we can write $u = e^{-\rho_1<D>^\kappa}(e^{-\delta_1<x>^\kappa}w)$, we get by use of Lemma 3.4 with $\epsilon'_0 = \epsilon_0/2 - \rho_1, \bar{\epsilon}_0 = \epsilon_0/2 - \delta_1, a_1 = a(x,\xi)e^{-\rho_1<\xi>^\kappa}$ and $a_2 = e^{-\delta_1<x>^\kappa}, \epsilon_{a_2} = 1$,

$$a(x,D)u(x) = a(x,D)(e^{-\rho_1<D>^\kappa}(e^{-\delta_1<x>^\kappa}w) = ((a(x,\xi)e^{-\rho_1<\xi>^\kappa}) \circ e^{-\delta_1<x>^\kappa})(x,D)w(x).$$

Noting that $a_1(x,\xi) := (e^{(\delta_1-\delta)<x>^\kappa}e^{(\rho_1-\rho)<\xi>^\kappa}) \circ (a(x,\xi)e^{-\rho_1<\xi>^\kappa}) \circ e^{-\delta_1<x>^\kappa} \in A^\kappa_{0,0}$, we obtain

(3.11)
$$\|au\|_{H^\kappa_{\rho_1-\rho,\delta_1-\delta}} = \|a_1(x,D)w\|_{L^2} \leq C\|w\|_{L^2} = C\|u\|_{H^\kappa_{\rho_1,\delta_1}}$$

for any $u \in H^\kappa_{\epsilon_0/2,\epsilon_0/2}$. Since $H^\kappa_{\epsilon_0/2,\epsilon_0/2}$ is dense in $H^\kappa_{\rho_1,\delta_1}$ from Lemma 3.2, we get the following theorem.

Theorem 3.5 Let $a \in A^\kappa_{\rho,\delta}(|\rho|,|\delta| \leq \epsilon_0/4), |\rho_1|,|\delta_1| < \epsilon_0/4$, where ϵ_0 are given in Proposition 2.3. Then $a(x,D)$ maps from $H^\kappa_{\rho_1,\delta_1}$ to $H^\kappa_{\rho_1-\rho,\delta_1-\delta}$ and satisfies the following inequality

$$(3.12) \qquad \|au\|_{H^\kappa_{\rho_1-\rho,\delta_1-\delta}} \leq C\|u\|_{H^\kappa_{\rho_1,\delta_1}}$$

for any $u \in H^\kappa_{\rho_1,\delta_1}$.

For $a \in A^\kappa_{\rho,\delta}$, we difine

$$(3.13) \qquad a^t(x,\xi) = os - \int\int e^{iy\eta} a(x+y,\xi+\eta) dy d\bar\eta,$$

and $a^*(x,\xi) = a^t(\bar x,\xi)$. Then we can prove the following lemma, by the same way as that of the proof (i) of Proposition 2.4.

Lemma 3.6. Let $a \in A^\kappa_{\rho,\delta}$ and $|\rho|,|\delta| \leq \epsilon_0$. Then $a^t(x,\xi)$ defined in (2.29) belongs to $A^\kappa_{\rho,\delta}$. Moreover it holds

$$(3.14) \qquad (a^t(x,D)u,\varphi)_{L^2} = (u, a(\bar x, D)\varphi)_{L^2},$$

$$(a^*(x,D)u,\varphi)_{L^2} = (u, a(x, D)\varphi)_{L^2},$$

for any $u, \varphi \in H^\kappa_{\epsilon_0,\epsilon_0}$.

The relation (3.14) and the inequality (3.12) yield

$$|(a^t u, \varphi)| \leq \|u\|_{H^\kappa_{\rho-\rho_1,\delta-\delta_1}} \|\bar a\varphi\|_{H^\kappa_{\rho_1-\rho,\delta_1-\delta}} \leq C\|u\|_{H^\kappa_{\rho-\rho_1,\delta-\delta_1}} \|\varphi\|_{H^\kappa_{\rho_1,\delta_1}},$$

if $|\rho|,|\delta| \leq \epsilon_0/4$ and $|\rho_1|,|\delta_1| < \epsilon_0/4$. Therefore taking account that $H^\kappa_{\epsilon_0/2,\epsilon_0/2}$ is dense in $H^\kappa_{\rho_1,\delta_1}$, we get from (3.14)

$$(3.15) \qquad \|a^t u\|_{H^\kappa_{-\rho_1,-\delta_1}} \leq C\|u\|_{H^\kappa_{\rho-\rho_1,\delta-\delta_1}},$$

for any $u \in H^\kappa_{\rho_1,\delta_1}$. Thus we get the following proposition.

Propostion 3.7. Let $a \in A^\kappa_{\rho,\delta}$ and $|\rho|,|\delta| \leq \epsilon_0/4$ and $|\rho_1|,|\delta_1| < \epsilon_0/4$. Then the pseudo differential operators $a^t(x,D)$ and $a^*(x,D)$ satisfy (3.15).

Noting that $(e^{\delta<x>^\kappa}e^{\rho<D>^\kappa})^t = e^{\rho<D>^\kappa}e^{\delta<x>^\kappa}$, we have for $u \in H^\kappa_{\rho,\delta}$

$$e^{\rho<D>^\kappa}e^{\delta<x>^\kappa}u(x) = (e^{\delta<x>^\kappa}e^{\rho<D>^\kappa})^t(e^{-\rho<D>^\kappa}e^{-\delta<x>^\kappa}e^{\delta<x>^\kappa}e^{\rho<D>^\kappa}u)(x)$$

$$= (e^{\delta \langle x \rangle^{\kappa}} e^{\rho \langle D \rangle^{\kappa}})^t \circ (e^{-\delta \langle x \rangle^{\kappa}} e^{-\rho \langle D \rangle^{\kappa}})^t e^{\delta \langle x \rangle^{\kappa}} e^{\rho \langle D \rangle^{\kappa}} u(x).$$

Moreover we can see from Proposition 2.4 that $(e^{\delta \langle x \rangle^{\kappa}} e^{\rho \langle \xi \rangle^{\kappa}})^t \circ (e^{-\delta \langle x \rangle^{\kappa}} e^{-\rho \langle \xi \rangle^{\kappa}})^t$ is in $A_{0,0}^{\kappa}$. Hence we obtain the fact below.

Proposition 3.8. *Let $|\rho|, |\delta| \leq \epsilon_0/4$. Then u belongs to $H_{\rho,\delta}^{\kappa}$ if and only if $u \in \tilde{H}_{\rho,\delta}^{\kappa}$.*

The following result on the multiple symbols of pseudodifferential operators is a special case of Lemma 2.2 of Chapter 7 in Kumanogo's book [12]. In [10] we gave the proof in the case of $d = 1$. We can show analogously in the case of $d > 1$.

Lemma 3.9. *Let $r_j(x, \zeta) \in \gamma^d S(1, g_0)(j = 1, 2, ..., v)$ satisfying*

$$|r_{j(\beta)}^{(\alpha)}(x, \xi)| \leq C_{r_j} \varepsilon_{r_j} |\alpha + \beta|!^d$$

and put

$$q_v(x, D) = r_1(x, D) r_2(x, D) \cdots r_v(x, D).$$

Then the symbol $q_v(x, \zeta)$ belongs to $\gamma^d S(1, g_0)$ and satisfies

(3.16)
$$|q_{v(\beta)}^{(\alpha)}(x, \zeta)| \leq C^v \prod_{j=1}^{v} C_{r_j} \bar{\varepsilon}_v^{-|\alpha + \beta|} |\alpha + \beta|!,$$

for $(x, \zeta) \in R^{2n}, \alpha, \beta \in N^n$, where C is independent of v and $\bar{\varepsilon}_v = min_{1 \leq j \leq v}\{\varepsilon_{r_j}/4\}, C_v = max_{1 \leq j \leq v}\{C_{r_j}\}$.

We can prove easily the following lemmas as a corollary of Lemma 3.9, by using the Neumann series method.

Lemma 3.10. *Let $r(x, \xi)$ be in $\gamma^d S(C_r, g_0)$. If $C_r > 0$ is sufficiently small, then there is the inverse $(I + r(x, D))^{-1}$ which is a pseudodifferential operator with its symbol contained in $\gamma^d S(1, g_0)$.*

Lemma 3.11. *Let $j(x, \xi) \in \gamma^d S(\varepsilon_1, g_1)$. Then if $\varepsilon_1 > 0$ is small enough, there are $k_1(x, \xi) \in \gamma^d S(\varepsilon_1 \langle x \rangle^{-1} \langle \xi \rangle^{-1}, g_1), \varepsilon_0 > 0$ independent of ε_1 and $r_{\infty}(x, \xi) \in A_{-\varepsilon_0, -\varepsilon_0}^{1/d}$ such that $(I + j(x, D))^{-1} = k(x, D) + k_1(x, D) + r_{\infty}(x, D)$, where $k(x, \xi) = (1 + j(x, \xi))^{-1}$.*

4 Fourier Integral Operators

For $\vartheta \in \gamma^d S(\rho_{\vartheta} \langle \xi \rangle^{\kappa} + \delta_{\vartheta} \langle x \rangle^{\kappa}, g_1)(\rho_{\vartheta}, \delta_{\vartheta} \geq 0)$, where $d\kappa \leq 1$, we denote

$$\phi(x, \xi) = x\xi - i\vartheta(x, \xi).$$

For $a \in A_{0,0}^{\kappa}$ we define a Fourier integral operator with a phase function $\phi(x, \xi)$ as follows,

$$(4.1) \qquad a_\phi(x,D)u(x) = \int_{R^n} e^{i\phi(x,\xi)} a(x,\xi)\hat{u}(\xi)\bar{d}\xi,$$

for $u \in H_{\epsilon_0,\epsilon_0}$. Putting $p(x,\xi) = a(x,\xi)e^{\vartheta(x,\xi)}$, we can see $p(x,\xi) \in A^\kappa_{\rho_\vartheta,\delta_\vartheta}$. Therefore we can regard $a_\phi(x,D)$ as a pseudo differential operator with its symbol $p = ae^\vartheta$ defined in §2 and consequently it follows from Theorem 3.5 that $a_\phi(x,D)$ acts continuously from $H^\kappa_{\rho,\delta}$ to $H^\kappa_{\rho-\rho_\vartheta,\delta-\delta_\vartheta}$. However in order to construct the inverse operator of $p(x,D)$ it is better to regard $p(x,D)$ as a Fourier integral operator. In paticular for $a = 1$ we denote

$$(4.2) \qquad I_\phi(x,D)u(x) = \int e^{i\phi(x,\xi)}\hat{u}(\xi)\bar{d}\xi,$$

$$(4.3) \qquad I^R_\phi(x,D)v(x) = \int e^{ix\xi}\bar{d}\xi \int e^{i\phi(y,\xi)}v(y)dy.$$

Theorem 4.1. *Let $a \in \gamma^d S(\langle x\rangle^m\langle\xi\rangle^\ell, g_1)$, $\vartheta \in \gamma^d S(\rho_\vartheta\langle\xi\rangle^\kappa + \delta_\vartheta\langle x\rangle^\kappa, g_1)$ and $\phi = x\xi - i\vartheta(x,\xi)$. Assume $d\kappa \leq 1$. Then if $\rho_\vartheta, \delta_\vartheta$ are sufficiently small, $\tilde{a}(x,D) = I_\phi(x,D)a(x,D)I^{-1}_\phi$ and $\tilde{a}'(x,D) = I_\phi(x,D)^{-1}a(x,D)I_\phi(x,D)$ are pseudodifferential operators of which symbols are given by*

$$(4.4) \qquad \tilde{a}(x,\xi) = p(x,\xi) + r(x,\xi),$$

$$(4.5) \qquad a'(x,\xi) = p'(x,\xi) + r'(x,\xi),$$

where

$$(4.6) \qquad p(x,\xi) - a(x - i\nabla_\xi\vartheta(x,\Phi),\xi + i\nabla_x\vartheta(x,\Phi)) \in \gamma^d S(\langle x\rangle^{m-1}\langle\xi\rangle^{\ell-1}, g_1),$$

$$(4.7) \qquad \bar{p}'(x,\xi) - a(x + i\nabla_\xi\vartheta(\Phi',\xi),\xi - i\nabla_x\vartheta(\Phi',\xi)) \in \gamma^d S(\langle x\rangle^{m-1}\langle\xi\rangle^{\ell-1}, g_1),$$

where $\Phi = \Phi(x,x,\xi)$ and $\Phi' = \Phi'(x,\xi,\xi)$ are given by (2.11) and (2.12) respectively and r,r' belong to $\gamma^d S(e^{-\epsilon_0(\langle\xi\rangle\langle x\rangle)^{\frac{1}{2d-1}}}, g_0)$ for an $\epsilon_0 > 0$ independent of ρ_ϑ.

This theorem is proved in [11] in the case of $d = \kappa = 1$. We can prove it by the similar way to that of [11]. Taking account of Proposition 2.1 and Lemma 2.2 and repeating the argument in the section 6 in [7] we can obtain the following Lemma 4.2-4.5.

Lemma 4.2. *Let $a(x,\xi) \in \gamma^d S(\langle x\rangle^m\langle\xi\rangle^\ell, g_1)$ and $\vartheta \in \gamma^d S(\rho_\vartheta\langle\xi\rangle^\kappa + \delta_\vartheta\langle x\rangle^\kappa, g_1)(\rho_\vartheta \geq 0, \delta_\vartheta \geq 0)$ and $d\kappa \leq 1$. Put $\phi = x\xi - i\vartheta(x,\xi)$ and $\tilde{a}(x,D) = a_\phi(x,D)I^R_{-\phi}(x,D)$. If $\rho_\vartheta + \delta_\vartheta$ is sufficiently small, then $\tilde{a}(x,\xi)$ belongs to $S(\langle x\rangle^m\langle\xi\rangle^\ell, g)$ and moreover satisfies*

$$(4.8) \qquad \tilde{a}(x,\xi) = \bar{p}(x,\xi) + r(x,\xi),$$

for $x, \xi \in R^n$, *and*

(4.9)
$$\tilde{p}(x,\xi) - \sum_{|\gamma|<N} \gamma!^{-1} D_y^\gamma \partial_\eta^\gamma \{a(x, \Phi(x,y,\eta)) J(x,y,\eta)\}_{y=x,\eta=\xi}$$

$$\in \gamma^d S(C^{1+N} N!^{2d-1} \langle x \rangle^{m-N} \langle \xi \rangle^{\ell-N}, g_1),$$

for any N, *where* $\Phi(x, y, \xi)$ *is a solution of the following equation,*

(4.10)
$$\Phi(x,y,\xi) - i\tilde{\nabla}_x \vartheta(x,y,\Phi(x,y,\xi)) = \xi,$$

(4.11)
$$\tilde{\nabla}_x \vartheta(x,y,\xi) = \int_0^1 \nabla_x \vartheta(y + t(x-y), \xi) dt,$$

$J(x,y,\xi) = \frac{D\Phi(x,y,\xi)}{D\xi}$ *is the Jacobian of* $\Phi, r(x,\xi) \in \gamma^d S(e^{-\varepsilon_0(\langle\xi\rangle\langle x\rangle)^{\frac{1}{2d-1}}}, g_0)$, *and* $C > 0, \varepsilon_0 > 0$ *are independent of* ρ_ϑ.

Lemma 4.3. *Let* $a(x,\xi)$ *and* ϑ *be satisfied with the same condition as one of Lemma 4.2. For* $\phi = x\xi - i\vartheta(x,\xi)$ *put* $a'(x,\xi) = I_{-\phi}^R(x,D) a_\phi(x,D)$. *Then if* ρ_ϑ *and* δ_ϑ *are sufficiently small,* $a'(x,\xi)$ *belongs to* $S(\langle\xi\rangle^m \langle x\rangle^\ell, g_1)$ *and moreover satisfies*

(4.12)
$$a'(x,\xi) = p'(x,\xi) + r'(x,\xi),$$

(4.13)
$$p'(x,\xi) - \sum_{|\gamma|<N} \gamma^{-1} D_y^\gamma \partial_\eta^\gamma \{a(\Phi'(y,\xi,\eta),\xi) J'(y,\xi,\eta)\}_{y=x,\eta=\xi}$$

$$\in \gamma^d S(C^{1+N} N!^{2d-1} \langle x \rangle^{m-N} \langle \xi \rangle^{\ell-N}, g_1),$$

for any non negative integer N, *where* $\Phi'(y,\xi,\eta)$ *is a solution of the equation*

(4.14)
$$\Phi'(y,\xi,\eta) - i\tilde{\nabla}_\xi \vartheta(\Phi'(y,\xi,\eta),\xi,\eta) = y,$$

(4.15)
$$\tilde{\nabla}_\xi \vartheta(y,\xi,\eta) = \int_0^1 \nabla_\xi \vartheta(y, \eta + t(\xi-\eta)) dt,$$

and $J'(y,\xi,\eta) = \frac{D\Phi'(y,\xi,\eta)}{Dy}$, *and* $r'(x,\xi) \in \gamma^d S(e^{-\varepsilon_0(\langle\xi\rangle\langle x\rangle)^{\frac{1}{2d-1}}}, g_0)(\varepsilon_0 > 0$ *is independent of* $\rho_\vartheta)$.

Lemma 4.4. *Let* $\vartheta(x,\xi) \in \gamma^d S(\delta_\vartheta \langle x \rangle^\kappa + \rho_\vartheta \langle \xi \rangle^\delta, g_1)$ *and* $d\kappa \leq 1, \rho_\vartheta \geq 0, \delta_\vartheta \geq 0$. *Assume that* $\vartheta(x,\xi) \geq \rho_0 \langle \xi \rangle^\kappa - \delta_0 \langle x \rangle^\kappa$. *If* $\rho_0 > 0$ *and* $\rho_\vartheta + \delta_\vartheta$ *and* $\rho_0 + |\delta_0|$ *are sufficently small, there is the inverse of* $I_\phi(x,D)$, *which maps continuously from* H_{ρ_1,δ_1} *to* $H_{\rho_1+\rho_0,\delta_1-\delta_0}$ *for* $|\rho_1|, |\delta_1|$ *small enough and satisfies*

(4.16)
$$I_\phi(x,D)^{-1} = I_{-\phi}^R(x,D)(I + j(x,D))^{-1} = (I + j'(x,D))^{-1} I_{-\phi}^R(x,D)$$

$$= I^R_{-\phi}(x, D)(k(x, D) + k_1(x, D) + r(x, D))$$

$$= (k'(x, \xi) + k_1'(x, D) + r'(x, D)))I^R_{-\phi}(x, D),$$

where $j(x, \xi) = J(x, 0, \xi) - 1 + r_1(x, \xi), j'(x, \xi) = J'(x, \xi, 0) - 1 + r_2(x, \xi), k(x, \xi) = J(x, 0, \xi)^{-1}, k'(x, \xi) = J'(x, \xi, 0)^{-1}$ and $k_1, k_1' \in \gamma^d S(\langle x \rangle^{-1} \langle \xi \rangle^{-1}, g_1)$ and $r, r' \in \gamma^d S(e^{-\varepsilon_0(\langle \xi \rangle \langle x \rangle)^{\frac{1}{2d-1}}}, g_0)$.

Lemma 4.5. *Let $a(x, \xi)$ and ϑ be satisfied with the same condition as one of Lemma 3,3. Let $\phi = x\xi - i\vartheta$. Then we have*

(4.17) $$\sigma(I_\phi(x, D)a(x, D))(x, \xi) = I_\phi \circ a(x, \xi) = e^{\vartheta(x, \xi)}(q(x, \xi) + r(x, \xi)),$$

(4.18) $$\sigma(a(x, D)I_\phi(x, D))(x, \xi) = a \circ I_\phi(x, \xi) = e^{\vartheta(x, \xi)}(q'(x, \xi) + r'(x, \xi)),$$

where r, r' is in $\gamma^d S(e^{-\varepsilon_0(\langle \xi \rangle \langle x \rangle)^{\frac{1}{2d-1}}}, g_0)$, if ρ_ϑ is sufficiently small, and q, q' satisfies

(4.19) $$q(x, \xi) - \sum_{|\gamma| < N} \gamma!^{-1} D^\delta_y \partial^\gamma_\eta \{a(x + y - i\bar{\nabla}_\xi \vartheta(x, \xi, \eta), \xi)\}_{y=\eta=0}$$

$$\in \gamma^d S(C^{1+N} N!^{2d-1} \langle x \rangle^{m-N} \langle \xi \rangle^{\ell-N}, g_1),$$

(4.20) $$q'(x, \xi) - \sum_{|\gamma| < N} \gamma^{-1} D^\gamma_y \partial^\gamma_\eta \{a(x, \xi + \eta - i\bar{\nabla}_x \vartheta(x, y, \xi))\}_{y=\eta=0}$$

$$\in \gamma^d S(C^{1+N} N!^{2d-1} \langle x \rangle^{m-N} \langle \xi \rangle^{\ell-N}, g_1),$$

for any positive integer N, and $C > 0$ and $\varepsilon_0 > 0$ are independent of ρ_ϑ, where $\bar{\nabla}_\xi \vartheta(x, \xi, \eta) = \int_0^1 \nabla_\xi \vartheta(x, \xi + t\eta) dt$ and $\bar{\nabla}_x \vartheta(x, y, \xi) = \int_0^1 \nabla_x \vartheta(x + ty, \xi) dt$.

Summing up Lemma 4.2-Lemma 4,5, we obtain Theorem 4.1.

5 Criterion to L^2−well posed Cauchy problem

For $T > 0$ let consider the following Cauchy problem,

(5.1) $$\partial_t u(t, x) - i\Delta u(t, x) - b(t, x, D)u(t, x) = 0,$$

(5.2) $$u(0, x) = u_0(x),$$

for $(t, x) \in (0, T) \times R^n$. We assume that $b(t, x, \xi)$ is in $C^0([0, T]; S^1_{1,0})$. Moreover we suppose that there are $C \in R, K > 0$ such that

(5.3) $$Reb(t, x, \xi) \le C,$$

for $x, \xi \in R^n$ with $|x|, |\xi| \ge K$ and $t \in [0, T]$. Then we can prove the following theorem by use of the same method as that of [3] and [8].

Theorem 5.1. *Assume that the above condition (5.3) is valid. For any $u_0 \in L^2$ and $f \in C^0([0, T]; L^2)$ there exists a unique solution $u \in C^0([0, T]; L^2) \cap C^1([0, T]; H^{-2})$ of the Cauchy problem (5.1)-(5.2).*

6   Proof of Theorem

Assume that $u(t, x)$ satisfies (1)-(2) in the introduction. We change the unknown function u to w as follows,

$$(6.1) \qquad w(t, x) = I_\phi(x, D)u(t, x),$$

where $\phi = x\xi - i\rho t\langle\xi\rangle^\kappa - i\epsilon\vartheta(t, x, \xi)$ and ϑ is given by

$$\vartheta(t, x, \xi) = \vartheta_0(x, \xi)\phi_0(\frac{\langle x\rangle}{M\langle\xi\rangle}) + t\langle\xi\rangle^{\sigma+\delta}(1 - \phi_0(\frac{\langle x\rangle}{M\langle\xi\rangle})),$$

$$\vartheta_0(x, \xi) = \frac{x \cdot \xi}{\langle x\rangle^{1-\sigma}\langle\xi\rangle^{1-\delta}}\phi_0(\frac{x \cdot \xi}{\langle x\rangle\langle\xi\rangle\varepsilon_1}) + \langle\xi\rangle^{\delta-\sigma}f(|x \cdot \xi|)[\phi_+(\frac{x \cdot \xi}{\langle x\rangle\langle\xi\rangle\varepsilon_1}) - \phi_-(\frac{x \cdot \xi}{\langle x\rangle\langle\xi\rangle\varepsilon_1})],$$

$$f(t) = \int_0^t (1 + s^2)^{\frac{\sigma-1}{2}}ds,$$

and $\phi_\pm(t) = \chi(\pm t), \phi_0(t) = 1 - \phi_+(t) - \phi_-(t)$ and $\chi(t) \in \gamma^d(R)$ such that $\chi(t) = 1$ for $t \geq 1, \chi(t) = 0$ for $t \leq 1/2, \chi'(t) \geq 0$ and $0 \leq \chi(t) \leq 1$. Then we can see that $\vartheta(t, x, \xi)$ belongs to $\gamma^d S(\langle x\rangle^\sigma\langle\xi\rangle^\delta, g_1)$ and that there are $\varepsilon_1 > 0, M > 0, K > 0, c_0 > 0$ such that ϑ satisfies

$$(6.2) \qquad (\partial_t + \xi \cdot \nabla_x)\vartheta(t, x, \xi) \geq c_0(\langle\xi\rangle^{2\delta}\langle x\rangle^{2\sigma-2} + \langle\xi\rangle^{\sigma+\delta} + \langle\xi\rangle\langle x\rangle^{\sigma+\delta-1}) - c_1,$$

for $x, \xi \in R^n$ with $|x|, |\xi| \geq K$ and $, |t| \leq T$.

It follows from Lemma 4.4 that if $|\epsilon|$ and $|\rho t|$ are sufficiently small, we have the inverse $I_\phi(x, D)^{-1}$. Therefore we get the following Cauchy problem of w from (1)-(2),

$$(6.3) \qquad \frac{\partial}{\partial t}w(t, x) = (\partial_t I_\phi)I_\phi(x, D)^{-1}w(t, x) + I_\phi(i\Delta + b(t, x, D))I_\phi(x, D)^{-1}w(t, x),$$

$$(6.4) \qquad w(0, x) = I_\phi(x, D)u_0(x).$$

Since $\rho t\langle\xi\rangle^\kappa + \vartheta(t, x, \xi) \in \gamma^d S(\langle\xi\rangle^\kappa + \langle x\rangle^\sigma\langle\xi\rangle^\delta + \langle\xi\rangle^\kappa, g_1)$ and $Im\nabla_x\phi(t, x, \xi) = \epsilon\nabla_x\vartheta(t, x, \xi) \in \gamma^d S(\langle x\rangle^\sigma\langle\xi\rangle^{\delta-1} + \langle\xi\rangle^{\kappa-1}, g_1)$, it follows from (4.10) that $\Phi(x, \xi) - \xi \in \gamma^d S(\langle x\rangle^{\sigma-1}\langle\xi\rangle^\delta, g_1)$. Hence we have from (4.6) in Theorem 4.1 and Lemma 2.3

$$(6.5) \qquad \sigma(I_\phi\Delta I_\phi^{-1})(x, \xi) = -|\xi - i\epsilon\nabla_x\vartheta(x, \Phi)|^2 + a_1(x, D),$$

$$= -(|\xi|^2 - \epsilon^2|\nabla_x\vartheta(t, x, \xi)|^2 - 2i\epsilon\xi \cdot \nabla_x\vartheta(t, x, \xi)) + a_1'(x, \xi)$$

where $a_1 \in S(\langle\xi\rangle\langle x\rangle^{-1}, g), a_1' \in S(\epsilon^2\langle x\rangle^{2\sigma-2}\langle\xi\rangle^{2\delta} + \langle\xi\rangle\langle x\rangle^{-1}, g_1)$ and moreover

$$\sigma(I_\phi b(t, x, D)I_\phi^{-1})(x, \xi) = b(t, x, \xi) + b_1(t, x, \xi),$$

$$\sigma(\partial_t I_\phi I_\phi^{-1})(x, \xi) = \sigma((\phi_t)_\phi I_\phi^{-1})(x, \xi) = -i\rho t\langle\xi\rangle^\kappa - i\epsilon\langle\xi\rangle^{\sigma+\delta}(1 - \phi_0(\frac{\langle x\rangle}{M\langle\xi\rangle})) + c(x, \xi),$$

where $b_1 \in S(\langle x\rangle^{\sigma-1}\langle\xi\rangle^\delta, g_1)$ and $c \in S(\langle x\rangle^{-1}\langle\xi\rangle^{2\kappa-1} + \langle x\rangle^{\sigma-1}\langle\xi\rangle^{\delta+\kappa-1}, g_1)$. Thus we obtain the equation of w from (6.3)-(6.4),

$$(6.6) \quad \frac{\partial w}{\partial t} = (i\Delta + b(t,x,D) + \rho\langle D\rangle^{\sigma+\delta} - \epsilon(\partial_t + \xi\cdot\nabla_x)\vartheta)(t,x,D)) + a_2(t,x,D))w(t,x),$$

$$(6.7) \qquad\qquad\qquad w(0) = I_{\phi(0)}(x,D)u_0(x),$$

where $a_2 = a_1' + b_1 + c \in S(\langle\xi\rangle^{2\delta}\langle x\rangle^{2\sigma-2} + \langle\xi\rangle\langle x\rangle^{-1} + \langle x\rangle^{\sigma-1}\langle\xi\rangle^\delta, g_1)$. Moreover taking account of the assumptions (5) in the introduction and of (6.2) , $\sigma+\delta = \kappa$, we can choose conviniently $K > 0, \epsilon$ and ρ such that we have

$$\rho\langle x\rangle^\kappa + Reb(t,x,\xi) - \epsilon(\partial_t + \xi\cdot\nabla_x)\vartheta)(t,x,\xi)) + Rea_2(t,x,\xi) \le 0,$$

for $x,\xi \in R^n$ with $|x|, |\xi| \ge K$,where $K > 0$ is sufficiently large. Therefore we can solve the Cauchy problem (6.6)-(6.7) by use of Theorem 5.1, since $w(0) = I_{\phi(0)}u_0$ belongs to L^2, and cosequently we get the solution $u = I_\phi(x,D)^{-1}w(t,x)$, which satisfies (6) from Lemma 1.2. In fact, taking account of $\rho t\langle\xi\rangle^\kappa + \varepsilon\vartheta(t,x,\xi) \ge \rho t\langle\xi\rangle^\kappa - \delta_0\rho t\langle x\rangle\langle^\kappa$ and $\rho t > 0$, we can see from Lemma 4.4 that $I_\phi(x,D)^{-1}$ maps from L^2 to $H^\kappa_{t\rho,-\delta_0}$. This completes the proof of Theorem.

References

[1] De Bouard A. Hayashi N. & Kato K. *Regularizing effect for the (generalized) Korteweg-de Vrie equations and nonlinear Schrödinger equations*, Ann. Inst. Henri Poincaré Analyse nonlinear vol. 12 pp. 673-725 (1995).

[2] Boutet de Monvel L. & Krée P. *Pseudo-differential operators and Gevrey classes*, Ann. Inst. Fourier Grenoble vol.17 pp. 295-323 (1967).

[3] Doi S. *Remarks on the Cauchy problem for Schrödinger type equations*, Comm. P.D.E. vol. 21 pp. 163-178 (1996).

[4] Hayashi N. & Saitoh S. *Analyticity and smoothing effect for Schrödinger equation*, Ann. Inst. Henri Poincaré Math. vol 52 pp. 163-173 (1990).

[5] Hayashi S., Nakamitsu K. & Tsutsumi M. *On solutions of the initial value problem for the nonlinear Schrödinger equations in one space dimension*, Math.Z. vol. 192 pp. 637-650 (1986).

[6] Jensen A. *Commutator method and a Smoothing property of the Schrödinger evolution group*, Math. Z. vol. 191 pp. 53-59 (1986).

[7] kajitani K. & Nishitani T. *The hyperbolic Cauchy problem*, Lecture Notes in Math. 1505,Springer-Verlag, Berlin,(1991).

[8] Kajitani K. *The Cauchy problem for Schrödinger type equations with variable coefficients* , Jour. Math. Soc. Japan vol.50 pp.179-202 (1998).

[9] Kajitani K. *Analytically smoothing effect for Schrödinger equations*, Proceedings of the International Conference on Dynamical Systems & Differential Equations in Southwest Missouri State University (1996).An added Volume I to Discrete and Continuous Dynamical Systems 1998, pp. 350-353 (1998)

[10] Kajitani K. & Baba A. *The Cauchy problem for Schrödinger type equations* , Bull. Sci. math. vol. 119 pp. 459-473 (1995)

[11] Kajitani K. & Wakabayashi S. *Analytically smoothing effect for Schrödinger type equations with variable coefficients*, Proceeding of Symposium of P.D.E. at University of Delaware 1997.

[12] Kato K. & Taniguti K. *Gevrey regularizing effect fot nonlinear Schrödinger equations*, Osaka J. Math. vol.33 pp. 863-880 (1996).

[13] Kato T.& Yajima K. *Some examples of smoothing operators and the associated smoothing effect*, Rev. Math. Phys. vol.1 pp. 481-496 (1989).

[14] Kumanogo H. *Pseudo-Differential Operators*, MIT Press (1981).

Semiclassical wavefunctions
and Schrödinger equation

MAURICE DE GOSSON
BLEKINGE INSTITUTE OF TECHNOLOGY,
SE 371 79 KARLSKRONA

AND

UNIVERSITY OF COLORADO AT BOULDER,
BOULDER CO 80 309

ABSTRACT.   It is well-known that the metaplectic representation allows the explicit construction of the solutions of Shrödinger's equation for all quadratic Hamiltonians. We generalise the metaplectic representation to the case of arbitrary physical Hamiltonians and obtain Feynman's formula as a particular case of our construction.

Au Professeur Jean Vaillant, avec respect

1.   INTRODUCTION

Consider Schrödinger's equation

$$i\hbar\frac{\partial\Psi}{\partial t} = -\frac{\hbar^2}{2m}\nabla_x\Psi + U\,\Psi \ , \ \ \Psi(x,0) = \Psi_0 \tag{1}$$

where $x = (x_1,...x_n)$, $m > 0$. It is well-known that ([5, 6, 4]) when the potential U is a quadratic form in the variables x_j this equation can be explicitly solved in integral form. Let in fact (f_t) be the flow determined by the Hamiltonian vector field $X_H = (\nabla_p H, -\nabla_x H)$ where

$$H = \frac{p^2}{2m} + U. \tag{2}$$

If U is quadratic (f_t) is a continuous one-parameter subgroup of the symplectic group $Sp(n)$ and can thus be lifted, in a unique way, to a continuous one-parameter subgroup (F_t) of the double covering $Sp_2(n)$ of $Sp(n)$. Identifying the latter with the metaplectic group $Mp^h(n)$, which will be defined below, and setting, for $\Psi_0 \in S(\mathbb{R}^n_x)$

$$\Psi(x,t) = F_t\Psi_0(x) \tag{3}$$

the function Ψ is a solution of Schrödinger's equation (1) with initial condition Ψ_0 (see, e.g., [2, 5, 10] for a proof on the Lie-algebraic level; we will prove this by other means below). This solution Ψ is given, except for at most isolated values t_k of t, by the formula

$$\Psi(x,t) = \left(\tfrac{1}{2\pi i\hbar}\right)^{n/2}\left[\det_{x,x'}\mathrm{Hess}(-W)\right]^{-1/2}\int e^{\frac{i}{\hbar}W(x,x';t)}\Psi(x')\,d^nx' \tag{4}$$

287

where W is the generating function of the flow (f_t); $\text{Hess}_{x,x'}(-W)$ matrix of second derivatives of $-W$.

In view of a famous result of Gronewold and van Hove (see [2, 6] for up-to-date expositions), this lifting procedure cannot be extended to solve Schrödinger's equation with an arbitrary potential. In fact, the construction of the solutions (3) relies on the following property of $Mp^h(n)$: for all polynomials H, K quadratic in (x_j, p_j) we have

$$\{H, K\} = 2\pi i\hbar[\widehat{H}, \widehat{K}] \tag{5}$$

where $\widehat{H}$ and $\widehat{K}$ are the operators obtained from H and K by Schrödinger's quantization rule

$$x \longmapsto \hat{x} = x \cdot \quad , \quad p \longmapsto \hat{p} = -i\hbar\nabla_x \quad , \quad px \longmapsto \tfrac{1}{2}(\hat{p}\hat{x} + \hat{x}\hat{p}).$$

Gronewold and van Hove show that it is not possible to construct a linear mapping $H \longmapsto \widehat{H}$ from the space $\mathcal{P}_k$ of polynomials in (x_j, p_j) with degree $k > 2$ in the space of Hermitian operators on $L^2(\mathbb{R}_x^n)$ in such a way that (5) is preserved (see [6] for a discussion of this result).

The purpose of this work is to study the properties of the functions (4) when W_t is the generating function determined by the flow (f_t) associated with a Hamiltonian of the type (2) with U arbitrary of class C^2. We will call these functions *semiclassical wavefunctions* (they are obtained from a classical object, the generating function W). We will see that for small values of t, these functions are close, "at the order t^2" of the exact solution of (1), and this fact will allow us, using an iterative method, to find the exact solution. In particular, our approach will allow us to recover Feynman's formula. (Our method has some similarities with that used by J. Vaillant [11] in another context.)

We will use the following standard notations: we denote by x^2 the scalar square $\langle x, x \rangle$ of $x = (x_1, ..., x_n)$; if A is a symmetric matrix we write Ax^2 instead of $\langle Ax, x \rangle$. We will assume that the potential U is a C^∞ function; most results will however hold under the weaker assumption that it is C^2.

2. Description of the Solutions: the Quadratic Case

Recall (see for example [1]) that $s \in Sp(n)$ is a free symplectic transformation if the equation $(x, p) = s(x', p')$ has, for given x and x', a unique solution (p, p'). Writing s in the canonical basis as

$$s = \begin{pmatrix} A & B \\ C & D \end{pmatrix}$$

(the blocks A, B, C, D being of order n), s is hence free if and only if $\det B \neq 0$. In this case there exists a generating function W, defined on $\mathbb{R}_x^n \times \mathbb{R}_x^n$ by

$$(x, p) = s(x', p') \iff \begin{cases} p = \nabla_x W(x, x') \\ p' = -\nabla_{x'} W(x, x') \end{cases} \tag{6}$$

and this function is given by

$$W(x, x') = \tfrac{1}{2} D B^{-1} x^2 - B^{-1} x \cdot x' + \tfrac{1}{2} B^{-1} A x'^2.$$

Conversely, if

$$W(x, x') = \tfrac{1}{2} P x^2 - L x \cdot x' + \tfrac{1}{2} Q x'^2. \tag{7}$$

is a quadratic form such that $P = P^T$, $Q = Q^T$ and $\det L \neq 0$, it is a generating function for the symplectic matrix

$$s = \begin{pmatrix} L^{-1}Q & L^{-1} \\ PL^{-1}Q - L^T & PL^{-1} \end{pmatrix}.$$

To every generating function (7) one associates the *two* generalized Fourier transforms:

$$S_{W,m} \Psi(x) = \left(\tfrac{1}{2\pi i \hbar} \right)^{n/2} \Delta(W) \int e^{\frac{i}{\hbar} W(x, x')} \Psi(x') \, d^n x'$$

where $\arg i = \pi/2$ and the quantity $\Delta(W)$ is defined by

$$\Delta(W) = i^m \sqrt{\left| \det \operatorname*{Hess}_{x,x'}(-W) \right|}$$

where $\operatorname{Hess}_{x,x'}(-W)$ is the matrix of second derivatives of $-W$. The integer m (the "Maslov index") corresponds to a choice of the argument of the determinant of $\operatorname{Hess}_{x,x'}(-W)$:

$$\arg \det \operatorname*{Hess}_{x,x'}(-W) = m\pi$$

and there are of course only possible choices for $m \bmod 2$. By definition, the metaplectic group $Mp^h(n)$ is the group of unitary operators generated by these generalized Fourier transforms; the projection $\Pi : S \longmapsto s$ of $Mp^h(n)$ on $Sp(n)$ defined by (6) is then a covering mapping of order 2 (for a proof, see Leray [7]). The formula (3) can be interpreted in the following way: there exists $\varepsilon > 0$ such that $f_t \in Sp(n)$ is free for $0 < |t| < \varepsilon$ (see [4], p.137). Choosing a generating function $W(x, x', t)$ for each f_t $(0 < |t| < \varepsilon)$ in such a way that the dependence $t \longmapsto W(x, x', t)$ is C^∞, the solution of Schrödinger's equation (1) is then given by formula $\Psi(x, t) = F_t \Psi_0(x)$ where the operator $F_t \in Mp(n)$ is defined by

$$F_t \Psi_0(x) = \left(\tfrac{1}{2\pi i \hbar} \right)^{n/2} \Delta(W) \int e^{\frac{i}{\hbar} W(x, x', t)} \Psi_0(x') \, d^n x' \tag{8}$$

the Maslov index $m(t)$ of $W(x, x', t)$ being chosen so that

$$\lim_{t \to 0} F_t \Psi_0(x) = \Psi_0(x)$$

(see [7, 3]). It can happen that for large values of t formula (8) no longer makes sense, since W is not in general defined for all t. This is not, however, a limitation of the method. In fact, since $F_t = (F_{t/N})^N$ for every integer N, we can write

$$\Psi(x, t) = (F_{t/N})^N \Psi_0(x)$$

allowing us to write the solution Ψ as an iterated integral, choosing N large enough, so that $F_{t/N}\Psi_0$ is given by formula (8) with t replaced by t/N. It is interesting to remark here that the integer N being arbitrary, we can write

$$\Psi(x, t) = \lim_{N \to \infty} (F_{t/N})^N \Psi_0(x). \tag{9}$$

Replacing W by the (crude) approximation

$$W_{Feyn} = m \frac{(x - x')^2}{2t} - U(x')t \tag{10}$$

we obtain the physicists Feynman's formula (see for instance [9]).

We will generalize this construction to the case of an arbitrary potential without using Feynman's formula. Our approach, which involves a precise study of the Hessian of the generating function , allows in fact to obtain a much more precise result.

3.   VAN VLECK'S DETERMINANT

The flow (f_t) determined by the Hamiltonian (2) has the following property: there exists $\varepsilon > 0$ such that f_t is a free symplectomorphism for $0 < |t| < \varepsilon$ ([4], p.137). Hence there exists a C^∞ function W such that

$$(x, p) = f_t(x', p') \Longleftrightarrow \begin{cases} p = \nabla_x W(x, x', t) \\ p' = -\nabla_{x'} W(x, x', t). \end{cases} \tag{11}$$

Moreover, for fixed x', the generating function determined by H satisfies the Hamilton-Jacobi equation

$$\frac{\partial W}{\partial t} + H(x, \nabla_x W) = 0 \tag{12}$$

(see [1, 4]). By definition, the determinant de van Vleck is the quantity

$$\rho(x, x', t) = \left(\frac{1}{2\pi i \hbar}\right)^n \operatorname*{Hess}_{x,x'}(-W)$$

that is à dire, vu (11):

$$\rho(x, x', t) = \left(\frac{1}{2\pi i\hbar}\right)^n \det \frac{\partial p'}{\partial x} = \left(\frac{1}{2\pi i\hbar}\right)^n \det \frac{\partial(x, x')}{\partial(p', x')}.$$

Van Vleck's satisfies an important continuity equation:

PROPOSITION 1. *The function* $(x, t) \longmapsto \rho(x, x'; t, t')$ *satisfies the equation*

$$\frac{\partial \rho}{\partial t} + \text{div}(\rho v) = 0 \tag{13}$$

where $v = p/m$ *is the velocity vector at* x *at time* t *of the trajectory passing through* x *and emanating from* x' *at time* t'.

Before we prove this let us recall the following result from the theory of differential equations (see [4, 8]). Consider the system

$$\dot{x}(t) = f(x(t), t) \quad , \quad x = (x_1, ..., x_n) \quad , \quad f = (f_1, ..., f_n) \tag{14}$$

where the f_j are real functions defined in an open set $U \subset \mathbb{R}^n$. Each solution $x_1, ..., x_n$ depends on n parameters $\alpha_1, ..., \alpha_n$. Set $\alpha = (\alpha_1, ..., \alpha_n)$ and write $x = x(\alpha, t)$. Suppose that

$$Y(\alpha, t) = \det\left(\frac{\partial x(\alpha, t)}{\partial(\alpha, t)}\right) \tag{15}$$

does not vanish for (α, t) in some open set $D \subset \mathbb{R}^n \times \mathbb{R}_t$. Then:

$$\frac{\partial Y}{\partial t}(\alpha, t) = Y(\alpha, t) \text{Tr}\left(\frac{\partial f}{\partial x}(x(\alpha, t))\right) \tag{16}$$

(Tr: the trace).

Set now $\rho(x, t) = \rho(x, x'; t)$ and consider the Hamilton equations

$$\begin{cases} \dot{x} = \nabla_p H(x, p, t) \\ \dot{p} = -\nabla_x H(x, p, t) \end{cases}$$

with initial conditions $x(0) = x'$ and $p(0) = p'$ (x' is fixed, and p' variable). The solutions $x(t)$ of the first equation are parametrized by p' since x' is fixed, and we can hence apply (16) with $f = \nabla_p H$, $\alpha = p'$ and

$$Y(p', t) = \det \frac{\partial x(p', t)}{\partial(p', t)} = \det\begin{pmatrix} \frac{\partial x}{\partial p'} & \frac{\partial x}{\partial t} \\ 0 & 1 \end{pmatrix}.$$

Setting $x = x(t)$ and $\dot{x} = \dot{x}(t)$, this function Y is given by

$$Y(p', t) = (2\pi i\hbar)^n \frac{1}{\rho(x, t)}$$

and formula (16) can hence be written

$$\frac{d}{dt}\left(\frac{1}{\rho(x, t)}\right) = \frac{1}{\rho(x, t)} \operatorname{Tr}\left[\frac{\partial}{\partial x}(\nabla_p H(x, p, t))\right]$$

that is

$$\frac{d}{dt}\rho(x, t) + \rho(x, t) \operatorname{Tr}\left[\frac{\partial}{\partial x}(\nabla_p H(x, p, t))\right] = 0. \tag{17}$$

Noting that the derivative of $t \longmapsto \rho(x(t), t)$ is

$$\frac{d\rho}{dt} = \frac{\partial\rho}{\partial t} + \nabla_x\rho \cdot \dot{x} = \frac{\partial\rho}{\partial t} + \nabla_x\rho \cdot \nabla_p H$$

and that we have

$$\operatorname{Tr}\left[\frac{\partial}{\partial x}(\nabla_p H)\right] = \nabla_x \cdot \nabla_p H \tag{18}$$

we finally get

$$\frac{\partial\rho}{\partial t} + \nabla_x\rho \cdot \nabla_p H + \rho\nabla_x \cdot \nabla_p H = 0$$

which is just (13).

LEMMA 1. *The following asymptotic expression of W holds for $t \to 0$:*

$$W(x, x', t) = \frac{m}{2t}(x - x')^2 - \overline{U}(x, x')t + O(t^3) \tag{19}$$

where we have set

$$\overline{U}(x, x') = \int_0^1 U(sx + (1 - s)x')\, ds \tag{20}$$

($\overline{U}(x, x')$ is thus the average value of the potential U on the segment $[x', x]$). We moreover have the following estimate for van Vleck's density:

$$\rho(x, x'; t) = \left(\frac{m}{2\pi i\hbar t}\right)^n (1 + O(t^2)). \tag{21}$$

PROOF. (For details see [4], Chapter 7). Let us prove (19). When $H = p^2/2m$ the generating function W is given by

$$W = W_f = m\frac{(x-x')^2}{2t}.$$

This suggests, in the general case, to write $W = W_f + R$. Inserting this expression in Hamilton-Jacobi's equation (12), we see that R has to be a regular of the following singular equation:

$$\frac{\partial R}{\partial t} + \frac{1}{2m}\left(\frac{\partial R}{\partial x}\right)^2 + \left(\frac{x-x'}{t-t'}\right)\frac{\partial R}{\partial x} + U = 0. \tag{22}$$

Looking for an asymptotic solution of the type

$$R = W_0 + W_1(t-t') + W_2(t-t')^2 + O\left((t-t')^3\right)$$

of this equation, where the W_j $(j = 0, 1, 2)$ only depend on x and x' one finds that $W_0 = 0$, and that W_1, W_2 must satisfy the conditions

$$\begin{cases} W_1 + (x-x')\dfrac{\partial W_1}{\partial x} + U = 0 \\[2mm] 2W_2 + (x-x')\dfrac{\partial W_2}{\partial x} = 0. \end{cases} \tag{23}$$

The general solution of the first of these equations is

$$W_1(x, x') = -\int_0^1 U(sx + (1-s)x')ds + \frac{k}{x-x'}$$

where k is an arbitrary constant. Since we want a regular solution, we must take $k = 0$, and hence

$$W_1 = -\overline{U}$$

where $\overline{U}$ is given by (20). Similarly, the only non-singular solution of the second equation (23) is $W_2 = 0$; the estimate 19 follows. Let us now prove (21). This can be done directly by using the continuity equation (13) satisfied by ρ, or alternatively by the following argument (see [4] for a rigorous justification). Since

$$W = W_f - \overline{U}t + O(t^3)$$

we have

$$\frac{\partial^2 W}{\partial x_i \partial x_j} = \frac{\partial^2 W_f}{\partial x_i \partial x_j} - \frac{\partial^2 \overline{U}}{\partial x_i \partial x_j} t + O(t^3).$$

Since

$$\frac{\partial^2 W_f}{\partial x_i \partial x_j} = \begin{cases} \frac{m}{t} & si \ i = j \\ 0 & si \ i \neq j \end{cases}$$

we have, by definition of ρ:

$$\begin{aligned} \rho(x, x'; t) &= \left(\frac{1}{2\pi i \hbar}\right)^n \det\left(\frac{m}{t} I_{n \times n} - \overline{U}''_{x,x'} t + O(t^3)\right) \\ &= \left(\frac{m}{2\pi i \hbar t}\right)^n \det\left(I_{n \times n} - \frac{1}{m}\overline{U}''_{x,x'} t^2 + O(t^3)\right) \\ &= \left(\frac{m}{2\pi i \hbar t}\right)^n (1 + O(t^2)) \end{aligned}$$

which is (21).

For instance, for the harmonic oscillator with Hamiltonian

$$H = \frac{1}{2m}(p^2 + m^2\omega^2 x^2)$$

$(n = 1)$ one finds that

$$W(x, x'; t, t') = m\frac{(x - x')^2}{2t} - \frac{m\omega^2}{6}(x^2 + xx' + x'^2)t + O(t^2).$$

Note that the approximation

$$W_{Feyn}(x, x'; t, t') = m\frac{(x - x')^2}{2t} - \frac{1}{2}m\omega^2 x'^2 t.$$

of (10) used in the usual Feynman formula is a very rough approximation of W; in fact $W - W_{Feyn} = O(t)$.

4.   THE SEMICLASSICAL WAVEFUNCTION

Assume, as above, that $0 < |t| < \varepsilon$, so that the generating function W determined by H exists, and set

$$G_{scl} = e^{\frac{i}{\hbar}W(x,x't)}\sqrt{\rho}(x, x', t) \tag{24}$$

where $\sqrt{\rho}$ is defined by

$$\sqrt{\rho}(x, x', t) = \begin{cases} \sqrt{|\rho|}(x, x', t) & (t > 0) \\ i\sqrt{|\rho|}(x, x', t) & (t < 0). \end{cases}$$

We then have:

PROPOSITION 2. *The function G_{scl} has the following properties:*

$$\lim_{t \to t'} G_{scl}(x, x'; t) = \delta(x - x') \tag{25}$$

and for fixed x' the function $(x, t) \mapsto G_{scl}(x, x'; t)$ satisfies

$$i\hbar \frac{\partial G_{scl}}{\partial t} = (\widehat{H} - Q)G_{scl} \tag{26}$$

where the function Q is given by

$$Q = -\frac{\hbar^2}{2m} \frac{\nabla_x^2 \sqrt{|\rho|}}{\sqrt{|\rho|}}. \tag{27}$$

PROOF. In view of Lemma 1 we have

$$\sqrt{\rho}(x, x'; t) = \left(\frac{m}{2\pi i \hbar t}\right)^{n/2} (1 + O(t^2)) \tag{28}$$

and

$$W(x, x'; t) = m \frac{(x - x')^2}{2t} + O(t)$$

for $t \to 0$, and consequently

$$G_{scl}(x, x'; t) = \left(\frac{m}{2\pi i \hbar t}\right)^{n/2} \exp\left[\frac{i}{\hbar}\left(m \frac{(x - x')^2}{2t}\right)\right] (1 + O(t))$$

for $t > 0$. Hence

$$\lim_{t \to 0^+} G_{scl}(x, x'; t) = \lim_{t \to 0^+} \left(\frac{m}{2\pi i \hbar t}\right)^{1/2} \exp\left[\frac{im}{\hbar} \frac{(x - x')^2}{2t}\right] = \delta(x - x').$$

By a similar argument, one finds that

$$\lim_{t \to 0^-} G_{scl}(x, x'; t) = \delta(x - x')$$

hence (25). Let us now show that G_{scl} verifies (26). Setting $a = \sqrt{|\rho|}$ we have

$$i\hbar \frac{\partial G_{scl}}{\partial t} = e^{\frac{i}{\hbar}W}\left(-\frac{\partial W}{\partial t}a + i\hbar\frac{\partial a}{\partial t}\right) \tag{29}$$

and also

$$i\hbar \frac{\partial G_{scl}}{\partial x_j} = e^{\frac{i}{\hbar}W}\left(-\frac{\partial W}{\partial x_j}a + i\hbar\frac{\partial a}{\partial x_j}\right) \quad (1 \le j \le n)$$

and hence

$$e^{-\frac{i}{\hbar}W}\widehat{H}G_{scl} = (\nabla_x W)^2 - \hbar^2\nabla_x^2 a + Ua - 2i\hbar\nabla_x a \cdot \nabla_x W - i\hbar a \nabla_x^2 W.$$

Since $\nabla_x W = p = mv$ it follows that

$$e^{-\frac{i}{\hbar}W}\left(i\hbar\frac{\partial G_{scl}}{\partial t} - \widehat{H}G_{scl}\right) =$$

$$-\frac{\partial W}{\partial t} - H\left(x, \nabla_x W, t\right) + \frac{\hbar^2}{2m}\nabla_x^2 a + \left(\frac{\partial a}{\partial t} + \nabla_x(av) + \frac{1}{2}a\nabla_x v\right).$$

that is, since W satisfies the Hamilton-Jacobi equation (12):

$$e^{-\frac{i}{\hbar}W}\left(i\hbar\frac{\partial G_{scl}}{\partial t} - \widehat{H}G_{scl}\right) = \frac{\hbar^2}{2m}\nabla_x^2 a + \left(\frac{\partial a}{\partial t} + \nabla_x(av) + \frac{1}{2}a\nabla_x v\right).$$

In view of the continuity equation (13) we have

$$\frac{\partial a}{\partial t} + \nabla_x(av) + \frac{1}{2}a\nabla_x v = 0$$

and hence

$$e^{-\frac{i}{\hbar}W}\left(i\hbar\frac{\partial G_{scl}}{\partial t} - \widehat{H}G_{scl}\right) = \frac{\hbar^2}{2m}\nabla_x^2 a$$

proving (26).

Proposition 2 implies that G_{scl} is the Green function of a certain integro-differential equation:

COROLLARY 1. *For each* $\Psi_0 \in \mathcal{S}(\mathbb{R}_x^n)$ *the function*

$$\Psi(x,t) = \int G_{scl}(x, x'; t)\Psi_0(x')\, d^n x' \tag{30}$$

is a solution of the Cauchy problem

$$i\hbar\frac{\partial\Psi}{\partial t} = (\widehat{H} - \widehat{Q})\Psi \quad , \quad \Psi(x,0) = \Psi_0(x) \tag{31}$$

where $\widehat{Q}$ is the operator $S(\mathbb{R}_x^n) \longrightarrow S(\mathbb{R}_x^n)$ defined by

$$\widehat{Q}\Psi(x,t) = \int Q(x,x';t)G_{scl}(x,x';t)\Psi_0(x')\,d^n x' \tag{32}$$

Q being the function (27).

PROOF. Differentiating under the integration sign in (30) we get

$$i\hbar\frac{\partial\Psi}{\partial t} - \widehat{H}\Psi = \int\left(i\hbar\frac{\partial G_{scl}}{\partial t} - \widehat{H}G_{scl}\right)\Psi_0 d^n x'$$

$$= \frac{\hbar^2}{2m}\int\frac{\nabla_x^2\sqrt{|\rho|}}{\sqrt{|\rho|}}G_{scl}\Psi_0 d^n x$$

hence

$$i\hbar\frac{\partial\Psi}{\partial t} = (\widehat{H} - \widehat{Q})\Psi.$$

That $\Psi(x,0) = \Psi_0(x)$ immediately follows from formula (25).

When the potential U is quadratic, one gets formula (8) yielding the solutions of Schrödinger's equation:

COROLLARY 2. *When U is a quadratic form in the coordinates x_j the function (30) is a solution of the Cauchy problem for Schrödinger's equation:*

$$i\hbar\frac{\partial\Psi}{\partial t} = \widehat{H}\Psi \quad , \quad \Psi(x,0) = \Psi_0(x). \tag{33}$$

PROOF. If U is quadratic, then the generating function W has the form (7), and Van Vleck's determinant is hence independent of the variables x, x'. It follows that in this case $Q = 0$, hence the corollary, using (31).

5. AN ALGORITHM FOR SOLVING SCHRÖDINGER'S EQUATION

Denote by G_t the operator which to Ψ_0 associates the function $\Psi(\cdot,t)$ defined by *(30)*. In general (that is when $Q \neq 0$) we have

$$G_t G_{t'} \neq G_{t+t'}. \tag{34}$$

This is hardly surprising, since equation (31) is then an integro-differential equation. The operators G_t allow us, however, to construct exact solutions to Schrödinger's equation (1). Here is how. Suppose, for example, that $t > 0$ and consider:

$$\Pi_t(N) = (G_{\Delta t})^N \ , \quad \Delta t = t/N.$$

One can easily prove that when $N \rightarrow \infty$ the sequence of operators $\Pi_t(N)$ converges towards a limit F_t (see [4]) and that we have

$$F_t F_{t'} = F_{t+t'}.$$

In fact, the following essential result holds:

PROPOSITION 3. *For every $\Psi_0 \in S(R_x^n)$ the function*

$$\Psi(x, t) = \lim_{N \rightarrow \infty} (G_{\Delta t})^N \Psi_0(x) \tag{35}$$

is a solution of the Cauchy problem

$$i\hbar \frac{\partial \Psi}{\partial t} = -\frac{\hbar^2}{2m} \nabla_x \Psi + U \Psi \ , \quad \Psi(x, 0) = \Psi_0. \tag{36}$$

PROOF. Making a second-order Taylor expansion at $t = 0$ of Ψ one gets, using the equation satisfied by Ψ:

$$\Psi(x, \Delta t) = \left[1 + \frac{\Delta t}{i\hbar} \left(-\frac{\hbar^2}{2m} \nabla_x^2 + U(x) \right) \right] \Psi_0(x) + O((\Delta t)^2) \tag{37}$$

(see [9]). Next note that the integral

$$G_{\Delta t} \Psi_0(x) = \int G_{scl}(x, x'; \Delta t) \Psi_0(x') \, d^n x'$$

satisfies the estimate

$$G_{\Delta t} \Psi_0(x) = \sqrt{\rho_f}(\Delta t) \int e^{\frac{i}{\hbar} W_f(x, x'; \Delta t)} \left(1 - \frac{i}{\hbar} \overline{U}(x, x') \Delta t \right) \Psi_0(x') \, d^n x'$$

$$+ O((\Delta t)^2)$$

where W_f is defined by

$$W_f(x, x'; \Delta t) = m \frac{(x - x')^2}{2\Delta t} \ , \quad \rho_f(\Delta t) = \left(\frac{m}{2\pi i\hbar \Delta t} \right)^n.$$

Applying the method of the stationary phase for $\Delta t \rightarrow 0$ we have

$$\int e^{\frac{i}{\hbar} W_f(x, x'; t)} \Psi_0(x') \, d^n x' = \left(\frac{2\pi i\hbar \Delta t}{m} \right)^{n/2} \left(1 + \frac{i\hbar \Delta t}{2m} \nabla_x^2 \right) \Psi_0(x) + O\left(\Delta t^2 \right)$$

and

$$\int e^{\frac{i}{\hbar} W_f(x,x';t)} \overline{U}(x,x') \Psi_0(x') \, d^n x' = \left(\frac{2\pi i \hbar \Delta t}{|m|} \right)^{n/2} \overline{U}(x,x,t') + O\left(\Delta t^2\right)$$

$$= \left(\frac{2\pi i \hbar \Delta t}{|m|} \right)^{n/2} U(x,t') + O\left(\Delta t^2\right)$$

from which follows that

$$G_{\Delta t} \Psi_0(x) = \left[1 + \frac{\Delta t}{i\hbar} \left(-\frac{\hbar^2}{2m} \nabla_x^2 + U(x) \right) \right] \Psi_0(x) + O((\Delta t)^2). \tag{38}$$

Comparing the estimates (37) and (38), we get

$$\Psi(x, \Delta t) - G_{\Delta t} \Psi_0(x) = O((\Delta t)^2). \tag{39}$$

Denoting by F_t the mapping which to $\Psi_0(x)$ associates $\Psi = \Psi(x,t)$ let us show that

$$F_t = \lim_{N \to \infty} (G_{\Delta t})^N. \tag{40}$$

In view of (39) we have

$$F_{\Delta t} = G_{\Delta t} + O((\Delta t)^2)$$

and hence

$$F_t = (F_{\Delta t})^N = (G_{\Delta t} + O((\Delta t)^2))^N.$$

One immediately checks that

$$(G_{\Delta t} + O((\Delta t)^2))^N = (G_{\Delta t})^N + N O((\Delta t)^2)$$

$$= (G_{\Delta t})^N + O(\Delta t)$$

hence (40).

REMARK. All the results above can be extended to the case where the potential appearing in the Hamiltonian depends on t, and even to the case where a vector potential is present, that is when H is of the type:

$$H = \frac{1}{2m} (p - A(x,t))^2 + U(x,t).$$

One must however in this case replace the flow (f_t) determined by H by the family of symplectomorphisms $(f_{t,t'})$ where $t \longmapsto f_{t,t'}(x',p')$ is the solution of Hamilton's equations with $x(t') = x'$ and $p(t') = p'$. One also has to replace the Green function $G = G(x, x'; t)$ by the function $G(x, x'; t, t')$ defined by

$$i\hbar\frac{\partial G}{\partial t} = -\frac{\hbar^2}{2m}\nabla_x G + U\,\Psi \quad, \quad \lim_{t\to t'} G = \delta(x - x').$$

Formula (35) then becomes the "ordered product"

$$\Psi(x, t) = \lim_{N\to\infty} \prod_{j=0}^{N-1} G_{t-j\Delta t, t-(j+1)\Delta t}\,\Psi(x')$$

where

$$G_{t,t'}\Psi(x) = \int G_{scl}(x, x'; t, t')\Psi(x')\, d^n x'.$$

REFERENCES

[1] V.I. Arnold, *Mathematical Methods of Classical Mechanics*, second edition, Graduate Texts in Mathematics, Springer-Verlag (1989)

[2] G. Folland, *Harmonic Analysis in Phase space*, Annals of Mathematics studies (Princeton University Press, Princeton, N.J., 1989).

[3] M. de Gosson, *Maslov Indices on Mp(n)*, Ann. Inst. Fourier, Grenoble, **40**(3) (1990), 537–555

[4] M. de Gosson, *The Principles of Newtonian and Quantum Mechanics: The Need For Planck's Constant ħ*, Imperial College Press/World Scientific, London (2001)

[5] V. Guillemin and S. Sternberg, *Geometric asymptotics*, Math. Surveys Monographs **14**, Amer. Math. Soc., Providence R. I. (1978)

[6] V. Guillemin and S. Sternberg, *Symplectic Techniques in Physics*, Cambridge University Press, Cambridge, Mass. (1984)

[7] J. Leray, *Lagrangian Analysis and Quantum Mechanics, a mathematical structure related to asymptotic expansions and the Maslov index*, MIT Press, Cambridge, Mass. (1981);

[8] V.P. Maslov and M.V. Fedoriuk, *Semi-Classical Approximations in Quantum Mechanics*, Reidel, Boston (1981)

[9] L.S. Schulman, *Techniques and Applications of Path Integrals* (Wiley, 1981).

[10] J.-M. Souriau, *Construction explicite de l'indice de Maslov*, Group Theoretical Methods in Physics, Lecture Notes in Physics, **50** (Springer-Verlag, 1975), 17–148.

[11] J. Vaillant, *Polynôme sous-caractéristique and dérivée de Lie de la forme élément de volume*, Comptes Rendus Acad. Sci. Paris **268**, 547–548 (1969)

Strong uniqueness in Gevrey spaces for some elliptic operators

F. COLOMBINI   Dipartimento di Matematica-Università di Pisa, Italy

C. GRAMMATICO   Dipartimento di Matematica-Università di Bologna, Italy

À Jean Vaillant, en témoignage d'amitié.

1   INTRODUCTION

We consider here the problem of strong uniqueness from the origin for particular smooth elliptic operators.

Let P be a second order, elliptic operator with smooth coefficients defined in an open neighborhood of the origin in $\mathbf{R}^2$. We denote by P_2 the principal part of P. Then if P has simple characteristic, as a consequence of the results in [1] and [2], we have the following:

1. if $P_2(0, D_x)$ is real, then P has the strong uniqueness from the origin,

2. if $P_2(0, D_x)$ is not real, then there exists a C^∞ function a flat at zero (more exactly in a suitable Gevrey class) and a function $u \in C^\infty$ flat at zero, not identically zero, such that

$$Pu + au = 0\,. \tag{1}$$

We emphasize the problem of strong uniqueness at the origin for u satisfying the inequality of the type

$$\left|\Delta^h u(x)\right| \le |W_0(x)|\,|u(x)| + |W_1(x)|\,|\nabla u(x)| + \cdots \tag{2}$$

$$+ \cdots + |W_h(x)|\left|\nabla^h u(x)\right| \qquad x \in \Omega,$$

301

where $h \in \mathbf{N}$ and Ω is a neighborhood of the origin in $\mathbf{R}^n$.

Let $C_b^\infty(\Omega)$ denote the space of functions in $C^\infty(\Omega)$ which are flat at the origin; we recall that $f(x)$ *is said to be flat at zero if* $D^\alpha f(0) = 0$ *for any* $\alpha \in \mathbf{N}^n$; we say that the relation (2) has the property of strong unique continuation at the origin, if the only function $u(x) \in C_b^\infty(\Omega)$ satisfying (2) is the zero function.

We recall that for a differential operator $P = P(x, y, t, D_x, D_y, D_t)$ of order m $(m \geq 2)$ with smooth coefficients, defined in a neighborhood of the origin in $\mathbf{R}^n$ $(n \geq 2)$ and of principal symbol $P_m(x, y, t, \xi, \eta, \tau)$, Alinhac [1] has shown, in particular, the following result (we denote the coordinates in $\mathbf{R}^n$ by (x, y, t) with $t \in \mathbf{R}^{n-2}$):

If $p_m(0, 0, 0; 1, \eta, 0)$ has two simple, non real and non conjugate roots, then there exist a neighborhood $\mathcal{V}$ of the origin, two functions $a, u \in C^\infty(\mathcal{V})$ both flat at zero, such that $Pu - au = 0$ in $\mathcal{V}$, and $\operatorname{supp} u$ is a neighborhood of the origin.

Hence, if we want to obtain unique continuation results in C^∞ for operators of order greater than two, it is not too restrictive to consider the case Δ^h $(h \in \mathbf{N})$.

This problem for the case $h = 1$ i.e. for inequalities of the type

$$|\Delta u(x)| \leq |W_0(x)| |u(x)| + |W_1(x)| |\nabla u(x)| \qquad x \in \Omega, \tag{3}$$

where $W_0(x)$, $W_1(x)$ are singular at the origin, has been studied for a long time.

The first results with singular $W_0(x)$ and $W_1(x)$ were obtained by Aronszajn, Krzywicki and Szarski [4] who examined potentials of the form: $|W_0(x)| \leq |x|^{-2+\varepsilon}$, $|W_1(x)| \leq |x|^{-1+\varepsilon}$ with $\varepsilon > 0$.

Afterwards a number of authors considered the problem of unique continuation, with $W_0(x)$ and $W_1(x)$ having singularities of type L^p (Hörmander [10], Barceló, Kenig, Ruiz, Sogge [5]).

In 1993 Regbaoui [15] proved a result of strong uniqueness for $W_0(x)$ and $W_1(x)$ satisfying the following estimates:

$$|W_0(x)| \leq C_1 |x|^{-2} \tag{4}$$

$$|W_1(x)| \leq C_2 |x|^{-1} \tag{5}$$

provided that $C_2 < \dfrac{1}{2}$.

Already in 1992, Pan [14] had shown a similar result, in $\mathbf{R}^2$, without any bound for the constant C_2 above, but for real valued functions u.

In 1997 the result of Regbaoui was improved (see [9]), by proving the strong unique continuation respectively for $W_0(x)$ and $W_1(x)$ as in (4) and (5), but with $C_2 < \dfrac{1}{\sqrt{2}}$. Later on Regbaoui in [16] extended this result to the case of operators $P(x, D) = \sum a_{ij}(x) D_i D_j$ and $P(0, D) = \triangle$.

Alinhac and Baouendi [2] have shown, in particular, that the strong unique continuation property holds for the relation (2) when $h = 2$ and $|W_l(x)| \le C |x|^{-4+l+\varepsilon}$, $l = 0, 1, 2$, for some $\varepsilon > 0$.

In 1999 we proved in [6] that the strong unique continuation holds for the inequality (2) with $|W_l(x)| \le C_l |x|^{l-2h}$, $l = 0, \ldots, h$, with bounds only on C_h constant.

In our recent work [6] we have proved, in particular, that

Given $h \in \mathbf{N}$ positive integer, if $u \in C_b^\infty(\mathbf{R}^2)$ verifies ($r = |x|$)

$$\left| \triangle^h u \right| \le \frac{C}{r^{2h}} \left| \partial_\theta^h u \right| \qquad x \in \mathbf{R}^2 \setminus \{0\}, \tag{6}$$

for a certain constant $C < (2h-1)!!$, then u is the zero function.

We have verified in [7] that the bounds for the constants in (6) are optimal. To this end we have the following result:

Let $h \in \mathbf{N}$ be a positive integer, then, for any $\varepsilon > 0$ we can find two functions $w \in C_b^\infty(\mathbf{R}^2)$ with supp $w \equiv \mathbf{R}^2$ and $a \in C^\infty(\mathbf{R}^2 \setminus \{0\})$ with $\|a\|_\infty \le (2h-1)!! + \varepsilon$ such that

$$\triangle^h w + \frac{a(x)}{r^{2h}} \partial_\theta^h w = 0 \qquad in \ \mathbf{R}^2 \setminus \{0\}. \tag{7}$$

More precisely
1) *if h is even we can take $|\Im a| \le \varepsilon$, $|\Re a| \le (2h-1)!! + \varepsilon$,*
2) *if h is odd we can take $|\Im a| \le (2h-1)!! + \varepsilon$, $|\Re a| \le \varepsilon$.*

In [6] we have also proved that the strong unique continuation property holds for functions verifying

$$\left| \triangle^2 w(x) \right|^2 \le \frac{\mathcal{M}}{|x|^6} |\nabla w(x)|^2 + \frac{\mathcal{C}}{|x|^4} \sum_{i,j=1}^n |D_i D_j w(x)|^2 \qquad x \in \mathbf{R}^n, \tag{8}$$

with $\mathcal{M}$ any positive constant and $\mathcal{C} < 9/4$. Whereas in [8] we have shown that

For every $\mathcal{C} > 9/4$ there exists a function $w \in C_b^\infty(\mathbf{R}^2)$ with supp $w \equiv \mathbf{R}^2$ satisfying the estimate (8) for a suitable constant $\mathcal{M}$.

Recently Le Borgne [12] has proved that for the inequality (8) it is possible to add derivatives of third order but with the potential $|x|^{-2+\varepsilon}$.

In these notes, we study the case in which P has Gevrey coefficients. From now on we assume by strong unique continuation the following

DEFINITION 1.1 *Let Ω be an open neighborhood of the origin in $\mathbf{R}^2$. We say that P has the strong unique continuation property if whatever $Pu = 0$ in Ω and $u \in C_b^\infty(\Omega)$ then $u \equiv 0$ in a neighborhood of zero.*

In 1981 Lerner [13] proved that if P is a second order, elliptic operator with simple characteristic and with Gevrey coefficients of order s, defined in neighborhood of the origin in $\mathbf{R}^2$, then the equation $Pu = 0$ has the strong unique property at zero if Gevrey's index s is smaller than a quantity depending on the cone $p(0, \mathbf{R}^2)$, where $p(x, \xi)$ is the principal symbol of P. Similar results are true in $\mathbf{R}^n$.

We shall expose, briefly, Lerner's results in $\mathbf{R}^2$ (see [13]). Then, we shall study particular operators P of fourth order with $P_4(0, D) = \triangle Q$, and Q being a second order, elliptic, homogeneous operator with real coefficients. We shall obtain the strong uniqueness if P has Gevrey coefficients of order smaller than a quantity depending on eigenvalues of the quadratic form associeted to Q (see theorem 3.1 below). We shall obtain this result in two different way: in the first case we shall give Carleman estimates for Q, in the second case as a consequence of the results in [13]. We hope that the first approach can be extended to $\mathbf{R}^n$.

2   SECOND ORDER CASE

Let Ω be an open neighborhood of the origin in $\mathbf{R}^2$. We denote by $\Re z$, $\Im z$ the real and imaginary part, respectively, of the complex number z.

Finally we denote by G^s the Gevrey space of order s.

We state first a classical Sjöstrand's lemma.

LEMMA 2.1 ([13], pag. 1174) *Let P be a second order, elliptic operator, defined in an open neighborhood of the origin in $\mathbf{R}^2$ and p its principal symbol.*

Then, either $p(0, \mathbf{R}^2 \setminus \{0\})$ is a convex cone of $\mathbf{C} \setminus \{0\}$, or it is $\mathbf{C} \setminus \{0\}$.

We state now the principal result in [13].

THEOREM 2.2 ([13], pag. 1165) *Let P be a second order, elliptic operator with Gevrey coefficients of order $s > 1$, defined in an open neighborhood of the origin in $\mathbf{R}^2$ and p its principal symbol.*

1. *If $p\left(0, \mathbf{R}^2 \backslash \{0\}\right)$ is a convex cone with angle 2ϕ [1], $0 \leq \phi < \dfrac{\pi}{2}$ and if*

$$s < 1 + \frac{1 - \sin \phi}{2 \sin \phi} \tag{9}$$

   then P has the strong unique continuation property at zero.

2. *If $p\left(0, \mathbf{R}^2 \backslash \{0\}\right) = \mathbf{C} \backslash \{0\}$ and if P has simple characteristic, then there exists a real number $\sigma_0 > 1$ depending only on $P\left(0, D_x\right)$ such that P has the strong unique continuation property at zero for any $s < \sigma_0$.*

We note that the operators as above are Gevrey hypoelliptic because their coefficients belongs to Gevrey class of order s, hence $Pu = 0$ implies $u \in G^s$.

We give two lemmas which we shall use later.

LEMMA 2.3 ([13], pag. 1166) *Let be $\nu > 0$ and $r^2(x)$ a positive quadratic form in $\mathbf{R}^n$; then the function $u(x) = exp\left(-r^{-\nu}\right)$ belongs to $G^{1 + \nu^{-1}}(\mathbf{R}^n)$.*

LEMMA 2.4 ([13], pag. 1166) *Let Ω be an open neighborhood of the origin in $\mathbf{R}^n$ and $u \in G^s(\Omega)$. If u is flat at zero, then there exists a function $v \in C^\infty(\Omega)$ flat at zero such that*

$$u = exp\left(-r^{-\nu}\right) v$$

provided $1 + \nu^{-1} > s$.

The heart of the proof in theorem 2.2 is that if P is a second order, elliptic operator with smooth coefficients defined in an open neighborhood of the origin in $\mathbf{R}^2$ and if it has simple characteristics then there esist two smooth, elliptic vector fields X_1, X_2 such that

$$P = X_1 X_2 + P_1 \tag{10}$$

where P_1 is a first order differential operator with coefficients C^∞.

---

[1]in this case, it is easy to see that P has simple characteristic

Now, let X be a smooth, elliptic vector field, we set

$$X_\gamma = e^{-\gamma\psi} X e^{\gamma\psi},$$

where $\psi = -\nu^{-1} r^{-\nu}$ $(\nu > 0)$ and $r = |x|$. If we choose ν such that

$$\nu + 2 > \frac{|X(0)|^2}{\min_{|x|=1} |\langle X(0), x\rangle|^2}, \qquad (11)$$

we can give a Carleman estimate for X. Then if we iterate Carleman estimates for X_1, X_2 fields in (10) we obtain a Carleman estimate for P by using $P_\gamma = e^{-\gamma\psi} P e^{\gamma\psi}$.

Let (x_1, x_2) be the coordinates in $\mathbf{R}^2$, we denote by (ξ_1, ξ_2) dual coordinates. Setting $p_0(\xi_1, \xi_2) = p(0, 0; \xi_1, \xi_2)$ we can write

$$p_0(\xi_1, \xi_2) = \alpha(\xi_1 + z\xi_2)(\xi_1 + w\xi_2) \qquad \alpha, z, w \in \mathbf{C};$$

from ellipticity of p_0 it follows that $\Im z, \Im w \neq 0$.

We can suppose, now, that $\alpha = 1$; we note that

$$\Re p_0(\xi_1, \xi_2) = \xi_1^2 + \xi_1\xi_2(\Re z + \Re w) + \xi_2^2(\Re z \cdot \Re w - \Im z \cdot \Im w);$$

it is easy to see that if $\Im z \cdot \Im w > 0$ then $\Re p_0$ vanishes and the imaginary axis is contained in the image of p_0. Hence $p(0, \mathbf{R}^2 \setminus \{0\}) = \mathbf{C} \setminus \{0\}$.

If $\Im z \cdot \Im w < 0$ the real part of p_0 can vanish or not vanish; in the second case the image of p_0 must be a convex cone with angle smaller than π.

Suppose that the image of p_0 is a convex cone with angle smaller than π. We can suppose also that $\Re p_0$ is a positive quadratic form; then there exists a linear map (rotation) $L : (s, t) \longmapsto (x_1, x_2)$ such that the principal symbol of $p_L(s, t; \sigma, \tau)$ evaluated at zero, becomes

$$p_L(0, 0; \sigma, \tau) = p_0(L(\sigma, \tau)) = e^{i\phi}\sigma^2 + e^{-i\phi}\tau^2$$

where 2ϕ is the angle of the cone $p(0, \mathbf{R}^2 \setminus \{0\})$.

When p verifies the cone property of angle 2ϕ, $0 \leq \phi < \dfrac{\pi}{2}$, we can find coordinates such that the relation (11) can be written, for any vector field, as

$$\nu > \frac{2\sin\phi}{1 - \sin\phi}. \qquad (12)$$

REMARK 2.1 *We note, taking into account lemma 2.4, that the assumption* $v_\gamma = e^{-\gamma\psi}u$ *for any* $\gamma > 0$ *and* $v_\gamma \in C_b^\infty$ *is satisfied if s is as in the theorem.*

We stress that the result holds again for function u such that:

1. $u = e^{\gamma\psi}v_\gamma$ for any $\gamma \gg 0$ and $v_\gamma \in C^\infty$ flat at zero,

2. $|Pu(x)| \leq C|x|^\varepsilon \left(\dfrac{|u(x)|}{|x|^2} + \dfrac{|\nabla u(x)|}{|x|} \right) \qquad x \in \Omega$ neighborhood of
   zero, for some constant $C > 0$ and for some $\varepsilon > 0$.

3   FOURTH ORDER CASE

Now, we consider two second order, elliptic operators R, S not proportional with real, constant coefficients.

After a change of coordinates, we can assume that

$$R = \triangle \quad e \quad S = \Gamma(D) = \lambda^2 \partial_{x_1}^2 + \mu^2 \partial_{x_2}^2, \quad \lambda > \mu > 0. \qquad (13)$$

Now, we can state our main result:

THEOREM 3.1 *Let* $P(x, D)$ *a fourth order, elliptic operator with Gevrey coefficients of order* $s > 1$, *defined in an open neighborhood* V *of the origin in* $\mathbf{R}^2$. *Suppose that*

$$P(x, D) = L(x, D)Q(x, D) + a(x) \qquad (14)$$

where L and Q are second order differential operators.
   We denote by L_2, Q_2 *the principal part of* L, Q *respectively, and assume that*

$$L_2(0, D) = \triangle \quad \text{and} \quad Q_2(0, D) = \Gamma(D)$$

with $\Gamma(D)$ *as in (13). If* $s < \dfrac{\lambda}{\lambda - \mu}$ *then P has the strong unique continuation property.*

REMARK 3.1 *Taking into account the results in* [1] *the operator in* (14) *has not in general the strong unique continuation property, if its coefficients are not Gevrey enough.*

As annonced in introduction we give two different proofs.

3.1   Proof I

We give Carleman estimates for Q operator and we note that similar estimates hold for L operator.

Proof—In a neighborhood of the origin in $\mathbf{R}^2$, by the assumption, there exist two smooth, elliptic vector fields X_1, X_2 such that

$$Q = X_1 X_2 + Q_1 \qquad (15)$$

and Q_1 is a first order differential operator with C^∞ coefficients.

Now, we consider the polar coordinates in the plane. We set, for $x \neq 0$ (see [9])

$$\frac{\partial}{\partial x_k} = \omega_{x_k} \frac{\partial}{\partial r} + \frac{1}{r} \Omega_{x_k}, \quad k = 1, 2,$$

where $\omega_{x_k} = \dfrac{x_k}{|x|}$ and Ω_{x_k} are suitable vector fields tangent to $\mathbf{S}^1$.

In polar coordinates, the vector fields in (15) becomes

$$X_k = (\alpha_k \omega_{x_1} + \beta_k \omega_{x_2}) \partial_r + \frac{\alpha_k}{r} \Omega_{x_1} + \frac{\beta_k}{r} \Omega_{x_2}, \qquad k = 1, 2, \qquad (16)$$

with α_k, β_k the coefficients of X_k fields. Moreover from the hypothesis $\Im \alpha_k (0) = \Re \beta_k (0) = 0$.

We give Carleman estimates for such vector fields, when applied to functions with support in $r \leq R_0$. Then, we can conclude in standard way.

Thus, let

$$X = (\alpha \omega_{x_1} + \beta \omega_{x_2}) \partial_r + \frac{\alpha}{r} \Omega_{x_1} + \frac{\beta}{r} \Omega_{x_2},$$

a smooth, elliptic vector field in the plain, with $\Im \alpha (0) = \Re \beta (0) = 0$.

For $u \in C_b^\infty$ having compact support in $r \leq R_0$ we consider

$$v_\gamma = e^{-\gamma \psi} u \qquad \forall \gamma > 0,$$

where $\psi = -\nu^{-1} r^{-\nu}$ $(\nu > 0)$ and $v_\gamma \in C_b^\infty$.

We can write

$$X = Y_1 + i Y_2,$$

with Y_1, Y_2 real smooth vector field defined in a neighborhood of the origin.

In polar coordinates

$$Y_1 = (\omega_{x_1} \Re \alpha + \omega_{x_2} \Re \beta) \partial_r + \frac{\Re \alpha}{r} \Omega_{x_1} + \frac{\Re \beta}{r} \Omega_{x_2},$$

while

$$Y_2 = (\omega_{x_1} \Im\alpha + \omega_{x_2} \Im\beta) \, \partial_r + \frac{\Im\alpha}{r} \, \Omega_{x_1} + \frac{\Im\beta}{r} \, \Omega_{x_2} \, .$$

Setting

$$X_\gamma = e^{-\gamma\psi} X e^{\gamma\psi} \, ,$$

we have

$$X_\gamma = Y_1 + \gamma \, r^{-\nu-2} \, \langle Y_1, x \rangle + i \left(Y_2 + \gamma \, r^{-\nu-2} \, \langle Y_2, x \rangle \right) \, ,$$

where $\langle Y, x \rangle = \sum a_k(x) \, x_k$ if $Y = \sum a_k(x) \dfrac{\partial}{\partial x_k}$.

From now on we write v instead of v_γ; we have

$$\|X_\gamma v\|^2 = \left\| Y_1 v + i \gamma \, r^{-\nu-2} \, \langle Y_2, x \rangle \, v \right\|^2 + \left\| i \, Y_2 v + \gamma \, r^{-\nu-2} \, \langle Y_1, x \rangle \, v \right\|^2 \quad (17)$$
$$+ \, 2 \, \Re \left(Y_1 v + i \gamma \, r^{-\nu-2} \, \langle Y_2, x \rangle \, v \, , \, i \, Y_2 v + \gamma \, r^{-\nu-2} \, \langle Y_1, x \rangle \, v \right),$$

where $\|\cdot\|$ and $(\cdot, \cdot)$ denote the norm and the inner product respectively in $L^2(\mathbf{R}^2)$.

Now we evaluate the scalar product in (17). At first

$$2 \, \Re \left(Y_1 v \, , \, \gamma \, r^{-\nu-2} \, \langle Y_1, x \rangle \, v \right) = 2 \, \Re \left((\omega_{x_1} \Re\alpha + \omega_{x_2} \Re\beta)^2 \, \partial_r v \, , \, \gamma \, r^{-\nu-1} \, v \right) \quad (18)$$
$$+ \, 2 \, \Re \left((\Re\alpha \, \Omega_{x_1} + \Re\beta \, \Omega_{x_2}) \, v \, , \, \gamma \, r^{-\nu-2} \, (\omega_{x_1} \Re\alpha + \omega_{x_2} \Re\beta) \, v \right) \, .$$

We set I_1, I_2 for the first and the second term respectively on the right hand side in (18). We have

$$I_1 = \gamma \left(\left(\nu \, (\omega_{x_1} \Re\alpha)^2 + \mathcal{O}(r) \right) v \, , \, r^{-\nu-2} v \right) \, ,$$

while

$$I_2 = 2 \, \Re \int \gamma \, r^{-\nu-1} \left((\Re\alpha \, \Omega_{x_1} + \Re\beta \, \Omega_{x_2}) \, v \, , \, (\omega_{x_1} \Re\alpha + \omega_{x_2} \Re\beta) \, v \right)_{L^2(\mathbf{S}^1)} dr \, ,$$

where $(\cdot, \cdot)_{L^2(\mathbf{S}^1)}$ denote the inner product in $L^2(\mathbf{S}^1)$.

Now taking into account the relations

$$\Omega_{x_1}^* = \omega_{x_1} - \Omega_{x_1} \quad \text{and} \quad \Omega_{x_1}(\omega_{x_1}) = 1 - \omega_{x_1}^2 \, ,$$

we have

$$2 \, \Re \left((\Re\alpha \, \Omega_{x_1} + \Re\beta \, \Omega_{x_2}) \, v \, , \, (\omega_{x_1} \Re\alpha + \omega_{x_2} \Re\beta) \, v \right)_{L^2(\mathbf{S}^1)} = (\mathcal{O}(r) \, v \, , \, v)_{L^2(\mathbf{S}^1)}$$
$$+ \, 2 \left((\Re\alpha)^2 \, \omega_{x_1}^2 \, v \, , \, v \right)_{L^2(\mathbf{S}^1)} - \left((\Re\alpha)^2 \, v \, , \, v \right)_{L^2(\mathbf{S}^1)}$$

Thus,

$$2\Re\left(Y_1 v\,,\,\gamma\, r^{-\nu-1}\,\langle Y_1, x\rangle\, v\right) = \gamma\left((\nu+2)\,(\Re\alpha)^2\,\omega_{x_1}^2\, v - (\Re\alpha)^2\, v\,,\, r^{-\nu-2}\, v\right)$$
$$+ \left(\mathcal{O}\left(r\right) v\,,\, v\right).$$

Similarly it is easy to see that

$$2\Re\left(i\,\gamma\, r^{-\nu-2}\,\langle Y_2, x\rangle\, v\,,\, i\, Y_2 v\right) = \gamma\left((\nu+2)\,(\Im\beta)^2\,\omega_{x_2}^2\, v - (\Im\beta)^2\, v\,,\, r^{-\nu-2}\, v\right)$$
$$+ \left(\mathcal{O}\left(r\right) v\,,\, v\right).$$

We choose, now, ν such that

$$\nu + 2 > \frac{\alpha^2(0) - \beta^2(0)}{\min_{\mathbf{S}^1}\left(\alpha^2(0)\,\omega_{x_1}^2 - \beta^2(0)\,\omega_{x_2}^2\right)}. \tag{19}$$

For ν as above and R_0 small enough, we can find $\varepsilon_0 > 0$ such that

$$\sigma \geq \varepsilon_0\,\gamma\left\|r^{-\frac{\nu}{2}-1} v\right\|^2 + 2\Re\left(Y_1 v\,,\, i\, Y_2 v\right),$$

where σ is the scalar product in (17). Now $2\Re\left(Y_1 v\,,\, i\, Y_2 v\right) = 2\Re\left(Z v\,,\, v\right)$, with Z a smooth vector field, hence

$$\sigma \geq \varepsilon_0\gamma\left\|r^{-\frac{\nu}{2}-1} v\right\|^2 + 2\Re\left(Z v\,,\, v\right).$$

From now on, we denote by $\mathcal{C}$ any positive constant and by B_R the ball which center is the origin and radius R.

If we take a constant $\mathcal{C} \leq \gamma R_0^{-\nu}$, from (17) we have

$$\|X_\gamma v\|^2 \geq \mathcal{C} R_0^\nu\gamma^{-1}\left\|Y_1 v + i\gamma r^{-\nu-2}\,\langle Y_2, x\rangle\, v\right\|^2$$
$$+ \mathcal{C} R_0^\nu\gamma^{-1}\left\|i Y_2 v + \gamma r^{-\nu-2}\,\langle Y_1, x\rangle\, v\right\|^2$$
$$+ \varepsilon_0\gamma\left\|r^{-\frac{\nu}{2}-1} v\right\|^2 + 2\Re\left(Z v\,,\, v\right).$$

So, if $r \leq R_0$ we have

$$\|X_\gamma v\|^2 \geq \mathcal{C}\left\|\gamma^{-\frac{1}{2}} r^{\frac{\nu}{2}} Y_1 v + i\gamma^{\frac{1}{2}} r^{-\frac{\nu}{2}-1}\left\langle Y_2, \frac{x}{r}\right\rangle v\right\|^2 \tag{20}$$
$$+ \mathcal{C}\left\|\gamma^{-\frac{1}{2}} r^{\frac{\nu}{2}} i Y_2 v + \gamma^{\frac{1}{2}} r^{-\frac{\nu}{2}-1}\left\langle Y_1, \frac{x}{r}\right\rangle v\right\|^2$$
$$+ \varepsilon_0\left\|\gamma^{\frac{1}{2}} r^{-\frac{\nu}{2}-1} v\right\|^2 - \gamma^{-1}\left\|r^{\frac{\nu}{2}+\frac{1}{2}} Z v\right\|^2 - \gamma\left\|r^{-\frac{\nu}{2}-\frac{1}{2}} v\right\|^2.$$

But, because X is an elliptic field, we deduce that there exists a constant $\alpha_0 \geq 0$ such that

$$\left\| r^{\frac{\nu}{2}+\frac{1}{2}} Z v \right\|^2 \leq \alpha_0 \left\{ \left\| r^{\frac{\nu}{2}+\frac{1}{2}} Y_1 v \right\|^2 + \left\| r^{\frac{\nu}{2}+\frac{1}{2}} Y_2 v \right\|^2 + \left\| r^{\frac{\nu}{2}+\frac{1}{2}} v \right\|^2 \right\}$$

$\forall\, v \in C_b^\infty (B_{R_0})$ having its support in B_{R_0}.

Because $r \leq R_0$, from (20), we have

$$\|X_\gamma v\|^2 \geq C \left\| \gamma^{-\frac{1}{2}} r^{\frac{\nu}{2}} Y_1 v + i \left\langle Y_2, \frac{x}{r} \right\rangle \gamma^{\frac{1}{2}} r^{-\frac{\nu}{2}-1} v \right\|^2 \tag{21}$$

$$+ C \left\| \gamma^{-\frac{1}{2}} r^{\frac{\nu}{2}} Y_2 v - i \left\langle Y_1, \frac{x}{r} \right\rangle \gamma^{\frac{1}{2}} r^{-\frac{\nu}{2}-1} v \right\|^2$$

$$+ \frac{\varepsilon_0}{2} \left\| \gamma^{\frac{1}{2}} r^{-\frac{\nu}{2}-1} v \right\|^2 + \frac{\varepsilon_0}{2} \left\| \gamma^{\frac{1}{2}} r^{-\frac{\nu}{2}-1} v \right\|^2$$

$$- \alpha_0 R_0 \left(\left\| \gamma^{-\frac{1}{2}} r^{\frac{\nu}{2}} Y_1 v \right\|^2 + \left\| \gamma^{-\frac{1}{2}} r^{\frac{\nu}{2}} Y_2 v \right\|^2 \right)$$

$$- R_0 \left\| \gamma^{\frac{1}{2}} r^{-\frac{\nu}{2}-1} v \right\|^2 .$$

Now, we note that

$$C \left\| \gamma^{-\frac{1}{2}} r^{\frac{\nu}{2}} Y_1 v + i \left\langle Y_2, \frac{x}{r} \right\rangle \gamma^{\frac{1}{2}} r^{-\frac{\nu}{2}-1} v \right\|^2 \geq \frac{\varepsilon}{2} \left\| \gamma^{-\frac{1}{2}} r^{\frac{\nu}{2}} Y_1 v \right\|^2 \tag{22}$$

$$- \varepsilon \left\| \left\langle Y_2, \frac{x}{r} \right\rangle \gamma^{\frac{1}{2}} r^{-\frac{\nu}{2}-1} v \right\|^2$$

for any $\varepsilon > 0$ and small enough.

Thus, if we take ε in (22) small enough and R_0 such that $\alpha_0 R_0 < \frac{\varepsilon}{2}$, we obtain the following estimate

$$\|X_\gamma v\|^2 \geq C \left(\left\| \gamma^{-\frac{1}{2}} r^{\frac{\nu}{2}} Y_1 v \right\|^2 + \left\| \gamma^{-\frac{1}{2}} r^{\frac{\nu}{2}} Y_2 v \right\|^2 + \left\| \gamma^{\frac{1}{2}} r^{-\frac{\nu}{2}-1} v \right\|^2 \right), \tag{23}$$

for some positive constant C.

Hence, we have proved that if R_0 is small enough, then there exists $C > 0$ such that (23) holds for any $v \in C^\infty$, flat at zero and having its support in B_{R_0}.

From (23), taking into account that X is an elliptic field, we have

$$\|X_\gamma v\|^2 \geq C \left(\gamma^{-1} \left\| r^{\frac{\nu}{2}} \partial_{x_1} v \right\|^2 + \gamma^{-1} \left\| r^{\frac{\nu}{2}} \partial_{x_2} v \right\|^2 + \gamma \left\| r^{-\frac{\nu}{2}-1} v \right\|^2 \right), \tag{24}$$

that is $\exists\, R_0 > 0$ and $\exists\, C > 0$ such that (24) holds $\forall\, \gamma \gg 0$ and $\forall\, v \in C^\infty$ flat at zero with support in B_{R_0}.

A similar estimate holds for $\|r^\alpha X_\gamma v\|$, that is $\exists\, R_0 > 0$ and $\exists\, C > 0$ such that the estimate

$$\|r^\alpha X_\gamma v\|^2 \;\geq\; C\left(\gamma^{-1}\left\|r^\alpha r^{\frac{\nu}{2}}\partial_{x_1}v\right\|^2 + \gamma^{-1}\left\|r^\alpha r^{\frac{\nu}{2}}\partial_{x_2}v\right\|^2\right) \qquad (25)$$

$$+\, C\,\gamma\left\|r^\alpha r^{-\frac{\nu}{2}-1}v\right\|^2,$$

holds for any $\gamma \gg 0$ and for any $v \in C^\infty$ flat at zero and having its support in B_{R_0}.

We return to Q operator and consider the operator Q_1. We can write

$$Q_1 = f(x)\,\partial_{x_1} + g(x)\,\partial_{x_2} + h(x)$$

with f, g, h smooth, complex-valued functions.

Now, if $v \in C_b^\infty(B_{R_0})$ with support in B_{R_0}, we have

$$\left\|e^{-\gamma\psi}Q_1\left(e^{\gamma\psi}v\right)\right\| \;\leq\; C\left\|e^{-\gamma\psi}\partial_{x_1}\left(e^{\gamma\psi}v\right)\right\|$$

$$+\, C\left\|e^{-\gamma\psi}\partial_{x_2}\left(e^{\gamma\psi}v\right)\right\| + C\,\|v\|,$$

hence

$$\left\|e^{-\gamma\psi}Q_1\left(e^{\gamma\psi}v\right)\right\| \;\leq\; C\left\|\frac{x_1}{r}\gamma r^{-\nu-1}v + \partial_{x_1}v\right\|$$

$$+\, C\left\|\frac{x_2}{r}\gamma r^{-\nu-1}v + \partial_{x_2}v\right\| + C\,\|v\|.$$

Taking into account that $r^{-\nu-1} \leq r^{-\nu-2}R_0$, it follows that

$$\left\|e^{-\gamma\psi}Q_1\left(e^{\gamma\psi}v\right)\right\| \leq C\gamma R_0\|r^{-\nu-2}v\| + C\,\|\partial_{x_1}v\| + C\,\|\partial_{x_2}v\|,$$

and hence

$$\left\|e^{-\gamma\psi}Q_1\left(e^{\gamma\psi}v\right)\right\| \leq CR_0\left(\gamma\|r^{-\nu-2}v\| + \|r^{-1}\partial_{x_1}v\| + \|r^{-1}\partial_{x_2}v\|\right). \quad (26)$$

We evaluate, now, $e^{-\gamma\psi}X_1X_2\left(e^{\gamma\psi}v\right)$; from (24) and (25) we have

$$\left\|e^{-\gamma\psi}X_1X_2\left(e^{\gamma\psi}v\right)\right\| \geq C\left(\gamma\|r^{-\nu-2}v\| + \|r^{-1}\partial_{x_1}v\| + \|r^{-1}\partial_{x_2}v\|\right), \quad (27)$$

with $C > 0$, provided that R_0 is small enough and the function v is as above.

Thus, from (26) and (27) we obtain that there exist $C, R_0 > 0$ such that

$$\left\| e^{-\gamma\psi} Q\left(e^{\gamma\psi} v \right) \right\| \geq C \left(\gamma \left\| r^{-\nu-2} v \right\| + \left\| r^{-1} \partial_{x_1} v \right\| + \left\| r^{-1} \partial_{x_2} v \right\| \right) , \qquad (28)$$

holds for any $\gamma \gg 0$ and for any $v \in C_b^\infty(B_{R_0})$ having its support in B_{R_0}.

To obtain Carleman estimates for P operator we make the following change of variables

$$\begin{cases} X = \sqrt{\lambda}\, x \\ Y = \sqrt{\mu}\, y, \end{cases}$$

so, the operators $\partial_x^2 + \partial_y^2$ and $\lambda^2 \partial_x^2 + \mu^2 \partial_y^2$ become, respectively, $\dfrac{1}{\lambda} \partial_X^2 + \dfrac{1}{\mu} \partial_Y^2$ and $\lambda \partial_X^2 + \mu \partial_Y^2$.

Now, for both new transformed operators, (19) becomes $\nu > \dfrac{\lambda - \mu}{\mu}$.

Finally, if we iterate the above Carleman estimate (28) we have

$$\left\| e^{-\gamma\psi} P\left(e^{\gamma\psi} v \right) \right\| \geq C\gamma^2 \left\| r^{-2\nu-4} v \right\| , \qquad (29)$$

for any $\gamma \gg 0$ and for any $v \in C_b^\infty(B_{R_0})$ having its support in B_{R_0}.

Let φ a real smooth function defined in $\mathbf{R}^+$ such that $(T < R_0)$

$$\varphi(r) = \begin{cases} 1 & \text{se } r < T/2 \\ 0 & \text{se } r > T. \end{cases}$$

Let u be the function of the theorem. Considering lemma 2.4, for ν as in (19) we can write $\varphi u = e^{\gamma\psi} \varphi v_\gamma$ for some $v \in C_b^\infty(B_{R_0})$, hence $e^{-\gamma\psi} P(\varphi u) = e^{-\gamma\psi} P\left(e^{\gamma\psi} \varphi v_\gamma \right)$. From (28) we have

$$\left\| e^{-\gamma\psi} P(\varphi u) \right\| \geq C \gamma^2 \left\| r^{-2\nu-4} \varphi v_\gamma \right\| ,$$

therefore, taking into account that $Pu = 0$ in a neighborhood of zero, it follows that

$$\gamma^2 \left\| r^{-2\nu-4} \varphi e^{-\gamma\psi} u \right\|_{\left(0, \frac{T}{2}\right)} \leq C \left\| e^{-\gamma\psi} P(\varphi u) \right\|_{\left(\frac{T}{2}, T\right)} , \qquad (30)$$

where $\| \cdot \|_{(a,b)}^2 = \displaystyle\int_{a \leq |x| \leq b} |\cdot|^2 \, dx$.

From (30), we obtain

$$\gamma^2 \left\| r^{-2\nu-4} u \right\|_{\left(0, \frac{T}{2}\right)} \leq C \left\| P(\varphi u) \right\|_{\left(\frac{T}{2}, T\right)}$$

for any γ large enough; if γ goes to infinity we deduce that $u \equiv 0$ in $B_{T/2}$. The proof is achieved.

3.2 Proof II

Proof—The theorem is an easy consequence of theorem 2.2 and of the following remark: we write the principal symbol of P evalued at zero

$$p\,(0,\xi) = p_1\,(0,\xi)\,p_2\,(0,\xi)$$

where p_1 and p_2 are principal symbols of $L = X_1 X_2 + T_1$ and $Q = X_3 X_4 + T_2$ respectively, while T_1 and T_2 are first order operators; moreover from the assumption we have

$$X_1\,(0) = \partial_{x_1} + i\partial_{x_2}\quad,\quad X_2\,(0) = \partial_{x_1} - i\partial_{x_2}$$

and also

$$X_3\,(0) = \lambda\partial_{x_1} + i\mu\partial_{x_2}\quad,\quad X_4\,(0) = \lambda\partial_{x_1} - i\mu\partial_{x_2}\,;$$

hence

$$p\,(0,\xi) = (\xi_1 + i\xi_2)\,(\xi_1 - i\xi_2)\,(\lambda\xi_1 + i\mu\xi_2)\,(\lambda\xi_1 - i\mu\xi_2)\,,$$

that we can write as

$$p\,(0,\xi) = \left(\lambda\xi_1^2 + \mu\xi_2^2 + i\xi_1\xi_2\,(\lambda - \mu)\right)\left(\lambda\xi_1^2 + \mu\xi_2^2 - i\xi_1\xi_2\,(\lambda - \mu)\right)\,. \quad (31)$$

We note that both factors in (31) have as image a cone in $\mathbf{C}$ with angle 2ϕ where $\sin\phi = \dfrac{\lambda - \mu}{\lambda + \mu}$.

We consider, now, the operator $\mathcal{P} = X_2 X_3$; applying the preceeding theorem to $\mathcal{P}$, we can choose new coordinates such that $\mathcal{P}$ becomes $\widetilde{\mathcal{P}} = \widetilde{X}_2\widetilde{X}_3$, and for $\widetilde{X}_2$, $\widetilde{X}_3$ the relation (12) becomes $\nu > \dfrac{\lambda - \mu}{\mu}$.

Return now to the operator P. In the new coordinates P becomes

$$\widetilde{P} = \widetilde{L}\widetilde{Q} + \widetilde{a}$$

where $\widetilde{L} = \widetilde{X}_1\widetilde{X}_2 + \widetilde{T}_1$, while $\widetilde{Q} = \widetilde{X}_3\widetilde{X}_4 + \widetilde{T}_2$.

We note that for the vector fields $\widetilde{X}_1$ and $\widetilde{X}_4$ the relation (12) becomes

$$\nu > \frac{\lambda - \mu}{\mu}\,.$$

Taking into account (25), we can iterate the estimate (28) for operators $\widetilde{L}$ and $\widetilde{Q}$; from the remark 2.1, it follows that P has the strong unique continuation property if s is as in the theorem.

References

[1] S.ALINHAC: *Non-unicité pour des opérateurs differentiels à caractéristiques complexes simples*, Ann. Sci. École Norm. Sup. **13** (1980), 385-393.

[2] S.ALINHAC, M.S.BAOUENDI: *Uniqueness for the characteristic Cauchy problem and strong unique continuation for higher order partial differential inequalities*, Amer. J. Math. **102** (1980), 179-217.

[3] S.ALINHAC, M.S.BAOUENDI: *A counterexample to strong uniqueness for partial differential equations of Schrödinger's type*, Comm. Partial Differential Equations **19** (1994), 1727-1733.

[4] N.ARONSZAJN, A.KRZYWICKI, J.SZARSKI: *A unique continuation theorem for exterior differential forms on Riemannian manifolds*, Ark.Mat. **4** (1962), 417-453.

[5] B.BARCELÓ, C.E.KENIG, A.RUIZ, C.D.SOGGE: *Weighted Sobolev inequalities for the Laplacian plus lower order terms*, Illinois J. Math. **32** (1988), 230-245.

[6] F.COLOMBINI, C.GRAMMATICO: *Some remarks on strong unique continuation for the Laplace operator and its powers*, Comm. Partial Differential Equations **24** (1999), 1079-1094.

[7] F.COLOMBINI, C.GRAMMATICO: *A counterexample to strong uniqueness for all powers of the Laplace operator*, Comm. Partial Differential Equations **25** (2000), 585-600.

[8] F.COLOMBINI, C.GRAMMATICO: *Strong uniqueness for Laplace and bi-Laplace operators in limit case*, in "Carleman estimates and applications to uniqueness and control theory" (F. Colombini and C. Zuily editors), Progress in Nonlinear Differential Equations and their Applications, Birkhäuser, Boston, **46** (2001), 49-60.

[9] C.GRAMMATICO: *A result on strong unique continuation for the Laplace operator*, Comm. Partial Differential Equations **22** (1997), 1475-1491.

[10] L.HÖRMANDER: *Uniqueness theorems for second order elliptic differential equations*, Comm. Partial Differential Equations **8** (1983), 21-64.

[11] D.JERISON, C.E.KENIG: *Unique continuation and absence of positive eigenvalues for Schrödinger operator*, Ann. of Math. **121** (1985), 463-494.

[12] P.Le Borgne: *Unicité forte pour le produit de deux opérateurs elliptiques d'ordre 2*, Indiana Univ. Math. J. **50** (2001).

[13] N.Lerner: *Résultats d'unicité forte pour des opérateurs elliptiques à coefficients Gevrey*, Comm. Partial Differential Equations **6** (1981), 1163–1177.

[14] Y.Pan: *Unique continuation for Schrödinger operators with singular potentials*, Comm. Partial Differential Equations **17** (1992), 953-965.

[15] R.Regbaoui: *Prolongement unique pour les opérateurs de Schrödinger*, Thèse, Université de Rennes, (1993).

[16] R.Regbaoui: *Strong unique continuation for second order elliptic differential operators*, J. Differential Equations **141** (1997), 201-217.

[17] T.Wolff: *A counterexample in a unique continuation problem*, Comm. Anal. Geom. **2** (1994), 79-102.

A remark on nonuniqueness in the Cauchy problem for elliptic operator having non-Lipschitz coefficients

DANIELE DEL SANTO Dipartimento di Scienze Matematiche, Università di Trieste, Via A. Valerio 12/1, 34127 Trieste, Italy

We are concerned here with the non-uniqueness of the solutions to the Cauchy problem for second order elliptic operators with real principal part.

It is well known that if the coefficients of the principal part of a second order elliptic operator are real and Lipschitz-continuous then a local uniqueness result holds for the solutions to the Cauchy problem belonging to $\mathcal{C}^2$ (see e. g. [1] or [2, Ch. XVII]).

On the other hand a classical example due to Pliś shows that the Lipschitz-continuity assumption (on the coefficients of the principal part) cannot be very much weakened, since the uniqueness property is not verified by the solutions of a particular second order elliptic operator which coefficients of the principal part are real and Hölder-continuous of exponent α, for all $\alpha < 1$ (see [3]).

Recently a sharp result on this subject has been obtained by Tarama (see [4]). In order to recall this result we introduce some notations. Let μ be a modulus of continuity, i. e. μ is a real valued continuous function defined on $[0,1]$, concave, strictly increasing and such that $\mu(0) = 0$. Let f be a complex valued function defined on Ω, open set of $\mathbf{R}^n$; f is μ-continuous (we will write $f \in \mathcal{C}^\mu(\Omega)$) if for each K compact set in Ω there exists $\varepsilon > 0$ such that

$$\sup_{x,y \in K,\ 0<|x-y|<\varepsilon} \frac{|f(x) - f(y)|}{\mu(|x - y|)} < +\infty.$$

It is easily seen that if for all $0 < \alpha < 1$,

$$\lim_{s \to 0^+} \frac{\mu(s)}{s^\alpha} = 0 \tag{1}$$

then $\mathcal{C}^{0,1}(\Omega) \subseteq \mathcal{C}^\mu(\Omega) \subseteq \bigcap_{\alpha<1} \mathcal{C}^{0,\alpha}(\Omega)$ where $\mathcal{C}^{0,1}(\Omega)$ and $\mathcal{C}^{0,\alpha}(\Omega)$ denote the space of Lipschitz-continuous and Hölder-continuous functions of exponent α respectively.

THEOREM 1 ([4, Th. 1.1]) Let μ be a modulus of continuity satisfying (1). Suppose that

$$\int_0^1 \frac{1}{\mu(s)}\, ds = +\infty. \tag{2}$$

Let Ω be an open neighborhood of a point z_0 in $\mathbf{R}^n$; let φ be a $\mathcal{C}^2$ real valued function on Ω such that $\nabla\varphi \neq 0$ on Ω and $\varphi(z_0) = 0$. Let P be a second order elliptic operator defined on Ω such that the coefficients of the principal part are real and μ-continuous and the other coefficients are complex valued and bounded.

Then P has the following local uniqueness property: there exits an open neighborhood Ω' of z_0 such that if $u \in \mathcal{C}^2(\Omega)$, $u = 0$ on $\{z \in \Omega : \varphi(z) \leq 0\}$ and $Pu = 0$ on Ω, then $u = 0$ on Ω'.

In the present note we prove that the condition (2) is necessary to the uniqueness in the Cauchy problem. This will be shown by a construction of a non-uniqueness example which is actually a slight modification of Pliś' one.

THEOREM 2 Let μ be a modulus of continuity satisfying (1). Suppose that

$$\int_0^1 \frac{1}{\mu(s)}\, ds < +\infty. \tag{3}$$

There exists a function $a \in \mathcal{C}^\mu(\mathbf{R})$ and there exist four functions $u, b_1, b_2, c \in \mathcal{C}^\infty(\mathbf{R}^3)$ such that $1/2 \leq a(t) \leq 3/2$ for all $t \in \mathbf{R}$, $\mathrm{supp}\,(u) = \{(t, x, y) \in \mathbf{R}^3 : t \geq 0\}$ and

$$\partial_t^2 u + \partial_x^2 u + a\partial_y^2 u + b_1 \partial_x u + b_2 \partial_y u + cu = 0 \quad \text{on } \mathbf{R}^3. \tag{4}$$

Proof: we follow very closely the proof of the Theorem 1 in [3], outlining the main points. We consider four $\mathcal{C}^\infty$ functions A, B, C and J defined on $\mathbf{R}$ such that

$$A(\tau) = 1 \quad \text{for } \tau \geq 2/3, \quad A(\tau) = 0 \quad \text{for } \tau \leq 1/2,$$

$$B(\tau) = 0 \quad \text{for } \tau \leq 0 \text{ or } \tau \geq 1, \quad B(\tau) = 1 \quad \text{for } 1/6 \leq \tau \leq 5/6,$$

$$C(\tau) = 0 \quad \text{for } \tau \geq 1/2, \quad C(\tau) = 1 \quad \text{for } \tau \leq 1/3,$$

$$J(\tau) = -2 \quad \text{for } \tau \geq 1/6 \text{ or } \tau \geq 5/6, \quad J(\tau) = 2 \quad \text{for } 1/3 \leq \tau \leq 2/3.$$

We introduce two sequences of positive real numbers $\{r_n\}, \{z_n\}$ such that

$$\sum_{n=1}^{+\infty} r_n < +\infty, \tag{5}$$

$$\{z_n\} \quad \text{is strictly increasing} \quad \text{and} \quad z_n > 1 \quad \text{for all } n \in \mathbf{N}. \tag{6}$$

Next we set

$$a_n = \sum_{k=1}^n r_k,$$

$$q_1 = 0 \quad \text{and} \quad q_n = \sum_{k=2}^{n} z_k r_{k-1},$$

$$p_n = (z_{n+1} - z_n)r_n,$$

and we suppose that

$$p_n > 1 \quad \text{for all } n \in \mathbf{N}. \tag{7}$$

We set finally

$$A_n(t) = A((t - a_n)/r_n), \quad B_n(t) = B((t - a_n)/r_n),$$

$$C_n(t) = C((t - a_n)/r_n), \quad J_n(t) = J((t - a_n)/r_n),$$

and

$$v_n(t, x) = \exp(-q_n - z_n(t - a_n)) \cos z_n x,$$

$$w_n(t, y) = \exp(-q_n - z_n(t - a_n) + J_n(t)p_n) \cos z_n y.$$

We define

$$u(t, x, y)$$

$$= \begin{cases} v_1(t, x) & \text{for } t \geq a_1, \\ A_n(t)v_n(t, x) + B_n(t)w_n(t, y) + C_n(t)v_{n+1}(t, x) & \text{for } a_{n+1} \leq t \leq a_n, \\ 0 & \text{for } t \leq 0. \end{cases}$$

It is possible to verify that supposing for all $K_1, \ldots, K_4 > 0$

$$\lim_{n \to +\infty} \exp(-q_n + K_1 p_n) z_{n+1}^{K_2} p_n^{K_3} r_n^{-K_4} = 0 \tag{8}$$

then the function u is in $\mathcal{C}^\infty$ (see [3, p. 97]). Moreover supp $(u) = \{(t, x, y) \in \mathbf{R}^3 : t \geq 0\}$.

We set

$$a(t) = \begin{cases} ((J_n'(t)p_n - z_n)^2 - J_n''(t)p_n)z_n^{-2} & \text{for } a_{n+1} \leq t \leq a_n, \\ 1 & \text{for } t \leq 0 \text{ or } t \geq a_1. \end{cases}$$

Since

$$a(t) = 1 - 2J_n'(t)p_n z_n^{-1} + (J_n'(t))^2 p_n^2 z_n^{-2} - J_n''(t)p_n z_n^{-2},$$

there exists a constant $C > 0$, depending only on the $\mathcal{L}^\infty$ norms of J and its first and second derivatives, such that the condition

$$\sup_{n \in \mathbf{N}} C(r_n^{-1} p_n z_n^{-1} + r_n^{-2} p_n^2 z_n^{-2} + r_n^{-2} p_n z_n^{-2}) < 1/2 \tag{9}$$

implies that $a(t) \in [1/2, 3/2]$ for all $t \in \mathbf{R}$. Moreover, for all $t \in [a_{n+1}, a_n]$,

$$|a'(t)| \leq C(r_n^{-2} p_n z_n^{-1} + r_n^{-3} p_n^2 z_n^{-2} + r_n^{-3} p_n z_n^{-2}), \tag{10}$$

where again C depends only on J and its derivatives. Since $p_n > 1$ for all $n \in \mathbf{N}$, we deduce from (10) that

$$|a'(t)| \leq C(r_n^{-2} p_n z_n^{-1} + r_n^{-3} p_n^2 z_n^{-2})$$

for all $t \in [a_{n+1}, a_n]$. From the concavity of μ we have that the function $\sigma \mapsto \sigma/\mu(\sigma)$ is increasing and consequently

$$\sup_{t,s\in[a_{n+1},a_n],\, t\neq s} \frac{|a(t) - a(s)|}{\mu(|t - s|)} \leq C \frac{r_n^{-1} p_n z_n^{-1} + r_n^{-2} p_n^2 z_n^{-2}}{\mu(r_n)}.$$

Then the μ-continuity of a will be implied by

$$\sup_{n\in\mathbf{N}} \frac{r_n^{-2} p_n z_n^{-1} + r_n^{-3} p_n^2 z_n^{-2}}{\mu(r_n)} < +\infty. \tag{11}$$

We finally define

$$b_1 = -\frac{(\partial_t^2 u + \partial_x^2 u + a\partial_y^2 u)\partial_x u}{u^2 + (\partial_x u)^2 + (\partial_y u)^2}, \qquad b_2 = -\frac{(\partial_t^2 u + \partial_x^2 u + a\partial_y^2 u)\partial_y u}{u^2 + (\partial_x u)^2 + (\partial_y u)^2}$$

and

$$c = -\frac{(\partial_t^2 u + \partial_x^2 u + a\partial_y^2 u)u}{u^2 + (\partial_x u)^2 + (\partial_y u)^2}.$$

If for all K_1, K_2 and $K_3 > 0$

$$\lim_{n\to+\infty} \exp(-p_n) z_{n+1}^{K_1} p_{n+1}^{K_2} r_{n+1}^{-K_3} = 0 \tag{12}$$

then the functions b_1, b_2 and c are in $\mathcal{C}^\infty$ and (4) holds. We refer to the cited paper of Pliś for the details. We choose

$$r_n = ((n + k)^2 \mu(1/(n + k)))^{-1}, \qquad z_n = (n + k)^3.$$

From (1) and (2) the conditions (5), (6), (7), (8), (9), (11) and (12) are verified if the integer k is sufficiently large. The proof is complete.

REFERENCES

1. L. Hörmander, On the uniqueness of Cauchy problem II, *Math. Scand.*, 7: 177–190 (1959).

2. L. Hörmander, "The Analysis of Linear Partial Differential Operators, III", Springer–Verlag, Berlin (1985).

3. A. Pliś, On non-uniqueness in Cauchy problem for an elliptic second order differential equation, *Bull. Acad. Pol. Sci.*, 11: 95–100 (1963).

4. S. Tarama, Local uniqueness in the Cauchy problem for second order elliptic equations with non–Lipschitzian coefficients, *Publ. Res. Inst. Math. Sci.*, 33: 167–188 (1997).

Sur le prolongement analytique de la solution du problème de Cauchy

YÛSAKU HAMADA

61-36 Tatekura-cho, Shimogamo, Sakyo-Ku, Kyoto, 606-0806, Japon

Dédié à Jean Vaillant

Résumé.

J. Leray et L. Gårding, T. Kotake et J. Leray ont étudié les singularités et des prolongements analytiques de la solution du problème de Cauchy dans le domaine complexe.

Nous étudions le prolongement analytique de la solution du problème de Cauchy pour l'opérateur différentiel à coefficients de fonctions entières.

Dans cet exposé, nous donnons une remarque sur le domaine d'holomorphie de la solution du problème de Cauchy pour certains opérateurs différentiels à coefficients polynomiaux. Ceci concerne les équations différentielles ordinaires de Darboux-Halphen, de J. Chazy et de la fonction modulaire.

1. INTRODUCTION ET RESULTATS

J. Leray [L] et L. Gårding, T. Kotake et J. Leray [GKL] ont étudié les singularités et un prolongement analytique de la solution du problème de Cauchy dans le domaine complexe.

[P], [PW] et [HLT] ont étudié des prolongements analytiques de la solution du problème de Cauchy pour l'opérateur différentiel à coefficients de fonctions entières ou polynomiaux.

Soit $x = (x_0, x') [x' = (x_1, \cdots, x_n)]$ un point de $\mathbf{C}^{n+1}$. On considère $a(x, D)$ un opérateur différentiel d'odre m, à coefficients de fonctions entières sur $\mathbf{C}^{n+1}$:
$$a(x, D) = \sum_{|\alpha| \leq m} a_\alpha(x) D^\alpha, D_i = \partial/\partial x_i, 0 \leq i \leq n.$$
Sa partie principale est noté $g(x, D)$. Nous supposons que
$$g(x; 1, 0, \cdots, 0) \equiv 1.$$

Soit S l'hyperplan $x_0 = 0$, non caractéristique pour g.

Étudions le problème de Cauchy

(1.1) $a(x, D)u(x) = v(x), D_0^h u(0, x') = w_h(x'), 0 \le h \le m - 1,$

où $v(x), w_h(x')$ sont des fonctions entières sur $\mathbf{C}^{n+1}$ et $\mathbf{C}^n$ respectivement.
D'après le théorème de Cauchy-Kowalewski, il existe une unique solution
holomorphe au voisinage de S dans $\mathbf{C}^{n+1}$. Jusqu'où est-ce que cette
solution locale peut se prolonger analytiquement? En général, il se passe
des diverses phénomènes compliquées.

Dans [H1], en appliquant un résultat de L. Bieberbach et P. Fatou, nous
avons construit un exemple tel que le domaine d'holomorphie de la solution
admet un point extérieur pour l'opérateur à coefficients de fonctions
entières. Dans [H2], nous avons donné un exemple tel qu' en disant
rudement, le domaine d'holomorphie de la solution ramifiée admet un point
extérieur pour un opérateur différentiel à coefficients polynomiaux. Dans
cet exposé, nous donnons un complément de [H2].

D'abord, pour l'expliquer, nous rappelons un résultat de [HLT].

THÉORÈME [HLT]. *Dans le problème* (1.1), *supposons que*
$g(x, D) = D_0^m + \sum_{k=1}^{m} L_k(x, D_{x'})D_0^{m-k}$, *où* $L_k(x, D_{x'}), 1 \le k \le m$, *est d'ordre*
k *en* $D_{x'}$ *et un polynôme en* x' *de degré* μk, μ *étant un entier* ≥ 0. *Alors il*
existe une constante $C(0 < C \le 1)$ *ne dépendant que de* $M(R)$ *telle que le*
problème (1.1) *possède une unique solution holomorphe sur*
$\left\{ x \in \mathbf{C}^{n+1}; | x_0 | \le C \min \left[(1+ \| x' \|)^{-\max(\mu-1,0)}, R \right] \right\}$,
où M(R) *est le module sur* $\{x_0; | x_0 | \le R\}$ *des coefficients des polynômes en*
x' *dans* $g(x, D)$ *et* $\| x' \| = \max_{1 \le i \le n} | x_i |$.

Donc au cas où $\mu = 0, 1$, la solution est une fonction entière sur $\mathbf{C}^{n+1}$. Ceci
a été déjà démontré dans [P], [PW] et [HLT].

Dans cet exposé, nous donnons des exemples tels que le domaine
d'holomorphie de la solution soit univalent et admette un point extérieur
pour l'opérateur différentiel à coefficients polynomiaux.

En fait, J. Chazy [C] a étudié les équations différentielles ordinaires du
troisème ordre et le système d'équations de Darboux-Halphen. (Aussi voir
[AF]). Nous employons ces résultats.

Considérons les problèmes de Cauchy

(1.2) $\quad \{D_0 + \sum_{i=1}^{3} H_i(x')D_i\}U_{1,j}(x) = 0,\ U_{1,j}(0, x') = x_j,\ 1 \le j \le 3,$

$\qquad [x = (x_0, x'), x' = (x_1, x_2, x_3)]$

où

$$H_1(x') = \frac{1}{2}[(x_2 + x_3)x_1 - x_2 x_3],$$

(1.3) $\qquad H_2(x') = \frac{1}{2}[(x_1 + x_3)x_2 - x_1 x_3],$

$$H_3(x') = \frac{1}{2}[(x_1 + x_2)x_3 - x_1 x_2].$$

Ceci concerne le système d'équations différentielles ordinaires de Darboux-Halphen ([C], [AF]).

Considérons les problèmes de Cauchy

(1.4) $\quad \{D_0 + x_2 D_1 + x_3 D_2 + (2x_1 x_3 - 3x_2^2)D_3\}U_{2,j}(x) = 0,$

$\qquad U_{2,j}(0, x') = x_j,\ 1 \le j \le 3.$

Ceci concerne l'équation différentielle ordinaire de Chazy. ([C], [AF]).

J. Leray [L] et L. Gårding, T. Kotake et J. Leray [GKL] ont étudié le problème de Cauchy, lorsque la surface initiale a des points caractéristiques. Dans [H2], nous avons étudié un cas exceptionel de [L] et [GKL]. Nous complétons ici des résultats de [H2].

Considérons les problèmes de Cauchy

(1.5) $\quad \{\sum_{i=0}^{3} A_i(x')D_i\}U_{3,j}(x) = 0,\ U_{3,j}(0, x') = x_j,\ 1 \le j \le 3,$

$\qquad [x = (x_0, x'), x' = (x_1, x_2, x_3)]$

où

$$A_0(x') = 2x_1^2(1 - x_1)^2 x_2,$$

(1.6) $\qquad A_1(x') = A_0(x')x_2,$

$$A_2(x') = A_0(x')x_3,$$

$$A_3(x') = 3x_1^2(1 - x_1)^2 x_3^2 - (1 - x_1 + x_1^2)x_2^4.$$

Ceci concerne l'équation différentielle ordinaire de fonction modulaire. ([C], [Hi], [AF]).

Nous avons alors

PROPOSITION 1.1. *Les domaines d'holomorphie* $\mathcal{D}_i, 1 \le i, j \le 3$, *des solutions* $U_{i,j}(x), 1 \le i, j \le 3$, *des problèmes* (1.2), (1.4) *et* (1.5) *sont des domaines univalents dans* $\boldsymbol{C}^4$. *Ils admettent un point extérieur dans* $\boldsymbol{C}^4$.

En utilisant une application birationnelle et une application algébrique, les problèmes (1.2) et (1.4) se transforment en les problèmes (1.5) avec les données transformées.

2. ESQUISSE DE LA PREUVE DE LA PROPOSITION 1.1

La fonction modulaire $w = \lambda(z)$ est holomorphe sur $\{z; \Im z > 0\}$ et son inverse $z = \nu(w)$ est holomorphe sur le revêtement universel $\mathcal{R}[C \setminus \{0,1\}]$ du domaine $C \setminus \{0,1\}$. $\lambda(z)$ a la frontière naturelle $\{z; \Im z = 0\}$. $W = \lambda\left(\frac{at+b}{ct+d}\right)$, a, b, c, d étant des constantes, $ad - bc = 1$, satisfait l'équation

$\{W; t\} = -R(W)\left(\frac{dW}{dt}\right)^2$, où $\{W; t\}$ est la dérivée de Schwarz:

$$\{W; t\} = \frac{d}{dt}\left(\frac{d^2W}{dt^2} \bigg/ \frac{dW}{dt}\right) - \frac{1}{2}\left(\frac{d^2W}{dt^2} \bigg/ \frac{dW}{dt}\right)^2,$$

$R(W) = (1 - W + W^2)/2W^2(1-W)^2$. ([Hi])

Les $x_1 = W, x_2 = dW/dt, x_3 = d^2W/dt^2$ satisfont donc

$$\frac{dx_1}{dt} = x_2, \quad \frac{dx_2}{dt} = x_3, \quad \frac{dx_3}{dt} = \frac{3x_3^2}{2x_2} - \frac{(1 - x_1 + x_1^2)x_2^3}{2x_1^2(1-x_1)^2}.$$

Avec les données $x'(0) = y'$, on a alors

$$x_1 = \lambda\left(\frac{a(y')t + b(y')}{c(y')t + d(y')}\right),$$

où les fonctions

$$b(y') = \nu(y_1)d(y'), \quad d(y') = \frac{\lambda'(\nu(y_1))^{1/2}}{y_2^{1/2}},$$

$$c(y') = \frac{\lambda''(\nu(y_1))y_2^{1/2}}{2\lambda'(\nu(y_1))^{3/2}} - \frac{\lambda'(\nu(y_1))^{1/2}y_3}{2y_2^{3/2}},$$

$$a(y') = \frac{1 + b(y')c(y')}{d(y')},$$

sont holomorphes au voisinage d'un point $y'^{(0)} = (y_1^{(0)}, y_2^{(0)}, y_3^{(0)}) \in (\mathbf{C} \setminus \{0,1\}) \times (\mathbf{C} \setminus \{0\}) \times \mathbf{C}$ et elles se prolongent analytiquement sur $\mathcal{R}[(\mathbf{C} \setminus \{0,1\}) \times (\mathbf{C} \setminus \{0\})] \times \mathbf{C}$.

Les solutions $U_{3,j}(x), 1 \leq j \leq 3$, des problèmes (1.5) sont alors holomorphes au voisinage d'un point $(0, x'^{(0)})$ de $\{x; x_0 = 0, x' \in (\mathbf{C} \setminus \{0,1\}) \times (\mathbf{C} \setminus \{0\}) \times \mathbf{C}\}$ et nous obtenons

$$U_{3,1}(x) = \lambda\left(\frac{a(x')x_0 - b(x')}{c(x')x_o - d(x')}\right), U_{3,2}(x) = -D_0U_{3,1}(x), U_{3,3}(x) = -D_0U_{3,2}(x).$$

En observant une ramification de $\nu(w)$, on voit que
$Q(x') = \Im[a(x')\overline{c(x')}], M(x') = \overline{a(x')}d(x') - b(x')\overline{c(x')}$ et
$N(x') = \Im[a(x')\overline{d(x')}]$ sont des fonctions analytiques de variables réelles
$\Re x', \Im x'$ au point $x'^{(0)}$ et elles sont uniformes et analytiques en $\Re x', \Im x'$ sur
$(\mathbf{C} \setminus \{0, 1\}) \times (\mathbf{C} \setminus \{0\}) \times \mathbf{C}$.
$P(x') = iM(x')/2Q(x'), R(x') = 1/2 \, | \, Q(x') \, |$ sont uniformes et analytiques
en $\Re x', \Im x'$ sur $\{x' \in (\mathbf{C} \setminus \{0, 1\}) \times (\mathbf{C} \setminus \{0\}) \times \mathbf{C}; Q(x') \neq 0\}$.

Définissons le domaine suivant:

$$
\mathcal{D}_3 = \left\{
\begin{array}{l}
x = (x_0, x') \in \mathbf{C} \times (\mathbf{C} \setminus \{0, 1\}) \times (\mathbf{C} \setminus \{0\}) \times \mathbf{C}; \\
| \, x_0 - P(x') \, | > R(x') \text{ pour } Q(x') > 0, \\
| \, x_0 - P(x') \, | < R(x') \text{ pour } Q(x') < 0, \\
\Im[M(x')x_0] - N(x') < 0 \text{ pour } Q(x') = 0
\end{array}
\right\}.
$$

Le domaine $\mathcal{D}_3$ est univalent et il admet un point extérieur dans $\mathbf{C}^4$.
D'après le théorème de Cauchy-Kowalewski et les representations des
solutions, en employant un technique des prolongements analytiques dans
[HLT] (Proposition 7.1 dans [HLT]), on voit que les domaines
d'holomorphie de $U_{3,j}(x), 1 \leq j \leq 3$, sont $\mathcal{D}_3$. Ceci prouve la Proposition
1.1 pour $U_{3,j}(x), 1 \leq j \leq 3$.

Ensuite, nous étudions $U_{1,j}(x), 1 \leq j \leq 3$.

Considérons l'application birationnelle de
$\{x' = (x_1, x_2, x_3) \in \mathbf{C}^3; x_1 \neq x_2, x_1 \neq x_3, x_2 \neq x_3\}$ sur
$\{X' = (X_1, X_2, X_3) \in \mathbf{C}^3; X_1 \neq 0, 1, X_2 \neq 0\}$:

$$X_1 = X_1(x') = (x_1 - x_3)/(x_1 - x_2),$$
$$X_2 = X_2(x') = (x_2 - x_3)(x_1 - x_3)/(x_1 - x_2) = (x_2 - x_3)X_1(x'),$$
$$X_3 = X_3(x') = (x_1 + x_2 - x_3)(x_2 - x_3)(x_1 - x_3)/(x_1 - x_2)$$
$$= (x_1 + x_2 - x_3)X_2(x'),$$

et donc nous avons

$$x_1 = x_1(X') = \frac{X_3}{X_2} - \frac{X_2}{X_1},$$

$$x_2 = x_2(X') = \frac{X_3}{X_2} + \frac{X_2}{1 - X_1},$$

$$x_3 = x_3(X') = \frac{X_3}{X_2} + \frac{X_2}{1 - X_1} - \frac{X_2}{X_1}.$$

Par cette application, les problèmes de Cauchy (1.2) se transforment en les problèmes de Cauchy suivants.

$$\{\sum_{i=0}^{3} A_i(X')D_{Xi}\}\hat{U}_{3,j}(X) = 0, [X = (X_0, X'), X' = (X_1, X_2, X_3)]$$

avec les données

$$\hat{U}_{3,1}(0, X') = \frac{X_3}{X_2} - \frac{X_2}{X_1},$$

$$\hat{U}_{3,2}(0, X') = \frac{X_3}{X_2} + \frac{X_2}{1 - X_1},$$

$$\hat{U}_{3,3}(0, X') = \frac{X_3}{X_2} + \frac{X_2}{1 - X_1} - \frac{X_2}{X_1},$$

où

$$U_{1,j}(x) = \hat{U}_{3,j}(x_0, X'(x')), 1 \le j \le 3.$$

Nous avons alors

$$\hat{U}_{3,1}(X) = \frac{U_{3,3}(X)}{U_{3,2}(X)} - \frac{U_{3,2}(X)}{U_{3,1}(X)},$$

$$\hat{U}_{3,2}(X) = \frac{U_{3,3}(X)}{U_{3,2}(X)} + \frac{U_{3,2}(X)}{1 - U_{3,1}(X)},$$

$$\hat{U}_{3,3}(X) = \frac{U_{3,3}(X)}{U_{3,2}(X)} + \frac{U_{3,2}(X)}{1 - U_{3,1}(X)} - \frac{U_{3,2}(X)}{U_{3,1}(X)}.$$

Posons

$$\mathcal{E}_1 = \left\{ \begin{array}{c} x = (x_0, x') \in \mathbf{C}^4; x' = (x_1, x_2, x_3); x_1 \ne x_2, x_2 \ne x_3, x_3 \ne x_1, \\ X_0 = x_0, \ (X_0, X'(x')) \in \mathcal{D}_3 \end{array} \right\},$$

alors $U_{1,j}(x), 1 \le j \le 3$, sont holomorphes sur $\mathcal{E}_1$.

Désignons par $\mathcal{D}_1 = (\overline{\mathcal{E}_1})^{(\circ)}$ l' intereur de l'adhérance $\overline{\mathcal{E}_1}$ de $\mathcal{E}_1$. Nous obtenons alors $\mathcal{D}_1 \setminus \{x_k = x_l, 1 \le k < l \le 3\} = \mathcal{E}_1$. D'autre part, d'après le théorème de Cauchy-Kowalewsk, lesi $U_{1,j}(x), 1 \le j \le 3$, sont holomorphes au voisinage de $S \cap \{x_k = x_l, 1 \le k < l \le 3\}$. D'après le théorème de Hartogs, $U_{1,j}(x), 1 \le j \le 3$, sont holomorphes sur $\mathcal{D}_1$. Nous pouvons voir facilement que les domaines d'holomorphie de $U_{1,j}(x), 1 \le j \le 3$, sont $\mathcal{D}_1$. $\mathcal{D}_1$ est un domaine univalent et il admet un poin extérieur dans $\mathbf{C}^4$. Ceci prouve la Proposition 1.1 pour $U_{1,j}(x), 1 \le j \le 3$.

Finalement nous étudions les problèmes (1.4).

Considérons l'application de $\mathbf{C}^3$ sur $\mathbf{C}^3$:

$$x_1 = x_1(X') = X_1 + X_2 + X_3,$$

$$x_2 = x_2(X') = \frac{1}{2}(X_1 X_2 + X_2 X_3 + X_3 X_1),$$

$$x_3 = x_3(X') = \frac{3}{2} X_1 X_2 X_3.$$

Soit $X_j(x'), 1 \leq j \leq 3$, les déterminations de la fonction algébrique définie par

$$\tau^3 - x_1 \tau^2 + 2 x_2 \tau - \frac{2}{3} x_3 = 0,$$

à un point $x'^{(0)}$ de $\{x' = (x_1, x_2, x_3) \in \mathbf{C}^3; \Delta(x') \neq 0\}$, où $\Delta(x')$ est le discriminant de cette équation algébrique.

Les $X_j(x'), 1 \leq j \leq 3$, se prolongent analytiquement sur leur surfaces de Riemann $\mathcal{R}_\tau$, c'est - à - dire, le revêtement du domaine $\{x' = (x_1, x_2, x_3) \in \mathbf{C}^3; \Delta(x') \neq 0\}$.
$X_j = X_j(\tilde{x}'), \tilde{x}' \in \mathcal{R}_\tau, 1 \leq j \leq 3$, applique $\mathcal{R}_\tau$ sur
$\{(X_1, X_2, X_3) \in \mathbf{C}^3; X_1 \neq X_2, X_2 \neq X_3, X_3 \neq X_1\}$. L'application
$x_0 = X_0, x_j = x_j(X'), 1 \leq j \leq 3$, transforment les problèmes (1.4) en les problèmes suivants

$$\{D_{X_0} + \sum_{i=1}^{3} H_i(X') D_{X_i}\} \mathcal{U}_{1,j}(X) = 0, \ 1 \leq j \leq 3,$$

avec les données

$$\mathcal{U}_{1,1}(0, X') = X_1 + X_2 + X_3,$$

$$\mathcal{U}_{1,2}(0, X') = \frac{1}{2}(X_1 X_2 + X_2 X_3 + X_3 X_1),$$

$$\mathcal{U}_{1,3}(0, X') = \frac{3}{2} X_1 X_2 X_3.$$

Nous avons alors , au voisinage d'un point de $(0, x'^{(0)})$ de $\{x_0 = 0, x' \in \mathbf{C}^3; \Delta(x') \neq 0\}$,

$$U_{2,j}(x) = \mathcal{U}_{1,j}(x_0, X'(x')), 1 \leq j \leq 3.$$

Nous obtenons donc, au voisinage du point $(0, x'^{(0)})$,

$$U_{2,1}(x) = \sum_{j=1}^{3} U_{1,j}(x_0, X'(x')),$$

$$U_{2,2}(x) = \frac{1}{2}\{\sum_{1 \leq j < k \leq 3} U_{1,j}(x_0, X'(x'))U_{1,k}(x_0, X'(x'))\},$$

$$U_{2,3}(x) = \frac{3}{2}U_{1,1}(x_0, X'(x'))U_{1,2}(x_0, X'(x'))U_{1,3}(x_0, X'(x')).$$

Soient x' un point arbitraire de $\{x'; \Delta(x') \neq 0\}$ et γ un chemin dans $\{x'; \Delta(x') \neq 0\}$ d'origine fixé $x'^{(0)}$ à x'. Prolongeons analytiquement tout $(X_i(x'), X_j(x'), X_k(x'))$, $(1 \leq i, j, k \leq 3, i \neq j, j \neq k, k \neq i)$, le long de γ et définissons le domaine suivant:

$$\mathcal{E}_2 = \left\{ \begin{array}{l} x; \ \Delta(x') \neq 0, (x_0, X_i(x'), X_j(x'), X_k(x')) \in \mathcal{D}_1, \\ 1 \leq i, j, k \leq 3, i \neq j, j \neq k, k \neq i \end{array} \right\}.$$

Désignons par $\mathcal{D}_2 = (\overline{\mathcal{E}_2})^{(\circ)}$ l' intérieur d'adhérance de $\overline{\mathcal{E}_2}$ de $\mathcal{E}_2$. Nous avons $\mathcal{D}_2 \setminus \{x'; \Delta(x') = 0\} = \mathcal{E}_2$. Pour chaque point x de $\mathcal{D}_2 \cap \{x'; \Delta(x') = 0\}$, il existe alors un voisinage $W(x)$ in $\mathcal{D}_2$ tel que les fonctions $U_{2,j}, 1 \leq j \leq 3$, sont holomorphes, uniformes et bornées dans $W(x) \setminus \{x'; \Delta(x') = 0\}$, et donc d'après le théorème de Riemann, elles sont holomorphes sur $\mathcal{D}_2$. Bien entendu , comme dans $U_{1,j}, 1 \leq j \leq 3$, nous pouvons le démontrer aussi, en utilisant les théorèmes de Cauchy-Kowalewski et de Hartogs. Nous pouvons voir facilement que le domaine d'holomorphie de $U_{2,j}, 1 \leq j \leq 3$, sont $\mathcal{D}_2$. $\mathcal{D}_2$ est un domaine univalent et il admet un point extérieur dans $\mathbf{C}^4$.

Ceci prouve la Poposition 1.1 pour $U_{2,j}, 1 \leq j \leq 3$.

On trouvera les démonstrations détaillées des résultats dans [H3].

RÉFÉRENCES

[AF] M. J. Ablowitz and A. S. Fokas, Complex Variables: Introduction and Applications, Cambridge Texts in Applied Mathematics, Cambridge University Press, 1997.

[C] J. Chazy, Sur les équations différentielles du troisième ordre et d'ordre supérieur dont l'intégrale générale a ses points critiques fixes, Acta Math. 34 (1911), 317-385.

[GKL] L. Gårding, T. Kotake et J. Leray, Uniformisation et développement asymptotique de la solution du problème de Cauchy linéaire à données holomorphes; analogue avec la théorie des ondes asymptotiques et approchées, Bull. Soc. Math. France 92 (1964). 263-361.

[G] R. C. Gunning, Introduction to Holomorphic Functions of Several Variables, Vol. I, Wadsworth & Brooks/Cole, 1991.

[HLT] Y. Hamada, J. Leray et A. Takeuchi, Prolongements analytiques de la solution du problème de Cauchy linéaire, J. Math. Pures Appl. 64 (1985), 257-319.

[H1] Y. Hamada, Une remarque sur le domaine d'existence de la solution du problème de Cauchy pour l'opérateur différentiel à coefficients des fonctions entières, Tôhoku Math. J. 50 (1998), 133-138.

[H2] Y. Hamada, Une remarque sur le problème de Cauchy pour l'opérateur différentiel de partie principale à coefficients polynomiaux, Tôhoku Math. J. 52 (2000), 79-94.

[H3] Y. Hamada, Une remarque sur le problème de Cauchy pour l'opérateur différentiel de partie principale à coefficients polynomiaux II, à paraître au Tôhoku Math. J.

[Hi] E. Hille, Ordinary Differential Equations in the Complex Domain, John Wiley,1976.

[L] J. Leray, Uniformisation de la solution du problème linéaire analytique de Cauchy près de la variété qui porte les données de Cauchy (Problème de Cauchy I), Bull. Soc. Math. France 85 (1957) 389-429.

[N] T. Nishino, Theory of Functions of Several Complex Variables [Tahensu Kansu Ron] (en japonais), Univ. of Tokyo Press, 1996.

[P] J. Persson, On the local and global non-characteristic Cauchy problem when the solutions are holomorphic functions or analytic functionals in the space variables, Ark. Mat. 9 (1971), 171-180.

[PW] P. Pongérard et C. Wagschal, J. Problème de Cauchy dans des espaces de fonctions entières, J. Math. Pures Appl. 75 (1996), 409-418.

On the projective descriptions
of the space of holomorphic germs

P. LAUBIN, University of Liège, Institute of Mathematics (B37), 4000 Liège, Belgium

Summary

A natural topology on the set $\mathcal{O}(S)$ of holomorphic germs near a subset S of a Fréchet space E is the locally convex inductive limit topology of the spaces $\mathcal{O}(\Omega)$ endowed with the compact open topology; here Ω runs over all open subset of E containing S. We discuss some known results concerning the projective characterization of this generally uncountable inductive limit. In particular, when S is a compact subset, Mujica gave a description of $\mathcal{O}(S)$ as the inductive limit of a suitable sequence of compact subsets. He used for this a set of intricate semi-norms. We give a projective characterization of this space using simpler semi-norms whose form is similar to the one used in the Whitney extension theorem for C_∞ functions. They are quite natural in a framework where extensions are involved. We also give a simple proof that this topology is strictly stronger than the topology of the projective limit of the non quasi-analytic spaces

Introduction

In this paper, we present some known results and some new advances concerning the topology of spaces of holomorphic functions in an open subset of a Fréchet space. The main problems which we will considered concern projective descriptions of spaces which appear as inductive limits.

We restrict ourselves to the compact open topology but give a projective characterization of holomorphic germs using norms whose form is similar to the one used in the Whitney extension theorem for C_∞ functions. They are quite natural in a framework where extension properties are involved.

1 NOTATIONS AND DEFINITIONS

It is interesting to consider the problem in a general Fréchet space. Let E be a complex Fréchet space and Ω an open subset of E. There are many equivalent definitions of holomorphy in this framework [1]. Let us use the following one.

DEFINITION 1 *A function $f : \Omega \mapsto \mathbb{C}$ is holomorphic if it is continuous and the complex valued of one complex variable $\mathbb{C} \ni z \mapsto f(x + zh)$ is holomorphic where it is defined for every fixed $x, h \in E$.*

The space of holomorphic functions in Ω, is denoted $\mathcal{O}(\Omega)$. If the E is finite dimensional, the assumption of continuity can be removed. In general, it can be replaced by the local boundedness.

Let us remind some basic facts concerning polynomials in a Fréchet space. A homogeneous polynomial p of degree k is a function

$$p : E \to \mathbb{C} : h \mapsto p(h) = u.h^{(k)} = u(h, \ldots, h)$$

where

$$u : E \times \ldots \times E \to \mathbb{C} : (h_1, , \ldots, h_k) \mapsto u(h_1, , \ldots, h_k)$$

is a k-linear symmetric mapping. It is said continuous if u is continuous. If $j \leq k$, we denote by $u.h^{(j)}$ the $(k - j)$-linear form on E defined by

$$u.h^{(j)}.(v, \ldots, v) = u.(h, \ldots, h, v, \ldots, v)$$

with j copies of h and $k - j$ copies of v. If $k = j$, this is just an element of $\mathbb{C}$.

If $A \subset E$ is bounded, let

$$\|p\|_A = \sup_{h \in A} |p(h)|.$$

It is well known [1] that if A is a bounded, convex and balanced subset of E then

$$\sup_{h_1, \ldots, h_k \in A} |u.(h_1, \ldots, h_k)| \leq \frac{k^k}{k!} \|p\|_A. \tag{1}$$

An holomorphic function has a local Taylor expansion

$$f(x + h) = \sum_{k=0}^{+\infty} \frac{1}{k!} D^k f(x).h^{(k)}$$

The symmetric polynomials $D^k f(x)$ are defined by

$$D^k f(x).(h, \ldots, h) = D_t^k [f(x + th)]_{|t=0}.$$

If $k \geq 0$, we also use the notation

$$\|D^k f\|_{K,A} = \sup_{x \in K} \sup_{h \in A} |D^k f(x).h^{(k)}|$$

if K is a compact subset of Ω and A is a bounded, convex and balanced subset of E.

We endow the space $\mathcal{O}(\Omega)$ of holomorphic functions in Ω with the usual compact open topology τ_0 defined by the semi-norms

$$p(f) = \sup_K |f|$$

where K runs over all compact subset of Ω. We restrict ourselves to this topology although some interesting problems also occurs for the τ_ω and τ_δ topologies.

Of course, Cauchy inequalities extend in this framework. If A is a convex and balanced subset of E such that $x + rA \subset \Omega$ then

$$\|D^k f(x)\|_A \leq \frac{k!}{r^k} \sup_{x+rA} |f|.$$

2 SOME KNOWN RESULTS

At least two natural inductive limits occur.

A) If $E = \mathbb{C}^n$ and S is an arbitrary subset of $\mathbb{R}^n$ we can consider

$$\mathcal{O}(S) = \text{ind}_{\Omega \supset S} \mathcal{O}(\Omega)$$

where Ω runs over all open subset of $\mathbb{C}^n$ containing the set S. This inductive limit defines on $\mathcal{O}(S)$ the strongest locally convex topology for which the inclusions $\mathcal{O}(\Omega) \to \mathcal{O}(S)$ are continuous. In general, this is a strongly uncountable limit. This is the space of germs of real analytic functions on S. It has be proved by Martineau [2] using abstract functional techniques that

$$\mathcal{O}(S) = \text{proj}_{K \subset S} \mathcal{O}(K).$$

Here the topology is the weakest locally convex one such that the inclusions $\mathcal{O}(\Omega) \to \mathcal{O}(S)$ are continuous.

For an open subset ω of $\mathbb{R}^n$, this can also be proved using the representation of the analytic functionals on $\mathcal{O}(\omega)$ by harmonic functions in $\mathbb{R} \setminus \{0\} \times \omega$ presented in [3].

B) If E is a general Fréchet space and K is any compact subset of E, we can also consider the inductive limit

$$\mathcal{O}(K) = \text{ind}_{\Omega \supset K} \mathcal{O}(\Omega).$$

where Ω runs over all open subset of E containing the set K. Mujica in [12] gave a projective description of this space using explicit semi-norms.

His result can be stated in the following way. The topology of $\mathcal{O}(K)$ is generated by semi-norms of the form

$$p(f) = \sum_{n=0}^{+\infty} \frac{\epsilon_n^n}{n!} \|D^n f\|_{K,M} \tag{2}$$

$$q(f) = \sup_k \sup_{1 \leq n \leq n_k} 2^n \left| \sum_{m=0}^n \frac{D^m f(x_k).a^{(k)}}{m!} - \sum_{m=0}^n \frac{D^m f(y_k).b^{(k)}}{m!} \right| \tag{3}$$

where

- M is any compact subset of E,

- $(\epsilon_n)_{\mathbb{N}}$ is any sequence of strictly positive numbers decreasing to 0,

- (x_k) and (y_k) are any sequences of K and (a_k), (b_k) are any null sequences such that

$$x_k + a_k = y_k + b_k$$

  for any k

- (n_k) is any sequence of positive integers.

The semi-norms of type (2) are the natural ones. It can easily been shown that the sum can be replaced by a sup.

The semi-norms of type (3) are formed to be able in the proof to glue together several Taylor expansions at points where some ambiguity between Taylor expansions can occur. In some way, they look two restrictive: they involve estimates on limited Taylor expansions approximating the function outside the compact set K. Such restrictions do not occur in Whitney extension theorem for the C^∞ framework.

Mujica also shows that for a locally connected compact subset K, the first type of semi-norms can be used alone. However if

$$K = \{1/m : m \in \mathbb{N}_0\} \cup \{0\} \subset \mathbb{C}$$

this is not the case. Indeed, consider the sequence $f_m \in \mathcal{O}\{z \in \mathbb{C} : \Re z \neq \frac{1}{m+1/2}\}$ with $f_m = 1$ if $\Re z < \frac{1}{m+1/2}$ and $f_m = 0$ if $\Re z > \frac{1}{m+1/2}$. It is bounded for the semi-norms of type (2) but not bounded in $\mathcal{O}(K)$.

The result of Mujica is also stated for a Riemann domain over a Fréchet space.

3 NEW RESULTS

3.1 Theorem

Consider the following semi-norms of Whitney type

$$\|f\|_{K,M,k} = \frac{\|D^k f\|_{K,M}}{k!}$$

$$+ \sup_{0 \le \ell < k, 0 < \rho < 1} \sup_{x, x+h \in K, h \in \rho M} \frac{1}{\ell! \rho^{k-\ell}} \Big\| D^\ell f(x+h) - \sum_{j < k-\ell} D^{j+\ell} f(x) . \frac{h^{(j)}}{j!} \Big\|_M.$$

Here $k \ge 0$ and M is any compact subset of E. These semi-norms are simpler than the ones used by Mujica and only constrain limited Taylor expansions between two points of K.

Note that $\|f\|_{K,M,k}$ is increasing with M and that $\|f\|_{K,\lambda M,k} = \lambda^k \|f\|_{K,M,k}$ if $\lambda > 0$. Using the Cauchy inequalities, one easily see that these semi-norms are continuous on $\mathcal{O}(K)$.

In a finite dimensional space, we can fix M equal to the closed unit ball and omit it in the notations. Then we obtain the semi-norms

$$\|f\|_{K,k} = \frac{\|D^k f\|_K}{k!} + \sup_{0 \le \ell < k} \sup_{x, x+h \in K} \frac{1}{\ell! \|h\|^{k-\ell}} \Big\| D^\ell f(x+h) - \sum_{j < k-\ell} D^{j+\ell} f(x) . \frac{h^{(j)}}{j!} \Big\|$$

with $\|D^k f\|_K = \sup_{x \in K} \sup_{\|h\| \leq 1} |D^k f(x).h^{(k)}|$.

The main result of [4] is the following projective description of $\mathcal{O}(K)$.

THEOREM 2 *If K is a compact subset of a Fréchet space E then the topology of $\mathcal{O}(K)$ is defined by the semi-norms*

$$p(f) = \sup_{k \in \mathbb{N}} \epsilon_k^k \|f\|_{K,M,k}$$

where M runs over all compact subsets of E and $(\epsilon_k)_{k \geq 0}$ over all sequences of real numbers decreasing to 0. Moreover, if K is locally connected then one can replace $\|f\|_{K,M,k}$ by $\|D^k f\|_{K,M}/k!$ in the definition of the semi-norms p.

An easy inspection of the proof of this theorem shows that in the definition of $\|f\|_{K,M,k}$, it is possible to replace $\ell!$ by any sequence a_ℓ satisfying $a_\ell \geq \ell!$. This is not a surprise since the control on the growth of the derivatives is already performed by the first part. However, the previous semi-norms are the natural ones since they correspond to the usual estimation of the error in the Taylor expansion of a holomorphic function.

It follows from the theorem 2 that, for any locally closed subset M of $\mathbb{R}^n$, the space $\mathcal{O}(M)$ is nuclear, complete, ultrabornological and a strictly webbed space. In fact, the first three properties are contained in the theorem 1.2 of [2]. Moreover, if K_j is a fundamental sequence of compact subsets of M, the sets

$$e_{n_1,\ldots,n_r} = \{f \in \mathcal{O}(M) : \|f\|_{K_j,k} \leq n_j^{k+1}, k \in \mathbb{N}, 1 \leq j \leq r\}$$

define a strict web of $\mathcal{O}(M)$. Using the projective description of theorem 2, this can be easily checked as in [5] where the case of an open subset of $\mathbb{R}^n$ is considered.

3.1 Another inductive limit

The next result shows that the set of sequences $(\epsilon_k)_{k \geq 0}$ cannot easily be relaxed.

Let $L = (L_k)_{k \in \mathbb{N}}$ be an increasing sequence of real positive numbers. If Ω is an open subset of $\mathbb{R}^n$, we denote by $C^{(L)}(\Omega)$ the set of all $f \in C_\infty(\Omega)$ such that for any compact subset K of Ω, we have

$$\|D^k f\|_K \leq A^{1+k} L_k^k$$

for every k and some $A > 0$. This space is endowed with the semi-norms

$$p(f) = \sup_{k \in \mathbb{N}} \epsilon_k^k \frac{\|D^k f\|_K}{L_k^k}$$

where K runs over all compact subsets of Ω and $(\epsilon_k)_{k \geq 0}$ over all sequences of real numbers decreasing to 0.

L or $C^{(L)}$ is said to be quasi-analytic if $u \in C^{(L)}(\Omega)$ and $D^\alpha u(x) = 0$ for every $\alpha \in \mathbb{N}^n$ and some $x \in \Omega$ imply $u = 0$. A classical theorem of Denjoy-Carleman says that $C^{(L)}(\Omega)$ is non quasi-analytic if and only if $\sum_{k=0}^{+\infty} 1/L_k < +\infty$. It is well known that, as a set,

$$\mathcal{O}(\Omega) = \bigcap_L C^{(L)}(\Omega)$$

where the intersection runs over all non quasi-analytic sequences L.

We have the following result.

THEOREM 3 *If Ω is a non void open subset of $\mathbb{R}^n$ then the topology of $\mathcal{O}(\Omega)$ is strictly stronger than the projective limit topology of the non quasi-analytic spaces $C^{(L)}(\Omega)$.*

Note that this projective limit topology is defined by the semi-norms

$$p(f) = \sup_{k \in \mathbb{N}} \eta_k^k \, \|D^k f\|_K$$

where K runs over all compact subsets of Ω and $(\eta_k)_{k \geq 0}$ over all sequences of real numbers decreasing to 0 and satisfying

$$\sum_{k=0}^{+\infty} \eta_k < +\infty.$$

This follows easily from the fact that, if the sequence η_k is summable, there is a sequence n_k converging to $+\infty$ such that $n_k \eta_k$ is still summable.

References

[1] S. Dineen, "Holomorphic germs on compact subsets of locally convex spaces", Functional analysis, holomorphy and approximation theory, Springer-Verlag Lectures Notes in Math., Vol. 843, 1981, 247-263.

[2] A. Martineau, "Sur la topologie des espaces de fonctions holomorphes", Math. Ann. 163, 62-88, 1966.

[3] L. Hörmander, "The analysis of linear partial differential operators I", Grundl. der math. Wiss. 256, Springer, 1983.

[4] P. Laubin, "A projective description of the space of holomorphic germs", Proc. of the Edinburgh Math. Soc., 2001, 44, 1-9.

[5] H.G. Garnir, M. De Wilde and J. Schmets, "Analyse fonctionnelle III, Espaces fonctionnels usuels", Math. Reihe 45, Birkhäuser, 1973.

[6] K.D. Bierstedt, "An introduction to locally convex inductive limits", in Functional Analysis and Its Applications, 35-133, World Scientific, 1988.

[7] S. Dineen, "Complex analysis in locally convex spaces", North-Holland Math. Studies, Vol. 57, 1981.

[8] L. Ehrenpreis, "Solution of some problems of division. Part IV", Amer. J. of Math., 82, 522-588, 1960.

[9] A. Hirschowitz, "Bornologie des espaces de fonctions analytiques en dimension infinie", Séminaire Pierre Lelong 1969/1970, Springer-Verlag Lectures Notes in Math., Vol. 205, 1970, 21-33.

[10] L. Hörmander, "An introduction to complex analysis in several variables", North-Holland Publ. Co., 1973.

[11] F. Mantovani and S. Spagnolo, " Funzionali analitici reali e funzioni armoniche", Ann. Sc. Norm. Sup. Pisa 18, 475-513, 1964.

[12] J. Mujica, "A Banach-Dieudonné theorem for germs of holomorphic functions", J. Funct. Anal. 57, 31-48, 1984.

[13] K. Rusek, "A new topology in the space of germs of holomorphic functions on a compact set in $\mathbb{C}^n$", Bull. Acad. Polon. Sci. 25, 1227-1232, 1977.

[14] P. Schapira, "Sur les ultra-distributions", Ann. Scient. Ec. Norm. Sup. 1, 395-415, 1968.

Microlocal scaling and extension
of distributions

M.K. Venkatesha Murthy

Dipartimento di Matematica Università di Pisa
Via Buonarroti 2 Pisa, Italy

We are concerned here with the following problem for extension of distributions defined in the complement of a submanifold to the whole manifold.

Suppose X is a paracompact C^∞ manifold and Y is a C^∞ submanifold of X. We assume the following standard geometric condition regarding the position of the singularity set of the distribution with respect to the submanifold Y. Given a distribution $u \in \mathcal{D}'(X \setminus Y)$ such that the closure of $WF(u)$ in $T^*(X)$ is orthogonal to the tangent bundle TY of Y, the question is concerned with determining conditions under which u admits an extension $\tilde{u} \in \mathcal{D}'(X)$. The extendability of a distribution naturally depends on its analytic behaviour as one approaches the submanifold Y.

The analytic behaviour of u near Y is expressed by the notion of *microlocal scaling degree*. We shall introduce the concept of *microlocal scaling degree* of a distribution along the submanifold Y, which classifies the class of extendible distributions on $X \setminus Y$. We obtain results analogous to those of Hörmander for homogeneous distributions defined on $\mathbb{R}^n \setminus 0$.

Scaling degree in the Euclidean space

We begin by recalling the notion of the scaling degree due to Steinman for distributions in the Euclidean space with respect to the origin. Roughly speaking the scaling degree of a distribution measures the strength of its singularity.

The $\mathbb{R}_+$ - action $(t, \varphi) \to \varphi^t$ defined by $\varphi^t(x) = t^{-n}\varphi(\frac{x}{t})$ on the space of test functions $\mathcal{D}(\mathbb{R}^n)$ induces by the adjoint action an $\mathbb{R}_+$ - action on the space of distributions $\mathcal{D}'(\mathbb{R}^n)$ defined by $\langle u_t, \varphi \rangle = \langle u, \varphi^t \rangle$.

Definition. – The scaling degree of a distribution $u \in \mathcal{D}'(\mathbb{R}^n)$ with respect to the origin 0 is given by

$$sd(u) = \inf\{k \in \mathbb{R}; \mathcal{D}' - \lim_{t \to 0} t^k u_t = 0\}$$

Thus $u \to sd(u)$ is a mapping on the space of distributions $\mathcal{D}'(\mathbb{R}^n)$ into $[-\infty, +\infty]$.

We shall need the following simple properties of $sd(u)$:

For C^∞ functions f we have $sd(f) \leq 0$

$sd(\partial_x^\alpha u) \leq sd(u) + |\alpha|$, for any multi-index α

$sd(x^\alpha u) \leq sd(u) - |\alpha|$, for any multi-index α

$sd(fu) \leq sd(u)$ for any C^∞ function f

$sd(u \otimes v) = sd(u) + sd(v)$

For the function $\exp(\frac{1}{x})$ which is not defined at the origin $sd(\exp(\frac{1}{x})) = +\infty$

For homogeneous distributions of degree l the scaling degree is $-l$

It is well known that all homogeneous distributions on $\mathbb{R}^n \setminus 0$ are not extendible (see Hörmander's book).

We have the following result on the extension of distributions from $\mathbb{R}^n \setminus 0$ to $\mathbb{R}^n$:

Theorem. – *(a) If a distribution $u \in \mathcal{D}'(\mathbb{R}^n \setminus 0)$ has scaling degree $sd(u) < n$ then there exists a unique $\tilde{u} \in \mathcal{D}'(\mathbb{R}^n)$ which extends u and has the same scaling degree.*

(b) If a distribution $u \in \mathcal{D}'(\mathbb{R}^n \setminus 0)$ has a finite scaling degree $sd(u) \geq n$ then there exists a (non unique) extension $\tilde{u} \in \mathcal{D}'(\mathbb{R}^n)$ having the same scaling degree and it is uniquely determined by its values on the subspace complementary to the space of test functions $\varphi \in \mathcal{D}(\mathbb{R}^n)$ which vanish at the origin upto order $[sd(u)] - n$.

Thus in the second case the extension is determined uniquely modulo a finite linear combination of derivatives of the Dirac distribution.

Definition. – The real number $h = sd(u) - n$ is called the degree of singularity of the distribution u.

If the degree of singularity is finite then the distribution can be extended to the whole space.

Micro-local scaling degree along a submanifold

In order to extend globally across a submanifold we need a microlocal version of the scaling degree which classifies the strength of the singularities of distributions defined in the complement of submanifolds of arbitrary codimension, which can be extended to the whole manifold. If Y is a submanifold of X and $u \in \mathcal{D}'(X \setminus Y)$ and if W is any submanifold of X which is transversal to Y then the restriction $u|_W$ of u to W is well defined. To classify the distributions which are extendible globally across Y we need a covariant extension of the scaling degree $sd(u)$ introduced above, namely the following concept of *microlocal scaling degree with respect to the conormal bundle $N^*Y \setminus 0$* or more genrally with respect to a conic subset Γ_0 of $N^*Y \setminus 0$.

First of all we recall the following standard facts on wave front sets of distributions (see Hörmanders book for details).

We denote by $\mathcal{D}'_\Gamma(X)$, for any closed conic set Γ of $T^*(X) \setminus 0$, the space of all distributions v on X whose wave front set $WF(v)$ is contained in Γ and $\mathcal{D}'_\Gamma(X)$ is provided with Hörmander's (pseudo -) topology:

A sequence u_j in $\mathcal{D}'_\Gamma(X)$ converges to u in $\mathcal{D}'_\Gamma(X)$ if

(i) $u_j \to u$ weakly in $\mathcal{D}'(X)$,

(ii) for any properly supported pseudo - differential operator A such that $\mu\mathrm{supp}\,(A) \cap \Gamma = \emptyset$, we have $Au_j \to Au$ in $C^\infty(X)$. (where $\mu\mathrm{supp}\,(A)$ is the projection of the wave front set $WF(k_A)$ of the Schwartz kernel k_A of A onto the second component.

(i.e. $\mu - supp(A) = \pi_2(WF(K_A))$, where $\pi_2 : T^*X \times T^*X \to T^*X$ is the projection onto to the second factor.)

We also need the following property on the restriction:

If M is a submanifold of X and $u \in \mathcal{D}'(X)$ are such that the wave front set $WF(u)$ does not intersect the conormal bundle N^*M of M then u admits a restriction $\gamma_M u$. Moreover, if Γ is any conic set such that $\Gamma \cap N^*M = \emptyset$, then the trace operator γ_M lifts to a sequentially continuous operator from $\mathcal{D}'_\Gamma(X)$ to $\mathcal{D}'(M)$.

Suppose X is a C^∞ manifold and Y is a C^∞ submanifold of X. Given a distribution $u \in \mathcal{D}'(X \setminus Y)$ such that the closure of $WF(u)$ in $T^*(X)$ is orthogonal to the tangent bundle TY of Y,

i.e. if $(y, \xi) \in WF(u)$ with $y \in Y$, $\xi \in T_y^* X$ then $\langle \xi, \vartheta \rangle = 0$ $\quad \forall \vartheta \in T_y Y$.

It follows, from the property of the trace operator recalled above, that under the hypothesis on u and Y, u admits a restriction to every submanifold C of X contained in a sufficiently small neighbourhood of Y in X intersecting Y transversally.

We may assume that X is provided with a Riemannian structure and Y is a totally geodesic submanifold of X. We take a star shaped neighbourhood of the zero section $Z(T_Y X)$ in TX and suppose that f denotes the local diffeomorphism $\mathcal{U} \to f(\mathcal{U}) \subset Y \times X$ defined by $f(x, \vartheta) = (x, exp_x \vartheta)$.

u^f denotes the distribution defined on $\mathcal{U}$ obtained as the pull back of $(1_Y \otimes u)$ by f and let $(u^f)_t(x, \vartheta) = u^f(x, t\vartheta)$ for $t \in \mathbb{R}_+$.

Consider a closed conic set Γ_0 in the conormal bundle of Y with the zero section removed, denoted by $N^*Y \setminus 0$.

Definition. – A distribution $u \in \mathcal{D}'(X)$ is said to have the microlocal scaling degree $\underline{k}$ along the submanifold Y with respect to the closed conic set Γ_0, denoted by $\mu sd_Y(u, ; \Gamma_0)$, if

(a) there exists a closed conic subset Γ of $T^*(T_Y X) \setminus 0$ with the properties

(i) $\Gamma_{TY} \perp f^*(Z(T^*Y) \times \Gamma_0)$ and

(ii) $WF((f^*(1_Y \otimes u))_t) \subset \Gamma$

(b) $\mu sd_Y(u; \Gamma_0) = \underline{k} = \inf\{k \in \mathbb{R}; \mathcal{D}'_\Gamma(T_Y X) - \lim_{t \to 0} t^k f^*(1_Y \otimes u)_t = 0\}$

in the sense of Hörmander's topology on $\mathcal{D}'_\Gamma(T_Y X)$.

Remark. – We may take for f in the above definition any local diffeomorphism satisfying certain obvious conditions:

$$\begin{cases} f(y, 0) = (y, y) & \text{for } y \in Y \\ f(TY \cap \mathcal{U}) \subset Y \times Y \\ f(y, \vartheta) \in \{y\} \times X & \text{for } y \in Y \text{ and} \\ d_\vartheta f(y, 0) = id_{T_y X} \end{cases}$$

Moreover,

$\mu sd(u, \Gamma_0)$ is independent of the choice of such an f.

This follows by making use of the stationary phase method.

The microlocal scaling degree has properties analogous to the ones satisfied by the scaling degree on the Eucleidean space

The other main properties of $\mu sd_Y(u; \Gamma_0)$ are the following:

1) The microlocal scaling degree $\mu sd(u)$ behaves well with respect to product of distributions when they are defined:

If $u \in \mathcal{D}'_\Gamma$ and $v \in \mathcal{D}'_\Sigma(X)$ are such that the zero section $Z(N^*Y)$ of the conormal bundle of Y, is not contained in $\Gamma \oplus \Sigma$ then the product $u.v$ is well defined in a neighbourhood of Y and

$$\begin{cases} \mu sd_Y(u.v, \Lambda) \le \mu sd_Y(u; \Gamma) + \mu sd_Y(v; \Sigma) \\ \text{where} \\ \Lambda = \Gamma \cup \Sigma \cup (\Gamma \oplus \Sigma) \end{cases}$$

2) Given a submanifold Y' of Y and the conic set $\Gamma_0 \subset N^*Y \setminus 0$ we have

$$\mu sd_{Y'}(u; \Gamma'_0) \le \mu sd_Y(u; \Gamma_0)$$

where $(\Gamma'_0 = \Gamma_0|_{Y'})$. This folllows using the sequential continuity of the restriction mapping.

For the extension property of distributions it is more convenient to make use of an equivalent concept of transversal microlocal scaling degree. For this purpose we decompose $T_Y X = TY \oplus E$ and the subbundle E defines in a neighbourhood of Y in X transversal submanifolds $\mathcal{C} = \{C_y; y \in Y\}$. Let $\mathcal{U}$ be a neighbourhood of the zero section $Z(T_Y X)$ in TX and $\pi_2 : Y \times X \to X$ be the second projection. Then $f_E = \pi_2 \circ f|_{E \cap \mathcal{U}}$ is a diffeomorphism onto a neighbourhood of Y in X (where $f(x, \vartheta) = (x, exp_x \vartheta)$). The images of the fibers of E define submanifolds transversal to Y at each of its points. Setting $(f_E^* u)(x, t\vartheta) = u_{t,\perp}(x, \vartheta)$ for $(x, \vartheta) \in E$ we define the transversal microscaling degree along Y by

$$\mu s d_{Y,\perp}(u, \Gamma_E) = \inf\{k \in \mathbb{R}; \lim_{t \to 0} t^k . u_{t,\perp} \doteq 0\}$$

the limit being taken in Hörmander's topology in $\mathcal{D}'_{\Gamma_E}$ where $\Gamma_E|_{Z(E)} = f_E^*(\Gamma_0)$ is a closed conic set in $T^*(E) \setminus 0$. (Here and else where in the following $Z(E)$ stands for the zero - section of the bundle E.)

As already mentioned

$$\mu s d_{Y,\perp}(u, \Gamma_E) = \mu s d_Y(u; \Gamma_0)$$

and this can be proved by using local frames and explicitly writing down the Fourier transforms.

Extension of distributions across submanifolds

Suppose $u \in \mathcal{D}'(X \setminus Y)$ is such that the closure in $T^*(X)$ of the wave front set $WF(u)$ is orthogonal to the tangent bundle TY. The microlocal scaling degree with respect to Y and the transversal microlocal scaling degrees of such distributions are defined in a natural way:

If $\zeta \in C^\infty(X)$ with $\operatorname{supp}\zeta \cap Y = \emptyset$ then $(f^*(1_Y \otimes \zeta) . f^*(u))_t$ is considered as a distribution in a neighbourhood of the zero section in $T_Y X$ and the micro local scaling and the micro-local transversal scaling degrees are defined as before using these distributions.

Choose a fibration $\mathcal{C}$ of the neighbourhood $\mathcal{U}$ of Y by transversal submanifolds C_y at $y \in Y$ defined by the subbundle E complementary to TY in TX. Then we have

$$\mu s d_Y(u; \Gamma_0) = s d(u|_{C_y})$$

with respect to the point $\{y\} = C_y \cap Y$ considered as the origin in $\mathbb{R}^d$, d being the dimension of the transversal manifold C_y.

In the following we restrict ourselves to the case of $\Gamma_0 = N^*Y \setminus 0$ and in which case we shall write $\mu s d_Y(u)$ for $\mu s d_Y(u; \Gamma_0) = \mu s d_Y(u; N^*Y \setminus 0)$.

Our main result is the following

Theorem. – *Suppose Y is a submanifold of X and $u \in \mathcal{D}'(X \setminus Y)$ satisfy the hypothesis that the closure in T^*X of its wave front set $WF(u)$ is orthogonal to TY and that $\mu s d_Y(u; N^*Y \setminus 0) < \infty$.*

(a) If $\mu s d_Y(u) < \operatorname{codim} Y$ then u admits a unique extension $\tilde{u} \in \mathcal{D}'(X)$ such that $\mu s d_Y(\tilde{u}) = \mu s d_Y(u)$.

(b) If $\operatorname{codim} Y \le \mu s d_Y(u; N^*Y \setminus 0) < \infty$ then u admits (nonunique) extensions $\tilde{u} \in \mathcal{D}'(X)$ with the same microlocal scaling degree as that of u such that $\tilde{u}$ is uniquely determined by its values on the subspace $\mathcal{F}$ consisting of smooth functions in $\mathcal{D}(X)$ such that

$$\mathcal{D}(X) = \mathcal{D}_{Y,h}(X) \oplus \mathcal{F}$$

where $\mathcal{D}_{Y,h}(X)$ denotes the subspace of $\mathcal{D}(X)$ consisting of functions φ which vanish on Y upto order $h = \mu s d_Y(u; N^*Y \setminus 0) - \operatorname{codim} Y$.

Brief sketch of the proofs

We shall only give a very rough idea of the proofs of the theorems.

We first consider the case of the euclidean space, namely, $X = \mathbb{R}^n$ and $Y =$ the origin.

(i) Suppose $u \in \mathcal{D}'(\mathbb{R}^n \setminus 0)$ and $sd(u) < n$. Let $\zeta \in \mathcal{D}(\mathbb{R}^n)$ be a cutoff function such that $\zeta(x) = 1$ in $|x| \le \epsilon$ for some fixed $\epsilon > 0$ and let $\zeta_t(x) = \zeta(tx)$, for $t \in \mathbb{R}_+$. Define the sequence of distributions $\{u_j\}$ on $\mathbb{R}^n$ by

$$u_j = (1 - \zeta_{2^j})u.$$

It is easily seen that $\{u_j\}$ is a Cauchy sequence in $\mathcal{D}'(\mathbb{R}^n)$ for the weak topology. By the weak sequential completeness of $\mathcal{D}'(\mathbb{R}^n)$ the sequence $\{u_j\}$ converges weakly to a limit $\tilde{u} \in \mathcal{D}'(\mathbb{R}^n)$. Then one can check $\tilde{u}$ is unique, also independent of the choice of the function ζ and defines the extension of u such that $sd(\tilde{u}) = sd(u)$.

(ii) In the case $n \leq sd(u) < +\infty$, let $h = sd(u) - n$. Consider the subspace
$$\mathcal{D}_h(\mathbb{R}^n) = \{\varphi \in \mathcal{D}(\mathbb{R}^n); \, \partial_x^\alpha \varphi(0) = 0, \forall |\alpha| \leq [h]\}$$
Then the subspace
$$\{v \in \mathcal{D}'(\mathbb{R}^n); \, \langle v, \psi \rangle = 0, \forall \psi \in \mathcal{D}_h(\mathbb{R}^n)\}$$
consists of distributions of the form
$v = \sum_{|\alpha| \leq [h]} c_\alpha \partial_x^\alpha \delta$, where c_α are complex numbers.
We can decompose any $\varphi \in \mathcal{D}(\mathbb{R}^n)$ uniquely as
$\varphi = \varphi_1 + \varphi_2$, where
$\varphi_1(x) = \sum_{|\alpha| \leq [h]} \gamma_\alpha(x)(\partial_x^\alpha \varphi)(0)$
and
$\varphi_2(x) = \sum_{|\alpha| = [h]+1} \psi_\alpha(x)x^\alpha$ with $\psi_\alpha \in \mathcal{D}(\mathbb{R}^n)$
We now define
$$\langle \tilde{u}, \varphi \rangle = \sum_{|\alpha| = [h]+1} \langle (x^\alpha u)\tilde{\,}, \psi_\alpha \rangle + \sum_{|\alpha| \leq [h]} (-1)^{|\alpha|} \langle \partial_x^\alpha \delta, \gamma_\alpha \rangle$$

Here, since
$sd(x^\alpha u)\tilde{\,} \leq sd(u) - |\alpha| \leq n + [h] - ([h]+1) < n$ by the case (i) above the unique extensions $(x^\alpha u)\tilde{\,}$ are well defined.

The general case

If $u \in \mathcal{D}'(X \setminus Y)$ with $WF(u) \perp TY$ then we take a submanifold C of X transversal to Y and we study the behaviour of $u|_C$. Roughly speaking, for any $y \in Y \cap C$, $u|_C$ looks like a distribution $u \in \mathcal{D}'(C_y \setminus \{y\})$) which can be identified with $\mathcal{D}'(\mathbb{R}^{\text{codim} Y} \setminus \{0\})$.

We can then use the result in the euclidean space to obtain an extension C_y across Y locally. In order to get an extension globally we need a microlocal argument. We also observe that even for the local extension we do not have a natural $\mathbb{R}_+$ - action on the sections C_y and on $C_y \setminus \{y\}$. For this purpose we take, in some sense, a tubular neighbourhood of Y in X.

More precisely, we can identify Y with the zero section $Z(T_Y X)$ in the tangent bundle TX. We can also assume that X is provided with a Riemannian metric so that we can define the flow in X along vector fields with origin at points on Y. Hence take a neighbourhood $\mathcal{U}$ of $Z(T_Y X)$ in TX and the local diffeomorphism

$f(y, \vartheta) = (y, \exp \vartheta) : \mathcal{U} \to Y \times X$.

Then

$f^*(1_Y \otimes u) = u^f \in \mathcal{D}'(\mathcal{U})$

and

$(u^f)_t(x, \vartheta) = u^f(x, t\vartheta)$, for $t \in [0, 1]$.

Some Remarks

1. Such an extension problem for distributions arises in a question of renormalization in quantum field theory where the Green's function associated to a massive Klein - Gordon field, that is , a distribution solution of the Klein - Gordon equation on a globally hyperbolic Lorentzian manifold (M, g):

$$(P_g + m^2 + \kappa R)\varphi = 0$$

where

$P_g = g^{\mu\nu}\nabla_\mu\nabla_\nu$, $\nabla = \nabla_g$ being the covariant derivation with respect to the connection defined by the metric g.

$m \geq 0$, $\kappa \in \mathbb{R}$ and R denotes the scalar curvature.

It is known, by the results of Mme. Choquet - Bruhat, that on a globally hyperbolic manifold (X, g) there exist uniquely determined global forward and backward fundamental solutions to the Klein - Gordon operator.

2. The Klein - Gordon operator P_g is of real principal type. The associated distinguished parametrix plays an essential role in the study of the so called Hadamard states ω_2 :

$$\omega_2((P_g + m^2 + \kappa R)\varphi \otimes \psi) = \omega_2((P_g + m^2 + \kappa R)\psi \otimes \varphi) = 0,$$

$$\forall \varphi, \psi \in \mathcal{D}(M)$$

The Hadamard states can be considered as operator valued distributions on $X \setminus Y$ where X is the product of M by itself and Y is the so called "small diagonal". In the four dimensional globally hyperbolic manifold (M, g), the parametrix has a singularity structure as that of the distribution defined by the Hadamard fundamental solution

$$F(x, y) = v_{-1}(x, y)\sigma(x, y)^{-1} + v_0(x, y) \cdot \log[\sigma(x, y)]$$

where $\sigma(x, y)$ denotes the signed square of the geodesic distance from x to y in the metric g and v_{-1} and v_0 are smooth functions of $(x, y) \in M \times M$ such that v_{-1} does not vanish where σ vanishes. Here signed means, we take the positive sign if the geodesic $\sigma(x, y)$ is space-like and the negative sign if it is time-like.

It was shown by Radzikowski that the Hadamard states can be characterized by the wave front set of the uniquely determined distinguished parametrix of P_g.

References

1. R. Brunetti, K. Fredenhagen, and M. Köhler, The microlocal spectrum condition and the Wick's polynomials of free fields, Comm. Math. Physics, vol. 180 (1996), 633 - 652

2. Y. Choquet - Bruhat, Seminaire de Physique Mathématique, College de France (1963).

3. Y. Choquet - Bruhat, Hyperbolic partial differential equations on manifolds, Battelles Rencontres, 1967 lectures in Mathematics and Physics, Ed. Cecile M. DeWitt and John A. Wheeler, Benjamin (1968).

4. L. Hörmander, Analysis of linear Partial differential operators, Vol I, Springer - Verlag (1983)

5. M. Radzikowski, Micro-local approach to the Hadamard condition in quantum field theory on curved space - time, Comm. Math. Physics, vol. 179 (1996), 529 - 553

6. M. Radzikowski, A local to global singularity theorem for quantum field theory on curved space - time, Comm. Math. Physics, vol 180 (1996) 1 - 22

7. O. Steinmann, Perturbation expansions in axiomatic field theory, Lecture Notes in Physics, 11 Springer - Verlag (1971)

A lower bound for atomic Hamiltonians and Brownian motion

A. DEBIARD and BERNARD GAVEAU
Laboratoire Analyse et Physique Mathématique,
Université Pierre et Marie Curie, Paris, France

A Monsieur Jean VAILLANT
En témoignage de profonde estime et d'amitié

Abstract: We give a lower bound for the ground states of atomic hamiltonians using an estimation of Feynman-Kac exponentials and the ergodic theorem for the brownian motion on the sphere.

1. A major problem of atomic or molecular physics is to determine the ground state of an atomic or a molecular system. The system is represented by a Hamiltonian operator H acting on a Hilbert space $\mathcal{H}$, and one wants to determine the lower bound of the spectrum of H on the Hilbert space $\mathcal{H}$, namely $\inf(\operatorname{spec} H)$. The traditional method uses the well-known variational principle of Rayleigh-Ritz, namely

$$\inf(\operatorname{spec} H) = \inf_{\substack{\|\Psi\| = 1 \\ \Psi \in \mathcal{H}}} \langle \Psi | H \Psi \rangle \ . \tag{1}$$

This method gives an upper bound of the ground state E_0. For any test function $\Psi \in \mathcal{H}$, one has from (1)

$$E_0 \leq \frac{\langle \Psi | H \Psi \rangle}{\langle \Psi | \Psi \rangle} \ . \tag{2}$$

For an atom, $\mathcal{H}$ is the space of electronic wave functions, namely the space of functions $\Psi\big((\vec{r}_1, \sigma_1), \ldots, (\vec{r}_N, \sigma_N)\big)$ where N is the number of electrons, $\vec{r}_1, \ldots, \vec{r}_N$ are their positions in $\mathbb{R}^3$, $\sigma_1, \ldots, \sigma_N$ are their spin indices ($\sigma_j = \pm 1$) and Ψ is L^2 and skew symmetric with respect to the exchange of particles (see [1]). H is given (in atomic units) by

$$H = -\frac{1}{2} \sum_{i=1}^{N} \Delta_i + V(\vec{r}_1, \ldots, \vec{r}_N) \tag{3}$$

where Δ_i is the Laplace operator in the $\mathbb{R}^3$ space of $\vec{r}_i$ and V is the Coulomb potential

$$V(\vec{r}_1, \ldots, \vec{r}_N) = -\sum_{i=1}^{N} \frac{Z}{r_i} + \sum_{1 \leq i \leq j \leq N} \frac{1}{|\vec{r}_i - \vec{r}_j|} \tag{4}$$

Z being the positive electric charge of the nucleus.

The difficulty here is the skew symmetric character of the wave function which is the mathematical expression of the Pauli principle. For $N = 2$, the ground state Ψ is symmetric with respect to $\vec{r}_1$ and $\vec{r}_2$ and skew symmetric with respect to σ_1, σ_2 and the problem is then reduced to the absolute ground state of the Hamiltonian (3) on $L^2(\mathbb{R}^6)$ and the spin indices can be forgotten. It is well known and easy to prove that the absolute ground state of a Hamiltonian like (3) on the whole space $L^2(\mathbb{R}^{3N})$ is indeed given by a symmetric function.

We shall consider, in this article, the absolute ground state of H on the space $L^2(\mathbb{R}^{3N})$, forgetting the Pauli principle and the spin indices. It is only for $N = 2$, that our results will have a physical meaning.

Our purpose is to find a lower bound for the absolute ground state of H, instead of the more classical upper bound given by the variational Rayleigh-Ritz principle.

2. In order to calculate a lower bound for the ground state of H, we consider the Cauchy problem for the heat equation

$$\begin{cases} \dfrac{\partial f}{\partial t} = -Hf \\ f_{|t=0} = u_0 \qquad \text{given} \end{cases} \tag{5}$$

Formally, $f = \exp(-tH)u_0$ and it is obvious that if u_o is not orthogonal to the ground state of H, one has

$$f(t, \vec{r}) \sim e^{-E_0 t}(u_0 | f_0) f_0(\vec{r}) \tag{6}$$

for large $t > 0$, where f_0 is the ground state of H, $Hf_0 = E_0 f_0$. As a consequence of (6), one has

$$\lim_{t \to \infty} \frac{1}{t} \log f(t, \vec{r}) = -E_0 . \tag{7}$$

One possibility to insure that the initial data u_0 of the Cauchy problem (5) is not orthogonal to the ground state f_0 is to take $u_0 \geq 0$, because f_0 does not vanish.

In order to obtain a lower bound for E_0, it is sufficient to calculate an upper bound for $f(t, \vec{r})$ for large times t.

3. We now need a representation for $f(t, \vec{r})$. This will be given by the Feyman-Kac formula (see e.g. [2] for the original reference).

Let us consider N 3-dimensional independent brownian motions $\left(\vec{b}_1(t), \ldots, \vec{b}_N(t)\right)$.

Then, the solution of the Cauchy problem (5) is given by

$$f(t, \vec{r}) = E \left\{ \exp\left(-\int_0^t V\left(\vec{b}_1(s), \ldots, \vec{b}_N(s)\right) ds \right) u_0\left(\vec{b}_1(t), \ldots, \vec{b}_N(t)\right) \right| $$

$$\left| \left(\vec{b}_1(0), \ldots, \vec{b}_N(0)\right) = (\vec{r}_1, \ldots, \vec{r}_N) \right\} \tag{8}$$

Here $\vec{r} = (\vec{r}_1, \ldots, \vec{r}_N)$ and in Eq. (8), the sign

$$E \left\{ \ldots \left| \left(\vec{b}_1(0), \ldots, \vec{b}_N(0)\right) = (\vec{r}_1, \ldots, \vec{r}_N) \right\} \right.$$

is an integral over all brownian paths run from time 0 up to time t, starting from $(\vec{r}_1, \ldots, \vec{r}_N)$ at time 0.

An elementary proof of Eq. (8), the well-known Feyman-Kac formula, is given in [3], starting from the Trotter formula to express the heat semi group

$$\exp(-tH) = \lim_{K \to \infty} \left(\exp\left(-\frac{t}{K}\Delta \right) \exp\left(-\frac{t}{K}V \right) \right)^K$$

using gaussian integrals to disentangle the $\exp\left(-\frac{t}{K}\Delta \right)$ from the $\exp\left(-\frac{t}{K}V \right)$.

4. So the problem reduces now to an upper estimation of the Wiener integral of Eq. (8). In general, this problem is extremely difficult. It is indeed an ergodic problem (see e.g., [4]). Here, we shall use the explicit form of V, in particular the fact that V is homogeneous of degree -1.

First, we notice the identity

$$\frac{1}{2}\Delta\varphi = -V \qquad \text{where } \varphi = \sum Zr_j - \frac{1}{2} \sum_{1 \leq i \leq j \leq N} |\vec{r}_i - \vec{r}_j| \tag{9}$$

and we use Ito's formula (see e.g. (5)) to write

$$\varphi\left(\vec{b}_1(t), \ldots, \vec{b}_N(t)\right) = \varphi\left(\vec{b}_1(0), \ldots, \vec{b}_N(0)\right) +$$

$$+ \int_0^t \sum_{j=1}^N \vec{\nabla}_j\left(\vec{b}_1(s), \ldots, \vec{b}_N(s)\right) . d\vec{b}_j(s) \tag{10}$$

$$+ \int_0^t \frac{1}{2} \sum_{j=1}^N (\Delta_j\varphi)\left(\vec{b}_1(s), \ldots, \vec{b}_N(s)\right) ds .$$

Ito's formula is a Taylor expansion, up to order 2, for a function of the brownian motion, using the fact that $d\vec{b}_i(s) \sim \sqrt{ds}$, $d\vec{b}_j(s)d\vec{b}_k(s) \sim \delta_{kj}ds$. From Eq. (9)–(10) we obtain

$$\exp\left(-\int_0^t V(\vec{b}_1(s),\dots,\vec{b}_N(s))ds\right)\exp\left(-\varphi(\vec{b}_1(t),\dots,\vec{b}_N(t))\right)$$

$$= \exp\left(-\varphi(\vec{b}_1(0),\dots,\vec{b}_N(0))\right)\exp\left(-\sum_{j=1}^{N}\int_0^t \vec{\nabla}_j\varphi(\vec{b}_1(s),\dots,\vec{b}_N(s))d\vec{b}_j\right) \tag{11}$$

So, in Eq. (8) we choose for u_0

$$u_0 = \exp(-\varphi) \tag{12}$$

which is a positive integrable function and we replace the integrand in the second member of Eq. (8) by the equivalent expression given by the second member of Eq. (11), so that

$$f(t,\vec{r}) = \exp(\varphi(\vec{r}))E\left\{\exp\left(-\int_0^t\sum_{j=1}^{N}\vec{\nabla}_j\varphi(\vec{b}_1(s),\dots,\vec{b}_N(s))d\vec{b}_j(s)\right)\right|$$

$$\left|(\vec{b}_1(0),\dots,\vec{b}_N(0)) = \vec{r}\right\}. \tag{13}$$

The estimation of Eq. (13) reduces to

$$f(t,\vec{r}) \leq \exp(\varphi(\vec{r}))E\left\{\exp\left(+\frac{pq}{2}\int_0^t\sum_{j=1}^{N}|\vec{\nabla}_j\varphi|^2(\vec{b}_1(s),\dots,\vec{b}_N(s))ds\right)\right|$$

$$\left|(\vec{b}_1(0),\dots,\vec{b}_N(0)) = \vec{r}\right\}^{1/q}; \tag{14}$$

here $q > 1$, p is the conjugate exponent $\dfrac{1}{p}+\dfrac{1}{q} = 1$.

This inequality is a consequence of the so called «exponential martingale» (see e.g. [5]) and of the Hölder inequality. Details are given in [6]. Now, the fundamental remark is that the function $\sum_{j=1}^{N}|\vec{\nabla}_j\varphi|^2$ is a homogeneous function of degree 0 and thus it is a function on the sphere S^{3N-1} of the Euclidean space $\mathbb{R}^{3N}$. Because the sphere is compact, one can apply the ergodic theorem for brownian motion (see (7)) to obtain for large t

$$\int_0^t\sum_j^{N}|\vec{\nabla}_j\varphi|^2(\vec{b}_1(s),\dots,\vec{b}_N(s))ds \sim \left\langle\sum_j^{N}|\vec{\nabla}_j\varphi|^2\right\rangle t \tag{15}$$

where $\langle\,\rangle$ denotes the average on the unit sphere S^{3N-1}. As a consequence, we see from Eqs. (14)-(15) that

$$\lim_{t\to\infty}\frac{1}{t}\log f(t,\vec{r}) \leq \frac{1}{2}\left\langle\sum_j^{N}|\vec{\nabla}_j\varphi|^2\right\rangle \tag{16}$$

and from Eq. (7), we deduce that

$$-\frac{1}{2}\left\langle \sum_j^N |\vec{\nabla}_j \varphi|^2 \right\rangle \leq E_0 \tag{17}$$

is the desired lower bound for the ground state of H on L^2 functions.

5. It remains to calculate the average value of $\left\langle \sum_j^N |\vec{\nabla}_j \varphi|^2 \right\rangle$ on the sphere S^{3N-1}. It is proved in [6] that

$$-\frac{1}{2}\left\langle \sum_j^N |\vec{\nabla}_j \varphi|^2 \right\rangle = \frac{N}{2}\left(Z^2 + \frac{N-1}{4} + \frac{2(N-1)Z}{\pi} \right). \tag{18}$$

The specialization to the case $N = 2$, $Z = 2$ which is the case of the helium atom gives, after coming back to atomic units the evaluation:

$$-\frac{1}{2}\left\langle \sum_j^N |\vec{\nabla}_j \varphi|^2 \right\rangle \simeq 80\,\mathrm{eV} \tag{19}$$

The experimental value of E_0 for the helium atom (namely the full ionization energy of the helium atom) gives $E_0 \simeq -79\,\mathrm{eV}$ (see e.g. (8)).

So, the lower bound of Eq. (17) is in excellent agreement with the experimental value.

6. The difficulty, now, is to extend this kind of method to any N-fermion systems. For $N \geq 3$, the absolute spatially fully symmetric ground state is not the physically relevant ground state. It is correct that $\left\langle \sum_j^N |\vec{\nabla}_j \varphi|^2 \right\rangle$ is still a lower bound of the physically relevant ground state, but it will not be a precise lower bound, because all the electronic correlations imposed by the Pauli principle have been neglected.

References

[1] L. Landau, L. Lifschitz: Mécanique quantique, Ed Mir, Moscou (1965)

[2] M. Kac: Probability and related topics in physical sciences, Interscience, New-York (1989)

   See also M. Kac: Integration in Function spaces and some of its applications, Lezioni Fermiane, Pisa (1980)

   and L. S. Schulman: Techniques and applications of Path integrals, Wiley, New-York (1981)

[3] B. Gaveau, L.S. Schulman : Grassmann valued processes for the Weyl and the Dirac equations, Phys. Rev. D 36 (1987), 1135-1140

[4] B. Gaveau: Estimation des fonctionnelles de Kac sur une variété compacte et première valeur propre de $\Delta + f$, Japan Acad. Sci 60 (1985), 361-364

also A. M. Berthier, B. Gaveau: Convergence des exponentielles de Kac et application en physique et en géométrie. J. Funct. Analysis 29 (1978), 416-424

and B. Gaveau, E. Mazet: Divergence des exponentielles de Kac et diffusion quantique. Publi. RIMS Kyoto 18 (1982), 365-377

[5] H. P. Mc-Kean: Stochastic integrals. Academic Press. New-York (1969)

[6] A. Debiard, B. Gaveau : The ground state of certain Coulomb systems and Feynman-Kac exponentials, Mathematical Physics, Analysis and Geometry 3 (2000), 91-100

[7] K. Itô, H. P. Mc Kean: Diffusion processes and their sample paths, Springer-Verlag, New-York (1964)

[8] I. N. Levine: Quantum chemistry, Prentice-Hall, Englewood (1991)

M. Karplus, R. N. Porter: Atoms and molecules, Benjamin Cummings, Reading (1970)

Acknowledgement: We thank neither the chiefs of CNRS, nor the French Minister of national education and the council of Paris 6, for having interrupted their very modest financial support during our work.

A compromised arrow of time

L. S. Schulman

Physics Department, Clarkson University
Potsdam, NY 13699-5820 USA
schulman@clarkson.edu

ABSTRACT

The second law of thermodynamics—the usual statement of the arrow of time—has been called the most fundamental law of physics. It is thus difficult to conceive that a single dynamical system could contain subsystems, in significant mutual contact, possessing opposite thermodynamic arrows of time. By examining cosmological justification for the usual arrow it is found that a consistent way to establish such justification is by giving symmetric boundary conditions at two (cosmologically remote) times and seeking directional behavior in between. Once this has been demonstrated, it is seen that entropy increase can be reversed and that the usual arrow is less totalitarian than previously believed. In the same vein, other boundary conditions, modeling shorter periods in the evolution of the cosmos, can be found that allow the simultaneous existence of two thermodynamic arrows, notwithstanding moderate interaction between the systems possessing those arrows. Physical consequences of the existence and detection of opposite-arrow regions are also considered.

1. Introduction

The thermodynamic arrow of time is so all-encompassing that, paradoxically, it can be invisible. The ancients did not explain friction, just as they did not think to try to explain why things fell *down*. There were myths for spring and myths for rain, but as far as I know, no myths for terrestrial gravity or friction. Nevertheless, once Newton had formulated reversible laws of mechanics (by studying the nearly frictionless motion of the planets) it was realized that the enormous asymmetries of our lives—breaking eggs, graying beards, heating brake pads—required an explanation that went beyond the laws of dynamics.

It is not always appreciated that this explanation involves two separate logical steps. The first is to go from reversible *microscopic* dynamics to irreversible *macroscopic* manifestations. A result of this kind is the Boltzmann H-theorem. The second step takes the first as given, but then asks, why do we have the particular arrow that we do, what is it that picks the actual direction? Is it a physical effect, like CP violation or the expansion of the universe? Or perhaps it's more subtle. Any sufficiently large system must have *some* arrow, so that the directionality question loses its meaning. One would still need to establish, however, that there must be but a single arrow. Background on both aspects can be found in [1], although emphasis in the present article and in [2,3,4] is on the second matter.

The study of irreversibility is closely related to ergodic theory and the tools in the present study are the tools of that field. In particular I use the model systems popular there, one of the most congenial of these being the "cat map." I will first use this to show how an arrow of time can *apparently* develop in a symmetric system and how this can be related to cosmology. Then a system with two arrows will be exhibited. Next I will treat the important physical question of whether we could hope to see this phenomenon. Finally I will take up one of the amusing aspects of this research, the possibility of causal paradoxes, very much like those that arise in time travel scenarios.

2. The cat map

The "cat map" is a mixing transformation of the unit square onto itself. As for "Schrödinger's cat," the whimsical name was acquired when a fundamental idea was illustrated by having unpleasant things happen to a cat. For $x, y \in I^2 =$ the unit square, the mapping $\phi : I^2 \to I^2$ is

$$
\begin{aligned}
x' &= \phi_1(x, y) \equiv x + y \mod 1 \\
y' &= \phi_2(x, y) \equiv x + 2y \mod 1
\end{aligned}
\quad \text{or} \quad
\begin{pmatrix} x' \\ y' \end{pmatrix} \equiv \begin{pmatrix} 1 & 1 \\ 1 & 2 \end{pmatrix} \begin{pmatrix} x \\ y \end{pmatrix} \mod 1 \qquad (1)
$$

The matrix on the right has determinant one, so that ϕ is measure preserving. If one thinks of I^2 as phase space, then this is a model of classical mechanics. Technically, ϕ is effective at producing apparently irreversible effects because the matrix in Eq. (1) has as its larger eigenvalue $\lambda \equiv \left(3 + \sqrt{5}\right)/2 \sim 2.6 > 1$.

But it is the non-technical illustration of the relaxation that earned the map its name. In a classic text [5], an initial pattern consisting of an image of a cat is shown. When ϕ is applied once, the image is stretched by a factor λ, squeezed transversely by $1/\lambda$, snipped (when overshooting I^2), and reassembled. Within two or three applications of ϕ, the poor cat is hardly recognizable.

Feline images will not be needed in this article. Rather I place several hundred points in I^2 and allow each to move independently under ϕ. This models a collection of non-interacting atoms, an "ideal gas," coming to equilibrium—or failing to do so, as the case may be.

To gain a quantitative handle on relaxation and to define an "entropy," it is necessary to introduce "coarse grains" in I^2. Some such device is always necessary (see [1]), since if one knows precise coordinates and microscopic dynamics there is no loss of information, hence no entropy increase. The coarse grains are here obtained by placing a grid over I^2 and taking as the *macroscopic* state *only* the number of points within each rectangle defined by the grid. For N ($\gg 1$) identical atoms in G coarse grains, the entropy is

$$
S = -\sum_{k=1}^{G} \rho_k \log \rho_k \qquad \text{with} \quad \rho_k = n_k/N \qquad (2)
$$

where n_k is the number of points in coarse grain (rectangle) #k.

In the next section the time dependence of the entropy will be illustrated. If all points are in a single rectangle, then $S = 0$; when the points are Poisson distributed $S \lesssim \log G$. The relaxation time to go from a single 0.1×0.1 grain to equilibrium is about 5. Using smaller grains, one can see that $S(t) \sim t \log \lambda$ (and $\log \lambda$ is the Lyapunov exponent).

3. Cosmology and two-time boundary value problems

Around 1960 Gold [6] argued that the thermodynamic arrow of time followed the expansion of the universe. In other words, your coffee cools because the quasar 3C273 recedes. Lest this reformulation test your credibility, let me give perspective. The expansion of the universe is another factual arrow of time, a "cosmological" arrow. Lacking any great asymmetry in the *laws* of nature, it is plausible that the two all-pervasive arrows, cosmological and thermodynamic, could have something to do with each other. To make the case you need to show how the distant and unnoticed-until-1930 cosmological arrow could induce the thermodynamic one. Gold argued that the expansion of the universe provides a sink for free energy, by accepting—due to its expansion—all radiation thrown off by stars, etc. He also showed how the influence of large scale processes reached to smaller scales, even to coffee at Café Epsilon. He clinched his argument with a *gedanken* experiment in which a star was put in an insulated box and held there till it equilibrated. At this point the contents of this large box no longer had a thermodynamic arrow. He then imagined that he would open a window in the box for a short time, during which radiation would *escape* to the universe at large, this escape (rather than its opposite) being the consequence of expansion (cf. Olber's paradox). With the box again closed and with some of its contents gone, the system within is no longer in equilibrium. The subsequent return to equilibrium is exactly what provides the sought-for arrow and the connection to the larger cosmological scenario. A consequence of Gold's ideas is that in a contracting universe (which occurs for some solutions of the equations of general relativity) the arrow would be reversed.

Back in 1972 I was totally enamored of this argument. But now one learns why it is useful for physicists to talk to mathematicians. I tried to explain Gold's argument to a mathematician colleague of mine, Andrew Lenard, and mid-story came to a halt. Trying to maintain his high standards of logic I realized that Gold's star-in-a-box argument was circular.

First a counterexample: suppose the radiation had *entered* the box (as in a contracting universe, a lá Olber). According to the tale above, there would *still* be an arrow in the box after the window closed, as the system recovered from the influx of radiation. It would need to recover from *any* change. So something is clearly fishy.

Now if Gold's arrow in his star-in-a-box parable does *not* arise from the efflux/influx distinction, where does it come from? The answer is, it is the arrow of the *narrator*. In listening to this story one automatically accepts that prior to the opening of the window there is no effect of the opening. *But that is already an assumption about the arrow of time!* If this same process of opening and closing had been observed by someone with an opposite arrow (a notion one must allow if one is engaged in establishing the existence of one or the other arrow) that observer would naturally have assumed that the effect of that process would have occurred on the other side of the interval within which the window was open, namely an epoch we (or the narrator) would consider to be *before* the opening.

How can this circularity be avoided? I proposed [7] that one should give boundary conditions at two remote times and look for the emergence of an arrow between them. For example, if our universe has both a big bang and a big crunch (expansion followed by contraction), then the following sort of boundary condition would fit this prescription. Take the matter and radiation distributions as roughly homogeneous at some time interval τ both after the big bang and before the big crunch. Such homogeneity appears to have obtained at the epoch of radiation-matter decoupling, with $\tau \sim 300,000$ years. Now include the expansion and contraction and try to derive that at first (after time-τ) there would be an arrow in one direction, then—as the big crunch becomes nearer in time than the big bang—the other. This does *not* mean that we must live in such a universe. It is only a framework within which Gold's idea can be logically explored.

As stated, the project is daunting. But I have taken the usual route of statistical mechanicians, namely to abstract the problem, retaining the essential conceptual features while making the mathematics tractable.

The model system is a "gas" of particles each of which has cat map dynamics in the phase space I^2. How to model the expansion of the universe? The essential feature (for our purposes) of a rapid expansion from a homogeneous configuration is that homogeneity is a far, far from equilibrium state when the dynamics of the system are dominated by gravity. Gravity makes things clumpier; it *enhances* density differences. So the homogeneous matter and radiation distribution that prevailed at τ—and which we will demand prevail at $T - \tau$ (with T the big crunch time)—is far from equilibrium, or it will become that as the universe expands [8].

For the cat map gas, the way to be out of equilibrium is to *not* be uniformly distributed, and for a given coarse graining this can be accomplished by putting all the points in a single grain. As observed earlier (following Eq. (2)) this gives the system entropy $S = 0$. But now we want *two*-time boundary conditions, and we want them symmetric. So we pick a pair of times, 0 and T, and demand the same sort of nonequilibrium state for the gas at both these times.

Solving two-time boundary value problems can be difficult. Imagine a similar problem for the gas in a 40 m^3 room. Demand that all the gas be in a single cubic meter at time-0 (so the rest of the room is in vacuum), and require that under particle-particle interactions alone (and with the room totally isolated) all the gas spontaneously find itself in that same cubic meter one hour later. It is extremely

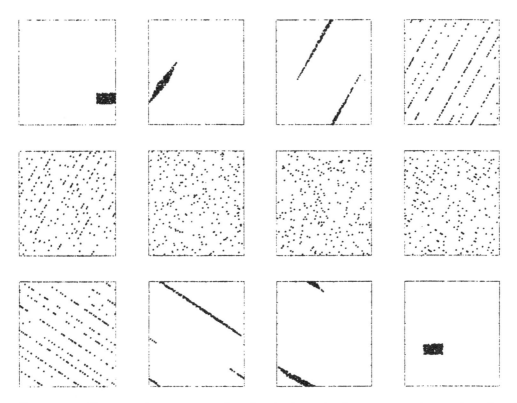

Figure 1. History of 250 points that begin in a single coarse grain at time-0 and end in a single coarse grain at time-19. Selected times are shown according to the following scheme: top row (left to right), [0 1 2 4]; second row, [5 8 11 14]; third row, [15 17 18 19].

unlikely that the exact initial data for the gas would lead to this eventuality, although *some* initial conditions do reach the desired time-T state. Finding them is another matter. If you randomly chose initial velocities and positions within the required cubic meter, a conservative estimate suggests that you would have less than one chance in $10^{10^{26}}$ to have all the particles arrive where you want them an hour later. For the ideal gas though the job is easy. I want a gas of 250 "atoms" and a coarse graining that divides I^2 into 50 rectangles. The trick is to randomly place 12,500 points in the desired initial rectangle and let these atoms evolve $T = 19$ time steps. The result is a square fairly uniformly covered with points. Then, *discard all points that do not fall into the desired final rectangle.* (For an ideal gas this does not affect the dynamics of those that remain.) This leaves you with about 250 successful points (1/50th of 12,500).

In Fig. 1 is a "movie" of the history of these points, with 12 selected time steps

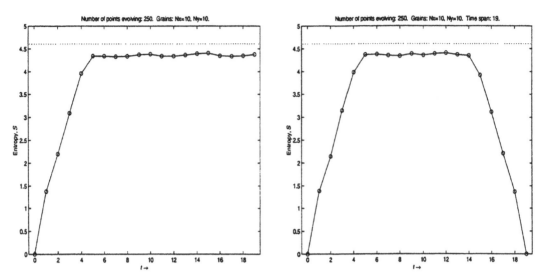

Figure 2. Entropy, S, as a function of time. The left figure (2a) is $S(t)$ *without* future conditioning. On the right (2b) is $S(t)$ for a two-time boundary condition solution, such as Fig. 1. Note the similarity of the initial time dependence. Also of interest is the reflection symmetry of Fig. 2b about its temporal midpoint.

shown. Note the first half of the movie: if I would show you the expansion of a gas with *no* future boundary condition you would not be able to tell the difference. To illustrate this, in Fig. 2a I show the entropy as a function of time for a gas of 250 particles under free expansion, and in Fig. 2b is the time dependence of the entropy for the actual "movie." Compare the initial increases in entropy for the two situations: they are indistinguishable, up to statistical fluctuations. For this *macroscopic* quantity, the patterns of entropy increase are essentially identical. A second point is that if Fig. 2b were flipped about its middle ($t = 9.5$) it would look the same (up to statistical fluctuations).

Now let's apply the message of this simulation to the grand issue of cosmology. Both boundary times, 0 and 19, are states of low entropy. This is analogous to the state of the universe long enough after "decoupling" (or before, coming back from the big crunch) for the homogeneous distribution no longer to represent equilibrium. For early times (in the cat movie) the entropy increases and indeed there is an arrow of time. As remarked, this arrow is indistinguishable from what you would have if there were _no_ future conditioning. Thus we have no way of knowing from this statistical information alone whether there is or is not a big crunch coming. The point of this demonstration is that the expansion of the universe creates a situation in which the homogeneous distribution, which is disordered—and likely—during one epoch, becomes extremely _unlikely_ at a later epoch (once gravity dominates). As the larger system proceeds to its more disordered state, as it *relaxes*, it is temporarily trapped in all sorts of metastable states, for example stars, which in turn relax and drive shorter term metastable states, such as people. Note too, from the symmetry about the midpoint in Fig. 2b, that the arrow toward the end is reversed.

4. Opposite arrows

So far we have talked about "the" thermodynamic arrow of time. Could it happen that within some large system there are *two* subsystems, each with a different arrow of time? To see how this might come about imagine that we do indeed live in a big bang-big crunch universe. Now get a well-insulated spaceship and put yourself in a cryogenic bath for a long time—until the universe has begun to contract. Now your alarm clock awakens you and you look out the porthole to see the world around you with an opposite-running arrow! Similarly I can imagine that there could be matter in regions that have been relatively isolated "since" the big crunch and which "survive" [9] into our epoch.

But could there really be two arrows at once, entropy increasing in one region, decreasing in another? Here is an argument against this possibility. From the standpoint of one of the observers (a sentient being in one component) the events in the other region are entropy *lowering*. An example of this would be if someone carefully arranged the positions and velocities of 11 billiard balls so they would come together, with 10 of them forming a triangle of balls at rest and the eleventh going away from them along one of the axes of the triangle (the reverse of a "break"). This would be a system with decreasing entropy; it becomes *more* ordered. Now if such a system is disturbed even slightly, someone coughing while the balls collided, it would ruin the enormous coordination required for the entropy lowering process. So for an observer in either region to see (and maybe yell in surprise at) the creation of order in the other, would be to destroy that order. It follows that opposite running arrows cannot coexist if systems interact.

The foregoing argument is wrong. It is given from the perspective of one observer, and implicitly assumes that the way to formulate a macroscopic problem is as an *initial* value problem in the time sense of that observer. It is true that for an initial value problem even small perturbations will disturb entropy lowering, but we should use arguments that do not already assume the validity of one arrow. As discussed, the way to do this is to give macroscopic boundary values at *two* times, and solve for the motion between.

This leads to the following formulation for producing simultaneous opposite arrows of time. Consider a system with two component subsystems; call them A and B. Give macroscopic boundary conditions such that A's entropy at time-0 is lower than it is at time-T, and B's entropy has the reverse specification. A and B *are* allowed to interact [10].

What sort of solution can be expected in the time interval $[0, T]$? If there were *zero* interaction, this would be equivalent to two separate problems. A's entropy

A, $t \longrightarrow$

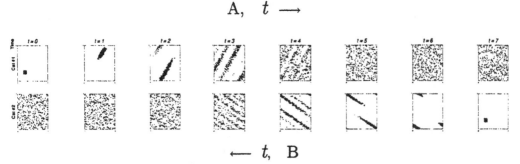

$\longleftarrow t,$ B

Figure 3. Images of a pair of interacting systems, each having a different arrow of time. For the upper row (system A) the boundary conditions are that all points are in a single coarse grain at time zero, with no condition set for time T ($= 7$). For the lower row (system B) all points are in a single coarse grain at time-T, with no requirement for time-0. The systems are coupled to each other with the parameter α (in the map ψ_α) taking the value 0.2.

would increase as a function of the time parameter t, while B's would decrease. This is a trivial consequence of the boundary conditions and the only anomaly is that B made a poor choice of time parameter. This emphasizes that "t" is a parameter for microscopic time evolution, with no a priori relation to thermodynamics.

Now allow interaction. Do the arrows survive? I investigated this with an extension of the cat map. Take two squares, A and B (copies of I^2), and allow N particles in each to evolve under the cat map. In addition, each point in A or B is associated with a particular point in the other box, and is influenced by the position of that other point. To describe the map efficiently introduce additional notation. Recall that the original cat map (Eq. (1)) is called ϕ. Now define a map, $\psi_\alpha(u, v) \equiv (u + \alpha v, v) \bmod 1$, depending on the real parameter α. Like ϕ, ψ is measure preserving, but the eigenvalues of the corresponding matrix are both unity, so it is not chaotic. For each time step, the combined motion of each pair of pairs $[(x_A, y_A), (x_B, y_B)]$ is a three-step process: 1) A-B interaction: $\psi_{\alpha/2}$ applied to (x_A, y_A) and (x_B, y_A) separately; 2) the usual cat map evolution: ϕ applied to (x_A, y_A) and (x_B, y_B) separately; 3) repeat step #1 [11].

The results of the simulation are shown in Fig. 3. Clearly the the goal of opposing arrows has been realized. It is interesting though to note the effects of the coupling. Without coupling the result of a single time step on the initial rectangle for system A would be a neat parallelogram, the image of that rectangle under the cat map (cf. Fig. 1, time step 1). In Fig. 3 it is clear that the parallelogram has been smudged. This is the effect of ψ and system B. Nevertheless, the overall expansion shows an arrow ($S(t)$ will be displayed momentarily). The effect of B on A is thus seen to be essentially noise, and correspondingly for the effect of A on B.

In Fig. 4L (left) is the time dependence of the entropy (but not for the same

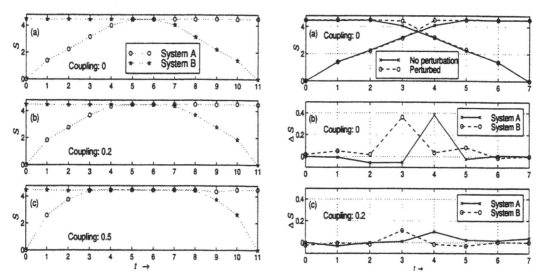

Figure 4. Entropy, S, as a function of time. The left figure (L) depicts $S(t)$ for three values of the parameter α. The uppermost (a), $\alpha = 0$ shows the non-interacting system and gives, as it trivially should, increasing entropy for A (as a function of the non-thermodynamic time parameter t), and decreasing entropy for B. Below that (b & c) are shown $S(t)$ for $\alpha = 0.2$ and 0.5, respectively. On the right (R) are shown entropy and entropy differences when the system is perturbed. The perturbation takes place effectively at time 3.5 and for A the changes in macroscopic behavior (whether for $\alpha = 0$ or 0.2) occur *later* by A's clock. For B the changes take place at lower t values, which are also later—by B's clock.

---

run as Fig. 3). Three values of α (in ψ_α) are shown. The first is $\alpha = 0$, for which as discussed, it is trivial that there should be two arrows. For $\alpha = 0.2$ increase and decrease of entropy in each system is quite similar, although looking carefully one can note that the relaxation is slightly quicker. This is consistent with the images of Fig. 3 and with the idea that the effect of each system on the other is noise. Finally the lowest image shows the behavior for the stronger coupling, $\alpha = 0.5$. Now the relaxation is rapid, for both systems, showing that with strong coupling the boundary values can force low entropy momentarily, but it is quickly lost.

For Fig. 4R (right) I take up another question: macroscopic causality. I take this to mean that macroscopic effect follows macroscopic cause, and distinguish it from microscopic causality, stated, e.g., in terms of field commutators. Defining a noncircular test of (macro) causality requires caution. If we were to use initial conditions, then the effect of a perturbation would *by definition* be subsequent.

In [1] appears a consistent test of causality in the framework of macroscopic two-time boundary conditions (our usual way of maintaining logical consistency in arrow-of-time questions). The system is required to be in particular coarse grains at times 0 and T, and is evolved microscopically from initial to final grains with a particular evolution law, yielding some macroscopic history. Next the system is evolved a second time, but now *with a perturbation*. The same boundary data are used for both runs. A "perturbation" consists of changing the evolution law on a single time step, some t_0, with $0 < t_0 < T$. In general the set of microscopic points satisfying the macroscopic boundary conditions without the perturbation will be different from those satisfying the (same) boundary conditions *with* the perturbation. So except for times 0 and T the points need not occupy the same coarse grains. Now we can state the test of macroscopic causality: if the *macroscopic* behavior is the same *before* the perturbation but not after, there is macroscopic causality. If it is the other way around, there is reverse causality. If the macroscopic behavior is different *both* before and after, there is no causality. Now apply this test to our elaborated cat map. The perturbation is that on a particular time step (t_0), instead of applying ϕ and ψ, another rule is used. That rule is that at t_0, instead of ϕ, we use a "faster" cat. That is, the matrix of the usual cat map Eq. (1), $\left(\begin{smallmatrix}1&1\\1&2\end{smallmatrix}\right)$, is replaced by $\left(\begin{smallmatrix}3&4\\2&3\end{smallmatrix}\right)$, which has a larger Lyapunov exponent.

In Fig. 4Ra an entropic history is shown for uncoupled systems, with, again, $T = 7$. The perturbation is nominally at $t_0 = 4$, but because entropy (S) values are calculated between steps, it is best to think of this as $t_0 = 3.5$. To better see the effect of the perturbation, in Fig. 4Rb I show only the entropy *change* due to the perturbation. For A the major difference occurs at 4, while for B it is at 3, consistent with causality. As indicated earlier, for uncoupled systems this result is trivial and only confirms our method. In Fig. 4Rc, coupling (0.2) is turned on and the same comparison made. Qualitatively causality persists, although the coupling (by inducing noise) reduces all deviations.

This demonstration of causality responds to another question the reader may have had. My evidence for arrows of time in the earlier discussion consisted solely of increase or decrease of entropy. This now shows that such increase or decrease corresponds to deep-rooted concepts of cause and effect.

5. Physical implications

The foregoing development shows that it is *possible* for there to be two opposite running arrows. But, *does it actually happen*? Could we look across the Milky Way and watch coffee unstir, stars unexplode, Humpty Dumpty come together?

This is a two-part question. First, is there opposite-arrow matter out there? Second, if it's there, could you observe it? Norbert Wiener thought you could not [12]. Thus I will devote a short subsection to checking this.

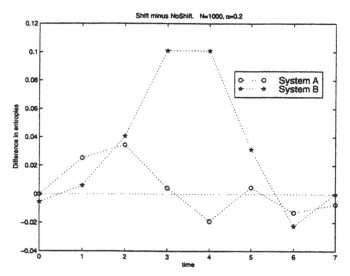

Figure 5. Difference in entropy between a run in which the evolution on time step #4 is the identity versus one in which points of A (alone) undergo a shift (details in the text).

5.1 Communication between opposite-arrow regions

Using the same modelling methods that established the possibility of contrary arrows, I will now show that a signal can pass from one system to the other. In [2] I considered electromagnetic radiation, and used the time-symmetric Wheeler-Feynman absorber theory [13] as a framework for studying electromagnetic communication between the regions. That demonstration has subtleties, particularly in connection with the origins of radiation reaction, that I now prefer to avoid. You need not resolve all the old issues of classical radiation in order to see that opposite arrow regions can have a macroscopic influence on one another.

Again we consider a two-time boundary value problem, and again A will be in a single grain at time-0 and B in a single grain at time-T. But in this experiment, at a particular intermediate time, t_0, I will wiggle A. By "wiggle" I mean that instead of the usual composite cat map step at t_0, the points of A will be shifted. That's all. On that same step, nothing happens to B. To be precise, in the example I worked out for Fig. 5, on time step 4, the points of B were mapped by the identity (2×2) matrix, while those of A underwent the following transformation: $x' = x + 0.3$, $y' = y + 0.5$, both mod 1. Of course this affects the points selected (to satisfy the two-time boundary value problem) at the microscopic level. But what is the macroscopic effect? To see this, I compare this run to one in which *neither* A *nor* B does anything on time step 4. (For all these the same boundary conditions were used.) This was done with more points than the earlier work in order to reduce statistical fluctuations. The entropic history of both runs is recorded, and Fig. 5

shows the *difference* between the entropies in the two cases.

First note that almost nothing happens to A. The shift is by integer multiples of 0.1 which is also the linear dimension of the coarse grains. But B does indeed respond to the "wiggling" of A. However primitive this may be as a form of communication, this shows that wiggling A causes a macroscopic response in B.

5.2 Possible physical origins of opposite-arrow regions

As discussed earlier, I believe the source of the thermodynamic arrow of time is the expansion of the universe [6]. Suggestions for the occurrence of opposite arrow regions will be developed in that same context. At the beginning of Sec. 4, I proposed one way to arrive at an opposite-arrow region that bucks the overall flow, namely a "spaceship," or isolation. Thus you could have a region that has relatively little interaction with its surroundings, so that at a time closer to (say) the big bang, its most important conditioning would be from the big crunch end of the universe. There is a mathematical question hidden in this possibility. Specifically, if you condition a process to take unlikely values at two widely separated times, then in between it will tend to move to more likely values, and will take the most likely values at the middle of its excursion. Its entropy will be a symmetric function of time (cf. [1]). By allowing spaceship survival and the like, I am introducing events for which a large excursion is moved away from the center. How unlikely is that? A much simplified way to examine this is to look at Brownian motion with diffusion constant D that is conditioned on starting from 0 and returning to 0 at time-T. It is easy to see [1] that the probability that it is at x at time-t is $1/[4\pi DTs(1-s)]\exp\left(-x^2/4DTs(1-s)\right)$, with $s = t/T$. Using this we can evaluate how unlikely various excursions are; for example, by integration it follows that $\Pr\left(x^2 > \rho^2 DT \text{ at } t\right) = 1 - \text{erf}\left(\rho/2\sqrt{s(1-s)}\right)$. If ρ is not large this function imposes little relative penalty on deviations being displaced from $T/2$. For example, for ρ as large as 5 (which is *extremely* unlikely—but *least* unlikely at $T/2$), it is only 3 times less unlikely at $0.4T$. (At $T/2$ the probability is about 10^{-12}.)

Starting therefore from the cosmo/thermo arrow connection, the only justification I can imagine for having opposite-arrow regions [14] is living in a big bang-big crunch universe in which some regions manage the isolation needed to pass through the epoch of maximal expansion with arrow intact [15].

5.3 Detection of opposite-arrow regions

What should this stuff look like, if it's there? By and large it should be extremely dull. Taking the opposite-arrow material to be "relics" from the future, as discussed in Sec. 5.2, one must first consider how far off the big crunch is. Of course this is completely unknown, since there is at present no evidence that there will *be* a big crunch. (On the contrary, evidence today favors an ever expanding universe—but bear in mind the caveat in [15]). Just to grab a number, suppose the big bang-big crunch time separation is 500 Gyr. We are a mere 10–15 Gyr from the big bang and still see stars burning. But anything that's 485 Gyr old would be unlikely to have hydrogen burning and would probably be invisible at optical—or any other easily seen—frequencies. So this stuff would have gravitational interactions but be pretty much invisible, characteristics attributed to "dark matter" [16]. Like other dark matter candidates, this would be picked up in microlensing observations, which by themselves however give no indication of statistical properties.

Nevertheless, there are ways to make a positive detection. One possibility is to see an "unexplosion." The idea is that even very old matter could have slowly developing processes that eventually lead to something dramatic—an explosion. From our perspective we would first see some fuzzy, relatively big region. It would gradually shrink in size, becoming brighter. Ultimately it would disappear.

A second method is to identify something that is *very* old. A way to achieve this is again to study the astrophysics of old objects, objects in which hydrogen burning has ceased, but in which other nuclear processes continue. Again these will be weak, for if they were strong they would long "ago" have burnt themselves out (unless a threshold for something traumatic is achieved, as in the last paragraph). If specific nuclei can be identified for these processes, then there may exist particular gamma emission lines that characterize matter of such great age.

A variant of this is the suggestion [17] that one could extend a technique now used to date old rocks, namely isotope abundance. Thus if there is evidence that a particular isotope with a lifetime of say 10^6 year was present at a rock's creation, then its absence today can be taken as evidence that the rock is many times that age. Similarly, isotopes with lifetimes much greater are known (even 10^{21} yr), and finding appropriate samples depleted of these isotopes could be another indicator.

The search for observational evidence is what I consider the most significant challenge now connected with the newly-raised possibility of opposite-arrow regions of the cosmos.

5.4 Cosmological implications

The central problem in the modern study of cosmology is, is the universe open or closed? Will it continue to expand forever, or will it turn around and collapse? Finding an opposite-arrow region would be strong evidence for a big crunch. As indicated, it is difficult to think of any other source of opposite-arrow material.

You could have a big crunch without a reversal of the thermodynamic arrow, and indeed early work on cosmology considered oscillating universes that warmed on each successive pass. Arguing from elegance, I consider that unlikely: if the geometry is symmetric and can nevertheless provide a thermodynamic arrow, why demand extra asymmetry (... notwithstanding Pauli's remark: "Elegance is for tailors").

As a small digression, I would like discuss other statistical ways to discern cosmological structure, supplementing material in [1]. Consider a world with a big bang and big crunch, a world that must meet boundary conditions at both ends, boundary conditions that are far from equilibrium (or extremely unlikely) with respect to time periods interior to the time interval in question. To clarify, suppose the big bang and big crunch are at $\pm \mathcal{T}/2$ and we study an interval $\pm T/2$, with $\mathcal{T} > T$ and $(\mathcal{T} - T)/2$ the duration of the interval from the big bang to the decoupling time. The homogeneous distribution of matter and its low metallicity at decoupling are natural when one works from an initial value close to the big bang. However, if one needs to meet this condition when *emerging* from a gravity-dominated universe and heading *toward* a big crunch, I expect it to be extremely *un*likely. Things must smooth out and you must get rid of all those big nuclei that formed along the way. So this is *not* an easy boundary condition to satisfy. One thing that would help would be to minimize the amount of processing of hydrogen, the amount that is caught up in burning stars. A state of the universe in which matter mostly did not accumulate in big masses, but stayed in bodies much smaller than the sun, would eliminate quite a bit of heavy element production, so that satisfying the boundary conditions we have described favors the existence of what would otherwise seem to be a disproportionate amount of nonluminous baryonic matter, for example, in the form of brown dwarfs.

Thus the prevalence of nonluminous baryonic matter, perhaps in the form of brown dwarfs, is consistent with the assumption of a geometrically and statistically symmetric universe. Could this "consistency" be upgraded to evidence for a big crunch? Not quite, because at this point one really doesn't know what the abundance of such matter *ought* to be, reflecting for example uncertainties in just how many brown dwarfs should be created in an episode of galactic star formation.

6. Paradoxes and causal loops

In this section we take up the causal loops and paradoxes that could arise if opposite-arrow material is present and if communication between opposite running regions occurred. These paradoxes have been used [18] to argue against Gold's idea, so it is worth confronting them here.

Consider the following tale of two observers with opposite-running thermodynamic arrows of time. We rename system A, Alice, and system B, Bob. The tale begins with Alice, who at 8 a.m. her time, sees rain coming in Bob's window (say this is 5 p.m. his time). With an hour's delay she informs Bob. He receives the signal at 4 p.m. his time and closes the window before the rain starts (say at 4:40p.m. his time). Did Alice see the rain coming in or not?

The resolution of this paradox is similar to that in [19] and I have dealt with it at length in [3]. The idea is that making the paradox precise requires setting up boundary conditions, in this case at two separate times, such that it would appear that no consistent intermediate motion could occur. Having done that, there are two possible resolutions. The first is that indeed your boundary conditions have no solution. From the perspective of the Setter of the boundary conditions this would mean that the world history He or She desired just doesn't happen. The second possible resolution uses assumptions of continuity and boundedness. In [3] it is shown that in reasonable situations, including that suggested above in the carpet paradox, there will be some intermediate motion that *does* satisfy the boundary conditions.

For the soggy carpet paradox such resolution could take the following form. You set up Alice at 7:30 a.m. her time with instructions to send a message to Bob about his window, should that be necessary. You "start" Bob at 3:30 p.m. his time with an open window. The self-consistent solution has Alice noting a *slightly* open window at 8 a.m. her time. Seeing this, Alice is unsure whether to send a message, and finally, at 9 a.m. her time, waffles, "It will rain in a bit, you might think about closing your window." Bob gets this message and balances fresh air versus dry carpet by mostly closing the window, but nevertheless leaving it open a crack—which is just what Alice sees at 8 a.m. her time! This kind of self-consistent loop is favored by (what I consider) the better science fiction writers in their dealing with time-travel paradoxes [20].

Acknowledgment

This article is dedicated to Prof. Jean Vaillant on the occasion of his retirement. Much of this work was done at the Technion-Israel Institute of Technology. I am grateful to P. Facchi, E. Mihokova, A. Ori, S. Pascazio, A. Scardicchio, L. J. Schulman, and M. Roncadelli for helpful discussions. My research is supported in part by the United States National Science Foundation grant PHY 97 21459.

References

[1] L. S. Schulman, *Time's Arrows and Quantum Measurement* (Cambridge University Press, Cambridge, 1997).

[2] L. S. Schulman, Opposite Thermodynamic Arrows of Time, Phys. Rev. Lett. **83**, 5419 (1999).

[3] L. S. Schulman, Resolution of causal paradoxes arising from opposing thermodynamic arrows of time, preprint.

[4] L. S. Schulman, Phys. Rev. Lett. **85**, 897 (2000).

[5] V. I. Arnold and A. Avez, *Ergodic Problems of Classical Mechanics* (Benjamin, New York, 1968).

[6] T. Gold, The Arrow of Time, Am. J. Phys. **30**, 403 (1962).

[7] L. S. Schulman, Correlating Arrows of Time, Phys. Rev. D **7**, 2868 (1973).

[8] "Equilibrium" is not the right term here, since systems acting only under Keplerian forces do not have an equilibrium. All that is needed for our argument

however, is that by becoming less homogeneous the system enters ever larger regions of phase space.

[9] The sense of the words "since" and "survive" in this sentence is with respect to the clock of the isolated region.

[10] In terms of our cosmological perspective, this time interval $[0, T]$ is far from both the big bang and the big crunch. The physical appearance of simultaneous arrows would be most likely to occur near the epoch of maximal expansion.

[11] The alternation of ϕ and ψ enhances apparent time reversal symmetry. ϕ itself is not time symmetric, but for our purposes all that matters is that Lyapunov exponents for ϕ and ϕ^{-1} are equal (which is why Fig. 2b is symmetric). This is indicates [1] that microscopic T violation alone need not provide a thermodynamic arrow.

[12] N. Wiener, *Cybernetics* (M.I.T. Press, Cambridge, Mass., 2nd ed., 1961), p. 34.

[13] J. A. Wheeler and R. P. Feynman, Interaction with the Absorber as the Mechanism of Radiation Rev. Mod. Phys. **17**, 157 (1945); J. A. Wheeler and R. P. Feynman, Classical Electrodynamics in Terms of Direct Interparticle Action Rev. Mod. Phys. **21**, 425 (1949).

[14] An "opposite-arrow region" consists of ordinary matter, differing from everyday matter only in statistical properties. It is not (necessarily) antimatter.

[15] Of course, when upwards of 90% of the matter believed to be in the universe is in forms currently unknown to the human race, one must be humble in reaching conclusions from the absence of alternative explanations or in declaring the impossibility of a phenomenon. After all, the justification for the thermodynamic arrow that I find most appealing—expansion of the universe—could not have been guessed until that expansion was observed.

[16] This may explain some dark matter, but not much. By Gold's hypothesis opposite-running-arrow regions should be a small fraction of what's around us, while dark matter seems to be a *large* fraction.

[17] F. Avignone and R. Creswick, private communication.

[18] R. Penrose, Big Bangs, Black Holes and 'Time's Arrow', in R. Flood and M. Lockwood, eds., *The Nature of Time* (Blackwell, Oxford, 1986), pp. 41-42.

[19] L. S. Schulman, Tachyon Paradoxes, Am. J. Phys. **39**, 481 (1971).

[20] R. A. Heinlein, *The Door into Summer* (Ballantine Books, New York, 1956).

Printed and bound by CPI Group (UK) Ltd, Croydon, CR0 4YY

23/10/2024

01778246-0007